REVIEW OF THE 21ST CENTURY TRUCK PARTNERSHIP, SECOND REPORT

Committee to Review the 21st Century Truck Partnership, Phase 2

Board on Energy and Environmental Systems
Division on Engineering and Physical Sciences

NATIONAL RESEARCH COUNCIL
OF THE NATIONAL ACADEMIES

THE NATIONAL ACADEMIES PRESS
Washington, D.C.
www.nap.edu

THE NATIONAL ACADEMIES PRESS • 500 Fifth Street, NW • Washington, DC 20001

NOTICE: The project that is the subject of this report was approved by the Governing Board of the National Research Council, whose members are drawn from the councils of the National Academy of Sciences, the National Academy of Engineering, and the Institute of Medicine. The members of the committee responsible for the report were chosen for their special competences and with regard for appropriate balance.

This study was supported by Contract DENT-0006206 between the National Academy of Sciences and the U.S. Department of Energy. Any opinions, findings, conclusions, or recommendations expressed in this publication are those of the author(s) and do not necessarily reflect the views of the organizations or agencies that provided support for the project.

International Standard Book Number 13: 978-0-309-22247-1
International Standard Book Number 10: 0-309-22247-8

Copies of this report are available in limited supply, free of charge, from the Board on Energy and Environmental Systems, National Research Council, 500 Fifth Street, NW, Keck W934, Washington, DC 20001; (202) 334-3344.

Additional copies of this report are available for sale from the National Academies Press, 500 Fifth Street, NW, Keck 360, Washington, DC 20001; (800) 624-6242 or (202) 334-3313; http://www.nap.edu.

Printed in the United States of America

THE NATIONAL ACADEMIES

Advisers to the Nation on Science, Engineering, and Medicine

The **National Academy of Sciences** is a private, nonprofit, self-perpetuating society of distinguished scholars engaged in scientific and engineering research, dedicated to the furtherance of science and technology and to their use for the general welfare. Upon the authority of the charter granted to it by the Congress in 1863, the Academy has a mandate that requires it to advise the federal government on scientific and technical matters. Dr. Ralph J. Cicerone is president of the National Academy of Sciences.

The **National Academy of Engineering** was established in 1964, under the charter of the National Academy of Sciences, as a parallel organization of outstanding engineers. It is autonomous in its administration and in the selection of its members, sharing with the National Academy of Sciences the responsibility for advising the federal government. The National Academy of Engineering also sponsors engineering programs aimed at meeting national needs, encourages education and research, and recognizes the superior achievements of engineers. Dr. Charles M. Vest is president of the National Academy of Engineering.

The **Institute of Medicine** was established in 1970 by the National Academy of Sciences to secure the services of eminent members of appropriate professions in the examination of policy matters pertaining to the health of the public. The Institute acts under the responsibility given to the National Academy of Sciences by its congressional charter to be an adviser to the federal government and, upon its own initiative, to identify issues of medical care, research, and education. Dr. Harvey V. Fineberg is president of the Institute of Medicine.

The **National Research Council** was organized by the National Academy of Sciences in 1916 to associate the broad community of science and technology with the Academy's purposes of furthering knowledge and advising the federal government. Functioning in accordance with general policies determined by the Academy, the Council has become the principal operating agency of both the National Academy of Sciences and the National Academy of Engineering in providing services to the government, the public, and the scientific and engineering communities. The Council is administered jointly by both Academies and the Institute of Medicine. Dr. Ralph J. Cicerone and Dr. Charles M. Vest are chair and vice chair, respectively, of the National Research Council.

www.national-academies.org

COMMITTEE TO REVIEW THE 21ST CENTURY TRUCK PARTNERSHIP, PHASE 2

JOHN H. JOHNSON, Michigan Technological University, *Chair*
JOSEPH M. COLUCCI, NAE,[1] Automotive Fuels Consulting, Inc.
CORALIE COOPER,[2] Northeast States for Coordinated Air Use Management, Boston, Massachusetts
DAVID E. FOSTER, University of Wisconsin-Madison
LARRY J. HOWELL, Consultant, Royal Oak, Michigan
JOHN G. KASSAKIAN, NAE, Massachusetts Institute of Technology, Cambridge
DAVID F. MERRION, Consultant, Detroit Diesel Corporation (retired), Brighton, Michigan
THOMAS E. REINHART, Southwest Research Institute, San Antonio, Texas
BERNARD ROBERTSON, NAE, Chrysler Corporation (retired), Bloomfield Hills, Michigan
CHARLES E. SALTER, Consultant, Chambersburg, Pennsylvania
KATHLEEN C. TAYLOR, NAE, General Motors, Research and Development Planning Center (retired), Fort Myers, Florida
WALLACE R. WADE, NAE, Ford Motor Company (retired), Novi, Michigan

Project Staff

JAMES J. ZUCCHETTO, Senior Program Officer/Board Director, Board on Energy and Environmental Systems, Study Director
MADELINE WOODRUFF, Senior Program Officer
E. JONATHAN YANGER, Senior Project Assistant
LANITA JONES, Administrative Coordinator

[1] NAE = member, National Academy of Engineering.
[2] Resigned from the committee, January 18, 2011.

BOARD ON ENERGY AND ENVIRONMENTAL SYSTEMS

ANDREW BROWN, JR., NAE,[1] Delphi Corporation, Troy, Michigan, *Chair*
RAKESH AGRAWAL, NAE, Purdue University, West Lafayette, Indiana
WILLIAM F. BANHOLZER, NAE, The Dow Chemical Company, Midland, Michigan
MARILYN BROWN, Georgia Institute of Technology, Atlanta
MICHAEL L. CORRADINI, NAE, University of Wisconsin-Madison
PAUL DECOTIS, Long Island Power Authority, Albany, New York
CHRISTINE EHLIG-ECONOMIDES, NAE, Texas A&M University, College Station
WILLIAM FRIEND, NAE, Bechtel Group, Inc. (retired), McLean, Virginia
SHERRI GOODMAN, CNA, Alexandria, Virginia
NARAIN HINGORANI, NAE, Consultant, Los Altos Hills, California
ROBERT J. HUGGETT, Consultant, Seaford, Virginia
DEBBIE NIEMEIER, University of California, Davis
DANIEL NOCERA, NAS,[2] Massachusetts Institute of Technology, Cambridge
MICHAEL OPPENHEIMER, Princeton University, New Jersey
DAN REICHER, Stanford University, Palo Alto, California
BERNARD ROBERTSON, NAE, Chrysler Corporation (retired), Bloomfield Hills, Michigan
ALISON SILVERSTEIN, Consultant, Pflugerville, Texas
MARK H. THIEMENS, NAS, University of California, San Diego
RICHARD WHITE, Oppenheimer & Company, New York, New York

Staff

JAMES J. ZUCCHETTO, Senior Program/Board Director
JOHN HOLMES, Senior Program Officer and Associate Board Director
DANA CAINES, Financial Manager
ALAN CRANE, Senior Program Officer
JONNA HAMILTON, Program Officer
LANITA JONES, Administrative Coordinator
ALICE WILLIAMS, Senior Project Assistant
E. JONATHAN YANGER, Senior Project Assistant

---

[1] NAE = member, National Academy of Engineering.
[2] NAS = member, National Academy of Sciences.

Acknowledgments

The Committee to Review the 21st Century Truck Partnership, Phase 2, is grateful to the representatives of the 21st Century Truck Partnership, including the four government agencies—the U.S. Department of Energy, the U.S. Environmental Protection Agency, the U.S. Department of Transportation, and the U.S. Department of the Army—and to the representatives from companies and national laboratories who contributed significantly of their time and effort to this National Research Council (NRC) study by giving presentations at meetings or responding to committee requests for information. The committee also acknowledges the valuable contributions of other individuals who provided information and presentations at the committee's open meetings. (Appendix B lists all of those presentations.)

The committee offers its special appreciation to Ken Howden, Director, 21st Century Truck Partnership, U.S. Department of Energy, Office of Vehicle Technologies (formerly the Office of FreedomCAR and Vehicle Technologies), for his significant contributions in coordinating responses to its questions and in making presentations to the committee.

Finally, the chairman gratefully recognizes the committee members and the staff of the NRC Board on Energy and Environmental Systems for organizing and planning the committee meetings and gathering information and drafting sections of the report. Jim Zucchetto in particular has done an outstanding job of facilitating the work of the committee and helping it to write a focused and timely report.

This report has been reviewed in draft form by individuals chosen for their diverse perspectives and technical expertise, in accordance with procedures approved by the NRC's Report Review Committee. The purpose of this independent review is to provide candid and critical comments that will assist the institution in making its published report as sound as possible and to ensure that the report meets institutional standards for objectivity, evidence, and responsiveness to the study charge. The review comments and draft manuscript remain confidential to protect the integrity of the deliberative process. We wish to thank the following individuals for their review of this report:

- Andrew Brown, Jr. (NAE), Delphi Corporation
- Douglas Chapin (NAE), MPR Associates
- Duke Drinkard, 21st Century Driver and Truck Alliance
- Roger Fruechte, General Motors (retired)
- Trevor Jones (NAE), ElectroSonics Medical
- Drew Kodjak, International Council for Clean Transportation
- Adrian Lund, Insurance Institute for Highway Safety
- Dale Stein (NAE), Michigan Technological University (retired)
- Ward Winer (NAE), Georgia Institute of Technology
- John Woodrooffe, University of Michigan

Although the reviewers listed above have provided many constructive comments and suggestions, they were not asked to endorse the conclusions or recommendations, nor did they see the final draft of the report before its release. The review of this report was overseen by Lawrence Papay (NAE). Appointed by the National Research Council, he was responsible for making certain that an independent examination of this report was carried out in accordance with institutional procedures and that all review comments were carefully considered. Responsibility for the final content of this report rests entirely with the authoring committee and the institution.

Contents

Figures and Tables

FIGURES

TABLES

Summary

BACKGROUND AND INTRODUCTION

In July 2010, the National Research Council (NRC) appointed the Committee to Review the 21st Century Truck Partnership, Phase 2, to conduct an independent review of the 21st Century Truck Partnership (21CTP). This Phase 2 review follows on the original NRC Phase 1 review of the Partnership conducted in 2007 and resulting in the report issued in 2008 (NRC, 2008). That 2008 review is referred to hereafter as the NRC Phase 1 report. It contains recommendations to which the 21CTP has responded (see Appendix C in this volume for the responses).

The 21CTP is a cooperative research and development (R&D) partnership including four federal agencies (the U.S. Department of Energy [DOE], the U.S. Department of Transportation [DOT], the U.S. Department of Defense [DOD], and the U.S. Environmental Protection Agency [EPA]), and 15 industrial partners (Allison Transmission, ArvinMeritor, BAE Systems, Caterpillar, Cummins Inc., Daimler Trucks North America [which includes Freightliner], Detroit Diesel Corporation [DDC], Eaton Corporation, Honeywell International, Navistar, Mack Trucks, NovaBUS, Oshkosh Truck, PACCAR, and Volvo Trucks North America).

Since the Phase 1 review, the Partnership has evolved in the face of changing budgets and new initiatives, such as the America Recovery and Reinvestment Act of 2009 (ARRA; Public Law 111-5), which injected funds during 2009 and 2010 for technology R&D on heavy-duty vehicles. The main leadership in the Partnership resides with the DOE's Office of Vehicle Technologies, which manages a number of DOE-funded R&D programs directly related to medium- and heavy-duty vehicle (MHDV) technologies. The other 21CTP agencies associate their own existing programs that are relevant to the goals of the 21CTP under the 21CTP umbrella.

The other factor, besides changing budgets and new initiatives, that makes budgets and projects involved in the 21CTP challenging to review is that the different agencies receive their budget appropriations from different committees in Congress. Thus, there is no central, overall control over budgets and accountability. Department of Energy staff organize meetings and conference calls, maintain the information-flow infrastructure (such as websites and e-mail lists), and have led the discussions for and preparation of the updated 21CTP Roadmap and Technical White Papers (DOE, 2010a, 2011) laying out Partnership goals. The management of individual projects under the 21CTP umbrella rests with the individual federal agencies that have funded the work. These agencies use the 21CTP information-sharing infrastructure to coordinate efforts and to ensure that valuable R&D results are communicated and that any overlap of activities among their respective efforts is reduced. The NRC's Phase 1 review of the overall 21CTP helped communicate to the various stakeholders and to Congress the ongoing R&D efforts in the agencies and on the various projects (NRC, 2008). It is anticipated that the present, Phase 2 review and report will help extend that avenue of communication to all interested parties.

The purpose of the 21CTP is to reduce fuel consumption and emissions while increasing heavy-duty vehicle safety by supporting research, development, and demonstration that can lead to commercially viable products and systems. The strategic approach of the Partnership includes the following: (1) develop and implement an integrated vehicle systems R&D approach that validates and deploys advanced technology; (2) promote research for engine, combustion, exhaust aftertreatment, fuels, and advanced materials; (3) promote research focused on advanced heavy-duty hybrid propulsion systems; (4) promote research to reduce parasitic losses (now called vehicle power demands); (5) promote the development of technologies to improve truck safety; (6) promote the development and deployment of technologies that substantially reduce energy consumption and exhaust emissions during idling; and (7) promote the validation, demonstration, and deployment of advanced truck and bus technologies, and grow their reliability sufficient for adoption in the commercial marketplace (DOE, 2006).

The organization of this report is similar to that of the NRC Phase 1 report. The committee reviewed the major areas that the 21CTP is addressing. (See Chapter 1 for the committee's complete statement of task). The committee's work was aided by its review of written materials and through presentations by 21CTP government and industry partners on technical progress and accomplishments (see Appendix B). In addition, the series of white papers referred to above summarized technical information, barriers, and, in many cases, goals and milestones, for six major focus areas:

1. *Engine systems*—which also includes fuels, aftertreatment, and materials;
2. *Hybrid propulsion systems*;
3. *Vehicle power demands*—formerly called parasitic losses, which aim to reduce energy losses such as those from rolling resistance or aerodynamics;
4. *Idle reduction*—which aims to reduce the amount of energy used for truck engine idling;
5. *Vehicle safety*—to reduce fatalities and injuries in truck-involved crashes; and
6. *Efficient operations*—which is a new area and white paper with the aim of reducing fuel consumption in the U.S. truck freight-delivery system.

This Summary first presents the committee's overall findings and recommendations from the review of the 21CTP as a whole. It then presents the major findings and recommendations, selected from Chapters 2 through 9, for the following: management strategy and priority setting for the Partnership, the first five focus areas defined by the white papers (listed above), the SuperTruck program begun in 2010, and the new, sixth focus area on efficient operations. The findings and recommendations from the chapters retain their original numbering to help the reader gain context by going to the original discussions. The report chapters also contain findings and recommendations in addition to those in this Summary.

The new SuperTruck program is funding the development and demonstration of full vehicle systems integrating a number of technologies into Class 8 heavy-duty, long-haul trucks with the aim of reducing load-specific fuel consumption (i.e., gallons per ton-mile). This new effort follows on the NRC Phase 1 report that called for integrating new technologies, including advanced diesel engines, into vehicle systems.

OVERALL OBSERVATIONS

Overall Report Finding S-1. The key benefit of the 21st Century Truck Partnership is the coordination of research programs directed toward the goal of reducing fuel usage and emissions while increasing the safety of heavy-duty vehicles. Federal involvement is bringing stakeholders to the table and accelerating the pace of technological development. Given the federal regulatory requirements to reduce emissions and fuel consumption, it seems the sharing of research and development (R&D) costs between the government and U.S. manufacturers of trucks and buses or heavy-duty vehicle components is appropriate to develop new technologies. Thus, the 21CTP is providing access to the extraordinary expertise and equipment in federal laboratories, in addition to seed funding that draws financial commitment from the companies to push forward in new technology areas. The Partnership provides the United States with a forum in which the member agencies, in combination with industry, academia, and federal laboratories, can better coordinate their programs. The steady decline in research funding from FY 2003 through FY 2007 was threatening the attainment of program goals. The actual funding and need for R&D are discussed in Chapter 1. The funding level in the years prior to the availability of funding through the American Recovery and Reinvestment Act of 2009 (ARRA) was not in proportion to the importance of the goal of reducing the fuel consumption of heavy-duty vehicles and providing advanced technology for the industry to meet the 2014-2018 and later fuel consumption regulations. The ARRA funds provided by Congress in 2009-2010 have significantly enhanced the ability of the Partnership to meet and demonstrate the goals for reducing fuel-consumption and improving safety in prototype vehicles.

Overall Report Recommendation S-1. The 21CTP should be continued to help meet the nation's goal of reduced fuel consumption in the transportation sector. In addition, the Partnership needs to review whether additional partners—such as major truck and component manufacturers that are not currently members—that could contribute to the R&D program should be recruited. Research funding should be commensurate with well-formulated goals that are strategic to reducing the fuel consumption of heavy-duty vehicles while improving safety, and all projects should be prioritized so that the 21CTP R&D program can be implemented within the available budget.

Overall Report Finding S-2. The 21CTP leadership responded substantively to most of the recommendations of the National Research Council's (NRC) Phase 1 review, which helped to contribute to the improved program that was the subject of this Phase 2 review. The committee commends the leadership of the Partnership for this effort.

Overall Report Recommendation S-2. The 21CTP program goals should continue to be established, reviewed, updated, related to available funding, and clearly stated in measurable engineering terms. The white papers defining the various technical areas of R&D should be reviewed and revised, as appropriate, periodically and prior to any future NRC review of the 21CTP. Given the "virtual" nature of the Partnership among 4 agencies and 15 industrial partners, the projects that are considered to be part of 21CTP should be better defined and, if part of the Partnership, indicated by a 21CTP notation in any 21CTP documentation.

MANAGEMENT STRATEGY AND PRIORITY SETTING

The NRC Phase 1 report (Recommendation 2-2; NRC, 2008) recommended the creation of "a portfolio management process that sets priorities and aligns budgets among the agencies and industrial partners." In response, the Partnership stated that although this recommendation "will be considered ... the ability to directly align budgetary decisions across the agencies, however desirable, may be outside the scope of this voluntarily collaborative organization" (see Appendix C). Given the individual control and oversight of the four agencies, the committee concluded that, although indeed highly desirable, such a portfolio management process is simply not likely to happen with the decentralized nature of the Partnership.

Although prioritization of projects across agencies is unlikely to happen in any meaningful way, the DOE has focused much of its 21CTP effort going forward on three SuperTruck projects, two funded with ARRA funds and one receiving DOE appropriated funds. These projects are directed toward demonstrating feasibility, fuel efficiency, and emissions compliance with full vehicle hardware for Class 8 long-haul freight trucks, as recommended in the NRC Phase 1 report (see Chapter 8 "SuperTruck Projects" in this report). The committee applauds the prioritization of available ARRA and DOE funds on these projects. Although improved collaboration and coordination among agencies would be welcome, the committee judges overall 21CTP program management to have improved since the Phase 1 report.

Finding 2-1. The 21CTP is a virtual organization facilitating communication among four agencies, government laboratories, and industry, but it has no direct control over research activities or funding across the agencies or by its industry partners. The committee continues to believe that the lack of single-point 21CTP authority is far from optimal, although it recognizes that this is necessary because of the various congressional committees that the agencies report to and that provide their budgets.

Recommendation 2-1. The Department of Energy (DOE) is urged to continue to improve the functioning of the 21CTP "virtual" management structure in every way possible. Such improved functioning would include strengthening interagency collaboration (particularly that involving the Environmental Protection Agency [EPA] and the Department of Defense [DOD])[1] and documenting and publishing specific 21CTP activity within all four agencies.

Finding 2-2. The EPA, DOD, and Department of Transportation (DOT) did not have a well-defined list of the projects and associated budgets that were included under the 21CTP umbrella. This stems in part from the virtual nature of the Partnership and partly, particularly within the DOE, from the natural overlap in activities on batteries, hybrids, materials, and other areas between the activities for light-duty vehicles and the 21CTP. Many of these activities are reviewed at the annual DOE Merit Review and at Directions in Engine-Efficiency and Emissions Research (DEER) conferences, and the new SuperTruck projects include an annual reporting requirement, but there is no dedicated report for the 21CTP.

Recommendation 2-2. The DOE should issue a brief annual report documenting the specific projects within the 21CTP and the progress made. The annual report should provide references to published technical reports from the involved agencies. This would especially help outside groups, future review committees, the Congress, and others to understand the structure, activities, and progress of the Partnership.

ENGINE SYSTEMS AND FUELS

The NRC Phase 1 report includes 12 findings and recommendations regarding engines. The Partnership concurred with many of the recommendations and incorporated several of them in the SuperTruck contracts. It did not, however, concur with the two findings and recommendations (NRC, 2008, Findings 3-1 and 3-8 and Recommendation 3-1 and 3-8) about the 50 percent brake thermal efficiency (BTE) goal for 2010 and the 55 percent BTE goal for 2013. However, the committee notes that the 21CTP has now changed the year for meeting the 50 percent BTE goal to 2015 and that for the 55 percent goal to 2018.[2]

Combustion

Finding 3-1. The committee reviewed nine diesel engine programs that were funded at a total of more than $100 million by the DOE and industry and that included the High Efficiency Clean Combustion (HECC) program, the Waste Heat Recovery (WHR) program, and others. Some programs met or exceeded their goals, for example achieving a 10.2 percent improvement in brake thermal efficiency (BTE) versus a 10 percent goal, whereas others did not quite meet the goals of 5 percent or 10 percent improvement in BTE. By combining HECC and WHR, each demonstrating greater than 10 percent improvement in BTE, together with other technologies, it should be possible to improve BTE by 20 percent to achieve the original DOE target of 50 percent peak BTE. However, the DOE target of 50 percent peak BTE was not met by the original goal of 2010.

[1] Subsequent to the committee's review of 21CTP programs, the DOE and the DOD entered into the Advanced Vehicle Power Technology Alliance (AVPTA) partnership on July 18, 2011. See, for example, "DOE, Army Alliance Underlines Achieving Energy Security" by Chris Williams, available at http://www.army.mil/article/62727/. Accessed October 18, 2011.

[2] The 55 percent BTE goal in the 21CTP updated white paper, "Engines," (DOE, 2011) is for a prototype engine system in the laboratory by 2015, whereas in the DOE Multi-Year Program Plan, the goal is for 2018 in a prototype engine (DOE, 2010b).

Finding 3-2. The DOE-funded research in advanced engine combustion at the national laboratories, in industry, and at universities is well managed and addresses important aspects for achieving an integration of advanced combustion processes that should be important enablers for achieving the 55 percent BTE goal as well as providing ongoing improvements. There also appears to be good interaction between the researchers performing the work and the industry stakeholders. Efforts to achieve 55 percent BTE are going to require complex and expensive technologies. It will be some time before it becomes clear whether there is a production-feasible and cost-effective way to achieve the 55 percent BTE target. The committee believes that this target carries considerable risk, even at the test cell demonstration stage.

Recommendation 3-1. The 21CTP fundamental research program should continue to provide important enablers for the 55 percent BTE goal, and the DOE should continue to look for leverage opportunities with other government- and industry-funded projects.

Recommendation 3-2. The DOE should ensure that the engine R&D for the goal of 50 percent BTE at over-the-road cruise conditions and the stretch goal of 55 percent BTE in an engine in a laboratory that will now be carried out under the SuperTruck program receive the appropriate share of the SuperTruck funding and benefit extensively from the DOE-funded research programs in advanced engine combustion.

Fuels

Finding 3-7. In spite of efforts to reduce the fuel consumption of light-duty and heavy-duty vehicles and to develop biomass-derived fuels (an effort which, except for corn-based ethanol, has not progressed as much as had been expected), petroleum will remain the primary source of light-duty and heavy-duty vehicle fuel for many years to come. Whereas future U.S. gasoline demand is expected to be flat for the next 20 years, diesel fuel demand is expected to grow, necessitating changes in refinery operations.

Recommendation 3-4. The DOE should reinstate its program for advanced petroleum-derived fuels (they will be transportation's primary fuels for many years to come) with the objective of maximizing the efficiency of their use.

Finding 3-9. The DOE established three different sets of goals for the fuels program from 2008 to 2011, which made an assessment of progress against the goals difficult. In total, little progress has been made toward the achievement of these DOE goals, which were not specified goals of the 21CTP.

Recommendation 3-6. The DOE fuel goals should be re-evaluated in line with the FY 2012 budget and the recommendations of this report. Specific plans for achieving these goals should be established.

Aftertreatment Technologies

Finding 3-10. The research agenda of the 21CTP is focused on improving the oxides of nitrogen (NO_x) reduction performance of selective catalytic reduction (SCR) and lean-NO_x-trap systems, improving the efficiency of and reducing the fuel consumption associated with particulate matter (PM) filter regeneration, and improving the ability to model aftertreatment systems. The DOE Cross-cut Lean Exhaust Emissions Reductions Simulations (CLEERS) program does a good job of coordinating the aftertreatment research programs within the 21CTP and disseminating the results to the technical community at large.

Finding 3-11. The demands on the aftertreatment system and its performance are intimately linked to the combustion process taking place within the cylinder. Consequently, the aftertreatment system must be developed and its performance evaluated in conjunction with the combustion system. The 21CTP realizes this, and its new goals for the aftertreatment program specifically state this.

Recommendation 3-7. The aftertreatment program within the 21CTP should be continued, and the DOE should continue to support the activities of CLEERS that interface with the activities of the aftertreatment technical community at large.

Emissions and Related Health Effects

Finding 3-13. The Advanced Collaborative Emissions Study (ACES), the Collaborative Lubricating Oil Study on Emissions (CLOSE), and the project on Measurement and Characterization of Unregulated Emissions from Advanced Technologies are comprehensive and cooperative projects that are investigating important issues related to potential heavy-duty diesel engine health effects. Based on the activities reported, the committee finds a high degree of collaboration among government agencies, national laboratories, and industry stakeholders.

Recommendation 3-9. The DOE should continue funding the Advanced Collaborative Emissions Study, the Collaborative Lubricating Oil Study on Emissions, and the project on Measurement and Characterization of Unregulated Emissions from Advanced Technologies until results are finalized and reported for all three studies.

Propulsion Materials

Finding 3-14. The propulsion materials program is addressing a broad range of materials issues associated with heavy-truck propulsion systems. Many of the initiatives are funded as cooperative R&D agreements (CRADAs) with significant industry cost sharing, showing strong support by industry for this area of work.

Recommendation 3-10. The DOE should fund programs in the areas outlined in its "21st Century Truck Partnership White Paper on Engines and Fuels" (February 25, 2011) in the section "Approach to Reaching Goals" covering materials R&D for valve trains, major engine components, air-handling systems (turbochargers and exhaust gas recirculation [EGR] systems), and exhaust manifold sealing materials.

High Temperature Materials Laboratory

Perhaps just as important as the direct support of the 21CTP is the extensive benefit to the broader research and development community that comes from the research conducted at the High Temperature Materials Laboratory (HTML). This research covers a wide range of challenging problems for which solutions require the unique instrumentation at HTML as well as the expertise of the knowledgeable DOE researchers who oversee and operate the facility. The fact that many academic researchers, as well as industry research specialists, seek collaboration with HTML speaks to the value of the facility with respect to the advancement of knowledge on many fronts.

HTML, located at Oak Ridge National Laboratory, was established more than 20 years ago as a National User Facility. It was created to provide specialized, and in some cases one-of-a-kind, instruments for materials research and characterization of value not only to the 21CTP but also to other programs needing a fundamental understanding of materials properties.

Finding 3-15. HTML continues to be a valuable resource for materials research for the 21CTP, providing specialized and in many cases unique instrumentation and professional expertise. The expertise of those who oversee the laboratory, and therefore the value of HTML to all users, is enhanced by the participation of the HTML staff themselves in the research.

Recommendation 3-10. The DOE should continue to provide 21CTP researchers and other potential users access to HTML, and it should make every effort to maintain support for HTML and to maintain the cutting-edge capability of the facility. Moreover, the DOE should provide sufficient funding for HTML, and for the research specialists who oversee and operate the facility, to enable continued research collaboration with the academic community, other government laboratories, and industry. In particular, HTML support should not be reduced to a level that allows only maintenance of the equipment for paying users.

HYBRID VEHICLES

Finding 4-4. The EPA and DOT's National Highway Traffic Safety Administration (NHTSA) issued their final rules on September 15, 2011, for "Greenhouse Gas Emissions Standards and Fuel Efficiency Standards for Medium- and Heavy-Duty Engines and Vehicles." Although these standards contain test procedures for determining fuel consumption for heavy-duty hybrid trucks, a manufacturer still needs a certificate of conformity showing that a vehicle's internal combustion engine meets the EPA criteria emission standards for heavy-duty engines (a procedure that does not recognize hybrid heavy-duty trucks). The California Air Resources Board (CARB) is currently drafting vehicle-level test procedures for heavy-duty hybrid vehicles.

Recommendation 4-3. As partners of the 21CTP, EPA and DOT's NHTSA should work with CARB to develop test procedures for the certification process for criteria emissions so that the emissions benefits of hybridization will be recognized, allowing the reduction in size or simplification of the emission control system of hybrid heavy-duty vehicles to be realized.

Finding 4-6. Six new stretch technical goals have been established by the 21CTP for heavy-duty hybrid vehicles. The committee agrees with the 21CTP that these are indeed stretch goals. Specific plans for achieving these new goals, some of which were carried over from the previous three goals that had been set for hybrids, were not provided to the committee. Nor was the rationale provided for these new goals, although they are appropriately focused on fuel consumption reductions, cost reduction, and a 15-year design life for the technologies. They appear to be reasonable technical goals. The cost and design life objectives in the previous goals had been identified earlier by the 21CTP as being necessary for achieving commercially viable heavy-duty hybrid vehicles. It is expected that a significant budget would be required through the target dates specified in the new goals, and a significant increase from the zero budget for heavy-duty hybrid R&D over the past 3 years would be required.

Recommendation 4-5. The 21CTP should establish plans and develop realistic budgets for accomplishing the six new stretch goals for heavy-duty hybrid vehicles in accordance with the committee's findings, explain the rationale behind the new goals, and provide the current status of the applicable technology for each of the goals so that the magnitude of the tasks for each can be assessed.

VEHICLE POWER DEMANDS (FORMERLY "PARASITIC LOSSES")

Finding 5-7. There is no rolling resistance test procedure with interlaboratory correlation universally employed as an industry standard.

Recommendation 5-4. The 21CTP, strongly supported by the DOT and EPA (the latter through its SmartWay program), should conduct an authoritative study of the several barriers (e.g., related to tread life, truck stability in blowouts, run-flat tires, and other topics) to the widespread carrier adoption of next-generation wide base single (NGWBS) tires. The DOT should specifically support reduction of barriers to NGWBS tire acceptance by requiring the universal use by tire manufacturers of a rolling resistance test procedure like that in ISO (International Organization for Standardization) 28580, to ensure that comparative interlaboratory data exist.

Finding 5-13. Summarizing the committee's findings on vehicle power demands: Project prioritization by the 21CTP roughly follows the consumption ranking of the several heavy-duty truck operating loads in Table 5-1 (see Chapter 5 in this report) and technology risk. However, sometimes market forces provide considerable impetus for quite good development and implementation—for example, in tire rolling resistance and, to a lesser extent, trailer aerodynamic components. The DOE has identified a strong role in which technology development costs and risks are high, as in its vehicle systems simulation and testing activities for heavy-duty trucks. It has generally followed these principles, to address high costs and risks, in the vehicle power demand projects. The SuperTruck projects will provide a unique Partnership opportunity to provide both further high-risk technology results for certain vehicle power demand reductions and real-world validation of numerous integrated systems.

Recommendation 5-8. Although it is tempting to assume that the SuperTruck projects will address all of the technologies required to reduce tractor-trailer fuel consumption, in practice many technologies may be left behind, particularly those that are not yet very mature. The Partnership should carefully review the technologies that have been identified and determine whether any technologies to reduce vehicle power demand are not being adequately addressed by the SuperTruck program. The DOE should define projects and find funding to support the development of technologies beyond the scope of SuperTruck.

ENGINE IDLE REDUCTION

Finding 6-1. The DOE, EPA, and DOT have funded a wide variety of idle reduction projects focused on implementation. A consolidated list of the funding provided for these projects was not provided to the committee, however, and the effectiveness of these projects could not be evaluated. The national patchwork of anti-idling regulations is an impediment to broader use of anti-idling measures.

Recommendation 6-1. The DOE, EPA, and DOT should develop a consolidated list of the funding provided for the idle reduction projects, review the effectiveness of these projects, and formulate a coordinated and consistent plan to encourage the adoption of idle reduction technologies to meet the goal of reducing fuel use and emissions produced by idling engines by at least two-thirds by 2017. The EPA and DOT should work to find incentives for states to promulgate uniform anti-idling regulations.

Finding 6-3. The Delphi solid oxide fuel cell (SOFC) auxiliary power unit (APU) provides several advantages over diesel APUs, but it has significant issues in its current development status, including the following: low efficiency of 25 percent versus the DOE's goal of 35 percent, and low demonstrated output power of 1.5 kW versus 3.0 kW believed sufficient by Delphi and 5 kW of typical diesel APUs; limited demonstrated durability; 2- to 5-hour warm-up time to the 750°C operating temperature; and high cost. The 10-year funding for this program expires in 2011.

Recommendation 6-3. The DOE should reassess the viability of the SOFC APU, particularly for application to the SuperTruck program, considering the following: (1) SOFC APU is still in the laboratory, (2) the low efficiency of 25 percent versus the DOE goal of 35 percent, (3) the low 1.5 kW output compared to the typical 5 kW diesel APUs, (4) the disadvantages associated with the requirement for continuous operation at 750°C, and (5) the expiration of funding from the DOE Office of Fossil Energy and EERE Fuel Cell Technologies Program of the DOE Office of Energy Efficiency and Renewable Energy after 10 years of development. The DOE should coordinate more closely with DOD in its fuel cell APU developments to ensure that the best technology is being pursued for the 21CTP's Goal 7 in the engine idle reduction focus area; that goal relates to the development and demonstration of viable fuel cell APU systems for military and other users (see Chapter 6 for the full text of Goal 7). (This recommendation is a follow-on to Recommendation 6-8 in the NRC Phase 1 report.)

Finding 6-4. Idle reduction technologies could provide 6 percent reduction in overall fuel consumption for Class 8 long-haul trucks with sleeper cabs, which is nearly 30 percent of the 20 percent reduction in the fuel consumption required to meet the proposed EPA/NHTSA 2017 fuel efficiency standards.

Recommendation 6-4. The 21CTP should review and potentially revise its idle reduction plans and goals in view of the fact that the proposed 2017 fuel efficiency standards provide an incentive for the adoption of idle reduction technologies as a means for achieving these standards for Class 8 long-haul trucks with sleeper cabs.

SAFETY

Finding 7-3. The DOT has met its heavy-truck safety goals for the past 4 years. However, the committee observes that the Transportation Research Board's (TRB's) 2010 study

Achieving Traffic Safety Goals in the United States: Lessons from Other Nations has shown that other nations have established more aggressive initiatives and goals with impressive results, and those results suggest that even greater improvement in highway safety is possible in the United States. The committee also notes that overall improvements in highway safety also yield improvements in heavy-duty truck safety, as most heavy-duty truck fatal accidents involve a light-duty vehicle.

Recommendation 7-3. The DOT should evaluate the conclusions and recommendations of the TRB study Achieving Traffic Safety Goals in the United States: Lessons from Other Nations of highway safety in other nations, and consider the possibility of establishing more aggressive initiatives and goals for highway safety in general. The DOT should also consider establishing more aggressive goals for heavy-duty truck safety.

SUPERTRUCK PROGRAM

Three projects have been selected for awards under the DOE's SuperTruck program; they will focus on measures to improve the fuel efficiency of Class 8 long-haul freight trucks. These projects will receive $115 million in DOE funding to develop and demonstrate full vehicle system-level technologies by 2015. Two of the project teams (Cummins, Inc. and Daimler Trucks North America, LLC) received ARRA funding for their projects, and Navistar, Inc. will be funded from DOE appropriations:

- Cummins, Inc. (Columbus, Indiana): Develop and demonstrate a highly efficient and clean diesel engine, an advanced waste heat recovery system, an aerodynamic Peterbilt tractor and trailer combination, and a solid oxide fuel cell auxiliary power unit to reduce engine idling.
- Daimler Trucks North America, LLC (Portland, Oregon): Develop and demonstrate technologies including optimized combustion, engine downsizing, electrification of auxiliary systems such as oil and water pumps, waste heat recovery, improved aerodynamics, hybridization, and possibly a fuel cell auxiliary power unit to reduce engine idling.
- Navistar, Inc. (Warrenville, Illinois): Develop and demonstrate technologies to improve truck and trailer aerodynamics, combustion efficiency, waste heat recovery, hybridization, idle reduction, and reduced rolling resistance tires.

The objective of the three SuperTruck projects is to develop and apply technologies leading to a system-level demonstration of highly efficient and clean diesel-powered Class 8 trucks that:

- Achieve a 50 percent increase in vehicle freight efficiency measured in ton-miles per gallon, which translates to a 33 percent reduction in load-specific fuel consumption (gallons per 1,000 ton-miles).
- Achieve at least a 20 percent improvement through engine thermal efficiency development, and achieve 50 percent BTE under highway cruise conditions.
- Evaluate potential approaches to 55 percent BTE in an engine via modeling, analysis, and potentially also laboratory tests

Finding 8-1. The three SuperTruck projects will be the flagship projects under the 21CTP for FY 2011 through FY 2014; the goals are in concert with recommendations made in the 2008 NRC Phase 1 report. A large portion of the DOE 21CTP budget will be devoted to these three projects. Each SuperTruck project integrates a wide range of technologies into a single demonstration vehicle (engine, waste heat recovery, driveline, rolling resistance, tractor and trailer aerodynamics, idle reduction, weight reduction technologies, etc.), and the contractors are pursuing sufficiently different technical paths to avoid excessive duplication of effort. The results will help determine which fuel-saving technologies are ready and cost-effective for original equipment manufacturer (OEM)-level product development programs.

Finding 8-4. The committee believes that the SuperTruck project teams have developed plans that address the needs of the SuperTruck program and that have a reasonable chance for success. The keys to success include proper implementation of the plans along with the flexibility to adapt to new information and intermediate results during the course of the project.

Finding 8-5. The SuperTruck projects allow each team to design its own test duty cycle(s) within certain constraints. One negative consequence of this approach is that the three trucks may never be tested using a common cycle for comparison.

Finding 8-6. The SuperTruck projects go beyond the scope of previous 21CTP projects. Instead of relying entirely on simulations and laboratory testing, each of these projects will result in a drivable truck. The committee believes that it is important to take technologies that have been developed to date and implement them in a real vehicle. Often, the application of new technologies in real-world applications yields unexpected results, and these results must be explored before any new technology can be considered ready for production implementation.

Recommendation 8-2. The DOE and the SuperTruck contractors should agree on at least one common vehicle duty cycle that will be used to compare the performance of all three SuperTruck vehicles. In addition, fuel consumption improvements should be calculated on the basis of the EPA/NHTSA fuel consumption regulations.

EFFICIENT OPERATIONS

The 21CTP recently proposed "efficient operations" as a new area for work under the 21CTP. The proposal is laid out in a draft white paper titled "Reducing Fuel Consumption in U.S. Trucking—A DOE-DOT Joint Study and Whitepaper" (DOE-DOT, 2011). In this draft, the two agencies explore opportunities to improve the efficiency of trucking operations, focusing on two areas of opportunity: (1) joint R&D efforts between DOE and DOT and (2) modifications of regulations (primarily DOT regulations).

Besides the many technologies available for reducing the fuel consumption of trucks, there are other ways of saving fuel that do not require any changes to vehicle or engine technologies—involving, for example, the ways that vehicles are operated and maintained, or the nature of regulations that may constrain or promote technology implementation and efficient operations. Infrastructure is also important because it can affect fuel consumption through factors such as vehicle speed fluctuation and congestion. Electronic features can be added to a truck that modify the performance of the engine or vehicle in ways that can save fuel.

The committee identified a number of areas and developed a number of findings and recommendations on the following topics for the 21CTP to consider in its white paper in formulating goals in order to reduce fuel consumption. The topics are as follows:

- Improved aerodynamic and rolling resistance performance for trailers,
- Exploitation of the use of intelligent vehicle systems, and
- Assessment of the potential impact of high-productivity vehicles and providing of leadership in getting them into trucking operations.

The following major findings and a recommendation are the result of the committee's review of the draft DOE-DOT (2011) white paper on efficient operations. They describe what the committee believes should be added to or changed in the white paper to help the 21CTP promote and enable more efficient trucking operations. (See Chapter 9 for other findings and recommendations.)

Finding 9-1. The DOE-DOT draft white paper proposes "efficient operations" as a new direction for the 21CTP. The committee agrees that this is an important area for R&D under the umbrella of the 21CTP. It also agrees that cooperation among the DOE and DOT and other agencies would be beneficial, particularly for assessing the possible effects of removing regulatory barriers to the use of fuel-saving measures.

Finding 9-10. The DOE-DOT draft white paper on efficient operations in its current form does not include any goals that could be used to prioritize and drive R&D efforts on efficient operations.

Recommendation 9-7. Specific goals for efficient operations should be developed, with strong consideration given to exploiting the potential for intelligent transportation systems (ITS) to reduce fuel consumption. In addition, priorities should be set for the R&D, testing, and data collection needed to analyze the benefits, drawbacks, and potential unintended consequences of removing barriers, including regulatory barriers, to the application of fuel-saving features. The draft white paper on efficient operations should be rewritten to take the findings and recommendations of the committee into account. The 21CTP partners, trucking fleets, and major suppliers should be involved in setting goals and research priorities

REFERENCES

DOE (U.S. Department of Energy). 2006. 21st Century Truck Partnership Roadmap and Technical White Papers. Doc. No. 21CTP-003. December. Washington, D.C.: Office of FreedomCAR and Vehicle Technologies.

DOE. 2010a. Updated 21st Century Truck Partnership Roadmap and Technical White Papers. Doc. No. 21CTP-003. September 1. Washington, D.C.: Office of FreedomCAR and Vehicle Technologies.

DOE. 2010b. Multi-Year Program Plan, 2011-2015. Vehicle Technologies Program. December. Washington, D.C.

DOE. 2011. Updated 21st Century Truck Partnership Technical White Papers. Working draft. February 25, 2011. Washington, D.C.: Office of Vehicle Technologies.

DOE-DOT (DOE-U.S. Department of Transportation). 2011. Reducing Fuel Consumption in U.S. Trucking—A DOE-DOT Joint Study and Whitepaper. 21st Century Truck Partnership Efficient Operations White Paper Draft. February 15. Washington, D.C.

NRC (National Research Council). 2008. Review of the 21st Century Truck Partnership. Washington, D.C.: The National Academies Press.

1

Introduction and Background

INTRODUCTION

In July 2010, the National Research Council (NRC) appointed the Committee to Review the 21st Century Truck Partnership, Phase 2, to conduct an independent review of the 21st Century Truck Partnership (21CTP). The results of the review are presented in this report. This Phase 2 review follows on the original review of the Partnership by the NRC, conducted in 2007 and resulting in what is referred to in this report as the NRC Phase 1 report, issued in 2008 (NRC, 2008).[1] The Partnership's responses to the recommendations in the Phase 1 report are contained in Appendix C of the present report.

The 21CTP is a cooperative research and development (R&D) partnership including four federal agencies (the U.S. Department of Energy [DOE], the U.S. Department of Transportation [DOT], the U.S. Department of Defense [DOD], and the U.S. Environmental Protection Agency [EPA]), and 15 industrial partners (Allison Transmission, ArvinMeritor, BAE Systems, Caterpillar, Cummins Inc., Daimler Trucks North America [which includes Freightliner], Detroit Diesel Corporation [DDC], Eaton Corporation, Honeywell International, Navistar, Mack Trucks, NovaBUS, Oshkosh Truck, PACCAR, and Volvo Trucks North America). The Partnership was formed in 2000 and announced on April 21, 2001, in a press event in Romulus, Michigan.[2]

The Partnership is a means of coordinating ongoing activities at the various agencies and private-sector companies to contribute to national goals that are, in the broadest terms, to "reduce fuel usage and emissions while increasing heavy vehicle safety" (DOE, 2010a). The 21CTP vision is "that our nation's trucks and buses will safely and cost-effectively move larger volumes of freight and greater numbers of passengers while emitting little or no pollution and dramatically reducing the dependency on foreign oil" (DOE, 2010b).

> The Partnership addresses the following "national imperatives":
>
> (a) Transportation in America supports the growth of our nation's economy both nationally and globally. (b) Our nation's transportation system supports the country's goal of energy security. (c) Transportation in our country is clean, safe, secure, and sustainable. (d) America's military has an agile, well-equipped, efficient force capable of rapid deployment and sustainment anywhere in the world. (e) Our nation's transportation system is compatible with a dedicated concern for the environment (DOE, 2010b).

This report builds on the NRC Phase 1 review and report and also, as part of its charge, comments on changes and progress that have occurred since the Phase 1 report was issued in 2008. The strategic approach of the Partnership includes the following elements as laid out in the 2006 21CTP roadmap (DOE, 2006, 2010b):

- Develop and implement an integrated vehicle systems research and development approach that validates and deploys advanced technology necessary for both commercial and military trucks and buses to meet the aforementioned national imperatives.
- Promote research for engines, powertrains, combustion, exhaust aftertreatment, fuels, and advanced materials to achieve both significantly higher efficiency and lower emissions.
- Promote research focused on advanced heavy-duty hybrid propulsion systems that will reduce energy consumption and pollutant emissions.
- Promote research to reduce vehicle power demands (also

[1] The chair, John H. Johnson, of the NRC committee for the Phase 1 review testified before the Energy and Environmental Subcommittee of the U.S. House of Representatives, on March 9, 2009. This subcommittee of the Committee on Science and Technology developed H.R. 3246, the Advanced Vehicle Technologies Act of 2009, which passed the House on September 11, 2009. Senator Debbie Stabenow introduced a companion bill in the Senate on December 7, 2009, but the bill was not passed by the Senate.

[2] For further details of the history, see DOE (2006) and NRC (2000, 2008).

referred to as parasitic losses) to achieve significantly reduced energy consumption.

- Promote the development of technologies to improve truck safety, resulting in the reduction of fatalities and injuries in truck-involved crashes.
- Promote the development and deployment of technologies that substantially reduce energy consumption and exhaust emissions during idling.
- Promote the validation, demonstration, and deployment of advanced truck and bus technologies, and grow their reliability sufficient for adoption in the commercial marketplace.

As is discussed in more detail in this report, the Partnership has been evolving and making some changes since the Phase 1 review. For example, the 2006 roadmap has been revised and updated, and a series of technical white papers that supported the 2006 roadmap have also undergone revisions (DOE, 2010c, 2011). The committee reviewed these updated documents as part of the Phase 2 review.

NATIONAL CONCERNS

The federal government, including the DOE, has addressed in varying degrees the economic, energy security, and environmental aspects of energy supply, distribution, and use for many decades, and the focus of efforts has changed from time to time. In recent years all three areas have had increasing attention by the administration and the Congress, given the rapid rise in energy prices in the 2007-2008 period, the severe recession of the past few years, the involvement in wars in the Middle East and the importance of that region for global oil supplies, and the attention to the environmental issue of global climate change. In addition, because of concerns about air quality and human health, a number of regulations have been passed over the years leading to more stringent exhaust emissions standards for both light-duty vehicles (cars, vans, and light trucks) and medium- and heavy-duty vehicles (MHDVs).

The economic concerns related to energy supply and energy use are generally framed in the language of affordability for the individual consumer as well as the impact on the U.S. economy from high energy prices and/or shortages. In recent years, not only have high energy prices been experienced but also there seems to be increased volatility in energy prices. Although the recent global and U.S. economic slowdowns depressed global as well as U.S. oil demand, worldwide oil consumption in general has risen rapidly during the past decade, mainly owing to rapid economic growth around the world. Nevertheless, even though the recent recession has moderated U.S. demand for imported oil, the Energy Information Administration (EIA, 2010) forecasts that the nation will continue to be highly dependent on imported oil. If global oil prices rise rapidly again because of supply-and-demand imbalances, future prices of oil will likely continue to put a strain on the U.S. economy. BP's recent *Statistical Review of World Energy* also shows an increase in world oil consumption of 3.1 percent from 2009 to 2010, reaching a level of 87.4 million barrels per day (bbl/day), and an increase in U.S. oil consumption of 2 percent, reaching 19.1 million bbl/day (BP, 2011). As a consequence, the United States is pursuing alternative sources of fuel and attempting to increase efficiency in oil usage.

The issue of energy security with regard to petroleum not only entails the economic concerns noted above but also is framed in terms of the U.S. dependence on imported petroleum. Oil use in the United States has varied during the past few years, but it has been around 20 million bbl/day and was 18.8 million bbl/day in 2009 (EIA, 2010). Most of this petroleum is used in the transportation sector, and about 25 percent of that is used for MHDVs. Regarding gasoline consumption, EIA (2010) projects that total transportation fuel use will grow between 2009 and 2035 but that total U.S. gasoline consumption will remain at about 9 million bbl/day from 2009 to 2035: these projections include the phasing in of new fuel economy regulations for light-duty vehicles by 2016 as discussed in the next section (EIA, 2010; Newell, 2010; see Figure 1-1). Total U.S. diesel fuel consumption, much of which is consumed by MHDVs, is projected to change from about 3.42 million bbl/day in 2009 to almost 4.5 million bbl/day in 2035. Fuel consumption by heavy-duty vehicles is projected to increase substantially in the United States as well as worldwide, and consumption by heavy-duty vehicles (Classes 6, 7, and 8; see the section below on "Classes and Use Categories of Trucks and Buses") consumption is expected to increase between 2010 and 2035 by 40 percent.[3,4] Thus, in round numbers, assuming an oil price of $100/bbl, expenditures for diesel fuel alone would be on the order of $125 billion per year in the United States.

The 21CTP is focused on reducing the fuel usage of heavy-duty vehicles, which consume about 25 percent of the petroleum currently used in the transportation sector, and the expected 40 percent increase in consumption by heavy-duty vehicles between 2010 and 2035. That usage is in contrast to light-duty vehicle consumption, which is expected to remain relatively unchanged. EIA (2010), in its liquid fuels projections, includes increasing the use of biofuels, which somewhat dampens the demand for petroleum-based fuels. EIA (2010) projects net petroleum imports to change from about 8.97 million bbl/day in 2009 to 8.52 million bbl/day in 2035, whereas total liquid fuels use (including the use of

[3] P. Davis, DOE, "U.S. Department of Energy Vehicle Technologies Program Overview," Presentation to the committee, September 8, 2010, Washington, D.C.

[4] No U.S. fuel economy standards for medium- and heavy-duty vehicles are in effect for the current model year. A Notice of Proposed Rule Making was issued October 26, 2010. Final standards issued by the U.S. Environmental Protection Agency (EPA) and the U.S. Department of Transportation National Highway Traffic Safety Administration (NHTSA) on September 15, 2011, will apply to model year 2014 (EPA/NHTSA, 2011). EIA's *Annual Energy Outlook 2010* projections do not include such standards.

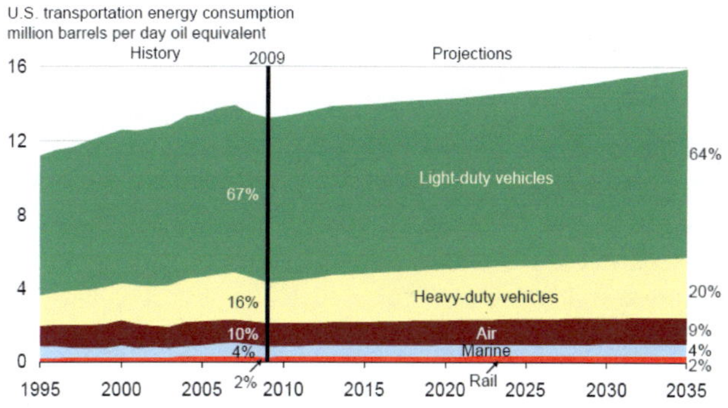

FIGURE 1-1 Trends in actual and projected U.S. transportation fuel use, 1995-2035. SOURCE: EIA (2010); Newell (2010).

biofuels) for transportation goes from about 13.5 million bbl/day in 2009 to 16.38 million bbl/day in 2035. Although the dependence on petroleum imports in these forecasts is somewhat ameliorated compared to recent trends, significant import dependence remains.

Added to the concern over high-priced oil and energy security is the concern regarding global warming. Nations around the world are beginning to exert more stringent control over human-made emissions, especially greenhouse gases (GHGs) such as carbon dioxide (CO_2). Numerous discussions have taken place in Congress about climate change, and many pieces of climate change legislation have been proposed. The consumption of petroleum in the U.S. transportation sector accounted for about 1.81 billion metric tons of CO_2-equivalent (CO_2-eq) emissions in 2009, about 33 percent of total U.S. CO_2-eq emissions from the burning of fuels. EIA (2010) forecasts that CO_2-eq emissions from petroleum in the transportation sector in 2035 will be about 2.065 billion metric tons, about 33 percent of the projected U.S. total emissions of 6.32 billion metric tons of CO_2-eq emissions from the burning of fuels. The EPA estimates that medium- and heavy-duty trucks and buses accounted for about 22 percent of CO_2-eq emissions from the transportation sector in 2008 (EPA, 2010). Consequently, the transportation sector and the significant portion of that sector that is composed of MHDVs are important to any policies that are aimed at reducing greenhouse gas emissions. Thus, for the foreseeable future, there will be pressure to control and reduce greenhouse gas emissions. One approach to addressing such emissions is to use fuel more efficiently; another is to use fuels that emit less CO_2 than petroleum-based fuels do.

Both the limited availability of oil and the additional pressures to reduce CO_2 emissions will have a profound impact on automotive vehicles worldwide. In addition, as briefly discussed in the next section, the United States is implementing policies on fuel economy that will directly impact the vehicle sector. These forces will pressure vehicle manufacturers to make renewed efforts to reduce both fuel consumption and exhaust emissions. Light-duty vehicle manufacturers have already made significant improvements in reducing fuel consumption and even more progress in reducing vehicle emissions. Emissions of oxides of nitrogen (NO_x) and particulate matter (PM) from heavy-duty vehicles have been significantly reduced by PM regulations that went into effect in 2007 and NO_x regulations that were phased in between 2007 and 2010. However, reductions in fuel consumption of the large commercial truck fleet have not been as impressive, partly because of the growth in the number of miles driven by large trucks during the past decade (NRC, 2008; NAS-NAE-NRC, 2009a).

Thus, with regard to economic considerations, energy security, and environmental reasons, the transportation sector is a key area for consideration and policy focus, and the medium- and heavy-duty vehicle component significant. The 21CTP can play an important role is this regard.

RECENT POLICY INITIATIVES

The economic and policy environment outside the DOE continues to change, and various external initiatives and policies can affect the DOE and specifically the Partnership: such developments that may affect the Partnership have continued to emerge since the publication of the NRC Phase 1 report (NRC, 2008). The control of emissions from the engines of heavy-duty trucks with gross vehicle weight (GVW) over 8,500 pounds (lb) began in 1973 in California and in 1974 in the United States as a whole. The federal standards were harmonized with California standards beginning with model year (MY) 2004, and stringent emissions standards for heavy-duty diesel engines came into effect, as noted in the previous section, in the 2007-2010 time period. These increasingly stringent engine emissions standards were an important driver for R&D for engine, emissions control, and fuels.[5] Reaching low emissions of hydrocarbons (HCs), carbon monoxide (CO), NO_x, and PM in terms of grams per brake horsepower-hour (g/bhp-h) was a significant challenge if fuel consumption was constrained to not increase. The changes in U.S. emission standards over time are presented in Figure 1-2 (DOE, 2010c, d).

With the increasing concern in Congress and the administration about energy security and greenhouse gas emissions, numerous actions are taking place that will create incentives for technology development for vehicles as well as improve the operational efficiency of managing the movement of freight in the United States (e.g., through driver education, longer combination vehicles, reducing congestion, etc.). Although the National Highway Traffic Safety Administration (NHTSA) has regulatory authority over fuel economy standards for vehicles, the EPA has the authority to regulate CO_2 and other greenhouse gases as air pollutants under the Clean Air Act, thus empowering the EPA to regulate vehicle

[5] A summary review of these emissions standards and changes can be found in the NRC Phase 1 report, Chapter 1 (NRC, 2008), as well as in references in this chapter (Ehlmann and Wolff, 2005; Johnson, 1988).

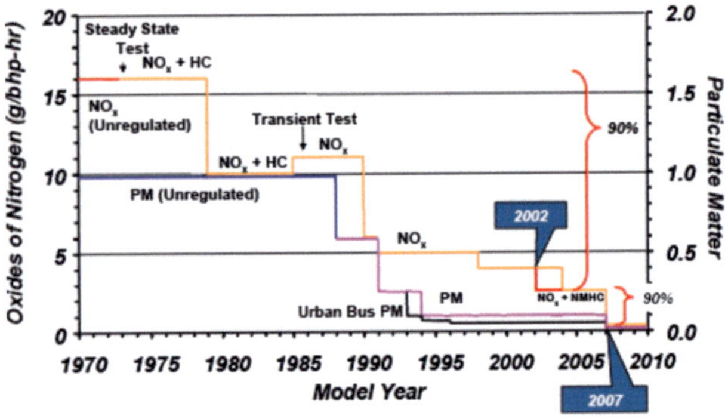

FIGURE 1-2 Historical trends in emission standards for new diesel engines, 1970-2010. SOURCE: DOE (2010c,d).

CO_2 emissions, which are directly related to vehicle fuel consumption measures. Following the announcement[6] in May 2009 of the President's National Fuel Efficiency Policy, the NHTSA and EPA have promulgated nationally coordinated standards for tailpipe CO_2-eq emissions and fuel economy for light-duty vehicles (LDVs)—which include both passenger cars and light-duty trucks.

The initial harmonized standards will affect MY 2012 LDVs, and compliance requirements will increase in stringency through MY 2016, building on the NHTSA's enacted corporate average fuel economy (CAFE) standard for MY 2011. The NHTSA has estimated the impact of the new CAFE standards and has projected that the proposed fleetwide standards for LDVs will increase fuel economy from 27.3 miles per gallon (mpg) in MY 2011 to 34.1 mpg in MY 2016, an average annual increase of 4.3 percent.[7] The NHTSA and EPA have also issued a Notice of Intent and Technical Assessment Report for fuel economy and GHG regulations for LDVs for the 2017-2025 time period to raise the fuel economy levels beyond these.[8]

With regard to MHDVs, the Energy Independence and Security Act (EISA) of 2007 (Public Law No. 110-140) directed the National Academy of Sciences (NAS) to conduct a study on the potential for technologies to reduce fuel consumption for such vehicles, and it directed the NHTSA to promulgate, for the first time ever, fuel economy standards for such vehicles. The NAS completed this task and submitted its report, *Technologies and Approaches to Reducing the Fuel Consumption of Medium- and Heavy-Duty Vehicles*, to the NHTSA and Congress in March 2010 (NRC, 2010a). In November 2010, the NHTSA and EPA proposed to regulate many of these vehicles, issuing a Notice of Proposed Rulemaking, namely, Greenhouse Gas Emissions Standards and Fuel Efficiency Standards for Medium- and Heavy-Duty Engines and Vehicles; Proposed Rule (40 CFR Parts 85, 86, 1036, et al.; and 49 CFR Parts 523, 534, and 535)(EPA/NHTSA, 2010).[9] Final standards issued by the EPA and NHTSA on September 15, 2011, will begin with model year 2014 (EPA/NHTSA, 2011). (See below in this chapter for additional details.)

The NRC (2010a) report and the NHTSA's and EPA's work have created a rich data and analysis base to support the DOE's efforts as well as providing background for the current review of the Partnership. In turn, the efforts of the DOE and the Partnership will be important in promoting technology development and helping to realize more efficient vehicles. Again, as with LDVs, these standards and attention to reducing fuel consumption and greenhouse gas emissions will stimulate R&D on advanced technologies for reducing fuel consumption for MHDVs.

Since the 1970s, Congress has supported legislation that requires increasing the production of fuels from renewable, bio-based sources and other alternative fuels as part of efforts to reduce petroleum-based fuel consumption. The EISA of 2007 includes a subtitle that amended the Renewable Fuels Standard contained in the Energy Policy Act of 2005 (EPAct 2005) and increased the volumes of renewable fuels to be phased in to the fuel supply substantially. The mandated volumes of renewable fuels to be used begin with 9 billion gallons in 2008 and reach 36 billion gallons in 2022. These fuels are anticipated to include corn-based ethanol, cellulosic-based ethanol, and biodiesel made from vegetable oils (e.g., from soybeans), animal fats, and cellulose. Much R&D is occurring to develop, demonstrate, and commercialize the advanced biofuels that would be made from cellulose, but costs and technology performance are still uncertain (NAS-NAE-NRC, 2009b).

Thus, the national landscape has shifted strongly toward addressing the nation's dependence on petroleum imports as well as emissions of greenhouse gases. The role of the public sector—through advanced R&D, and especially in partnering with the private sector, where the ultimate decisions will be made to deploy and commercialize new technology—is an important complement to these regulatory, market-pull requirements. In this vein, the Partnership's role in fostering technology that can reduce fuel consumption by medium- and heavy-duty vehicles has gained in importance since the NRC Phase 1 review.

[6] See President Obama's National Fuel Efficiency Policy at http://www.whitehouse.gov/the-press-office/president-obama-directs-administration-create-first-ever-national-efficiency-and-em.

[7] Note that this standard addresses CO_2-eq emissions so that manufacturers have some other means of receiving credits for reducing these emissions such as reducing hydrofluorocarbons in air-conditioning systems or using alternative fuels. If manufacturers only rely on reductions in vehicle fuel consumption, NHTSA estimates that manufacturers will have to comply with a 35.5 mpg standard by 2016. This is about a 40 percent increase in fuel economy over current (2010) standards.

[8] See http://www.nhtsa.gov/About+NHTSA/Press+Releases/2010/DOT+and+EPA+Announce+Next+Steps+toward+Tighter+Tailpipe+and+Fuel+Economy+Standards+for+Passenger+Cars+and+Trucks. Accessed October 15, 2010.

[9] See http://www.nhtsa.gov/staticfiles/rulemaking/pdf/cafe/CAFE_2014-18_Trucks_FactSheet-v1.pdf. Accessed December 15, 2010.

OVERVIEW OF THE 21ST CENTURY TRUCK PARTNERSHIP: AREAS OF INTEREST AND ORGANIZATION

As a means of providing focus and a set of goals and objectives for the Partnership as a whole, the Partnership developed a roadmap and supporting technical white papers (DOE, 2006). The roadmap with supporting background was reviewed during the NRC Phase 1 review and included a strategic vision, as well as a discussion of benefits of the program and the description of five main technical areas that the Partnership focused on, namely: (1) engine systems, (2) hybrid propulsion, (3) parasitic losses, (4) idle reduction, and (5) safety (DOE, 2006). The 21CTP established goals for each of these areas, and a fuller description of the five areas was included in the NRC Phase 1 report as well as an evaluation of progress toward the goals (NRC, 2008).

Since the NRC Phase 1 review, the Partnership has revised the roadmap and supporting white papers and goals (DOE, 2010c, 2011). The technical areas covered by the white papers are substantially the same, but there are some changes. The technical areas include (1) engine systems; (2) hybrid propulsion; (3) vehicle power demands, previously called parasitic losses, which included, for example, aerodynamics, tire rolling resistance, and other areas; (4) idle reduction; (5) safety; and (6) efficient operations, which is a new area that addresses the system of trucks and their operation for the efficient delivery of goods. These areas and the associated goals are discussed in further detail in the remaining chapters of this report. In addition, three major cost-shared contracts were awarded in the April through September 2010 time frame to carry out R&D and demonstrate for a complete tractor-trailer a freight efficiency improvement of 50 percent in ton-miles per gallon of fuel.

Activities of the Partners

The DOE's Office of Vehicle Technologies, which is within its Office of Energy Efficiency and Renewable Energy (EERE), has the primary role in the department for pursuing the development of advanced vehicle technologies both for LDVs and MHDVs. The LDV activities are included in what was the FreedomCAR and Fuel Partnership (FCFP), which is now the U.S. DRIVE (Driving, Research, and Innovation for Vehicle efficiency and Energy sustainability) program; the medium- and heavy-duty activities are included in the 21CTP. The FCFP and U.S. DRIVE programs include work on combustion and emissions control, fuel cells, hydrogen storage, batteries, lightweight materials, and power electronics. In terms of budgets, the LDV program activities have been much larger than the 21CTP, and in fact during the past decade the emphasis has increased on LDVs and declined on MHDV activities. For example, in FY 2010, the budget of the Office of Vehicle Technologies was about \$311 million, of which about \$38.5 million was devoted to MHDV applications.[10] There is some overlap between work that is done for LDVs and MHDVs—for example, in such areas as the understanding and modeling of advances in combustion, advances in lightweight materials, or advances in electrochemistry and battery technologies—and such overlapping areas are all managed under the Office of Vehicle Technologies to support both LDV and MHDV technologies, as appropriate. Consequently, advances made in technical areas that are characterized and budgeted as part of the U.S. DRIVE could benefit MHDVs.

The DOE also contracts work out to the private sector and involves the 21CTP industry partners. As noted in the previous section and discussed later in the report, for example, the DOE has awarded contracts to industry to address many aspects of reducing the fuel consumption of long-haul trucks. The EPA also has development work on hydraulic hybrid technologies for some classes of trucks and also funds work on combustion. It also works with the private sector and promotes and provides information on various technologies for the reduction of fuel consumption and of greenhouse gas emissions through its SmartWay program. The DOD also is very interested in improving the fuel efficiency and reducing the fuel consumption of its noncombat vehicles; for combat vehicles it is interested in increased power density and low heat rejection. The DOT is focused on safety issues, including the use of advanced technology and regulations that can improve highway safety, as well as the overall system and infrastructure for moving freight efficiently and economically.

Lines of Authority

The Partnership was originally under the leadership of the DOD (the U.S. Army Tank-Automotive Research and Development Command). In November 2002, that authority passed to the DOE (DOE, FCVT, 2006, pp. 4-7), specifically to the Office of FreedomCAR and Vehicle Technologies (which is now called the Office of Vehicle Technologies).

The other agencies associate their own existing programs that are relevant to the goals of the 21CTP under the 21CTP umbrella, so the DOE has little influence over the research programs of its DOT, DOD, and EPA partners. The other factor that makes budgets and projects involved in the 21CTP unclear is that the different agencies receive their budget appropriations from different committees in Congress. Thus, there is no central, overall management over 21CTP budgets and accountability. DOE staff organize meetings and conference calls, maintain the information-flow infrastructure (such as websites and e-mail lists), and have led the discussions for and preparation of the updated 21CTP Roadmap and Technical White Papers (DOE, 2010c) laying out Partner-

[10] P. Davis, DOE, "U.S. Department of Energy Vehicle Technologies Program Overview," presentation to the committee, September 8, 2010, Washington, D.C.

ship goals. The management of individual projects under the 21CTP umbrella rests with the individual federal agencies that have funded the work. These agencies use the 21CTP information-sharing infrastructure to coordinate efforts and to ensure that valuable research results are communicated and that any overlap of activities among their respective efforts is reduced.

According to the official Roadmap and Technical White Papers of the 21CTP (DOE, 2006, p. 6):

> DOE has been assigned to lead the federal R&D component of this program because of the close alignment of the stated 21st Century Truck Program goals and research objectives with DOE's mission "to foster a secure and reliable energy system that is environmentally and economically sustainable…." Since early 1996, DOE's FreedomCAR and Vehicle Technologies Program (and predecessor offices), in collaboration with trucking industry partners and their suppliers, has been funding and conducting a customer-focused program to research and develop technologies that will enable trucks, buses, and other heavy vehicles to be more energy-efficient and able to use alternative fuels while simultaneously reducing emissions. DOT brings to this program its mission-oriented intelligent transportation systems and highway transportation safety programs. DOD, as a major owner and operator of trucks, will define the military mission performance requirements and will fund appropriate dual-use and military-specific technologies so that national security will benefit by innovations resulting from this Program. R&D will be closely coordinated with EPA so that critical vehicle emissions control breakthroughs cost-effectively address the increasingly stringent future EPA standards needed to improve the nation's air quality.

Classes and Use Categories of Trucks and Buses

Industry classifies trucks and buses by weight, based on the vehicle's Gross Vehicle Weight Rating (GVWR), or on the maximum in-service weight set by the manufacturer, or—in the trucking industry—on the gross vehicle weight (GVW) plus the average cargo weight. The use categories of vehicles are not as well defined as the weight classes are; they depend on widely varying industry usage. For example, the same vehicle may be called heavy-duty by one segment of the industry and medium-duty by another.

Figure 1-3 gives an idea of the variety of medium- and heavy-duty vehicles to which developments in the 21CTP could be applied. It is based on the DOT classification system using a truck's GVWR. The DOE Transportation Energy Data Book (Davis et al., 2009, Table 5.7) developed information on vehicle weight classification, as did the NRC (2010a) report on medium- and heavy-duty vehicles. In general:

- Class 1 and 2 vehicles lighter than 10,000 lb are considered light trucks, such as pickups, small vans, and sport utility vehicles. They are generally spark-ignited, gasoline-fueled internal combustion engines, and more than 80 percent are for personal use. This class of vehicle up to about 8,500 lb comes under CAFE requirements for cars. Class 2 trucks with GVWR above 8,500 lb are similar to Class 3 trucks. Class 2B trucks (8,500 to 10,000 lb GVWR) include pickup trucks, sport utility vehicles, and large vans.
- Classes 3 through 6 are medium- and heavy-duty vehicles with single rear axles and use either gasoline-

Light-Duty		*Medium Heavy-Duty*				*Heavy-Duty*	
Class 1	Class 2	Class 3	Class 4	Class 5	Class 6	Class 7	Class 8
Less than 6,000 lb	6,000 to 10,000 lb	10,000 to 14,000 lb	14,000 to 16,000 lb	16,000 to 19,500 lb	19,500 to 26,000 lb	26,000 to 33,000 lb	Greater than 33,000 lb

FIGURE 1-3 Illustrations of typical vehicles in the various weight classes.

TABLE 1-1 Summary of Annual Miles and Fuel Use for Different Classes of Medium- and Heavy-Duty Vehicles in the United States Based on 2002 Survey Data

U.S. Medium- and Heavy-Duty Vehicle Population, Mileage and Fuel Use by Weight Class						
Vehicle Size	Population (millions)	Annual Miles (million miles)	Annual Fuel Use (million gallons)	% of Population	% of Annual Miles	% of Fuel Use
Class 2B	5.800	76,700	5,500	52.8	35.1	19.3
Class 3	0.691	9,744	928	6.3	4.5	3.3
Class 4	0.291	4,493	529	2.6	2.1	1.9
Class 5	0.166	1,939	245	1.5	0.9	0.9
Class 6	1.710	21,662	3,095	15.6	9.9	10.0
Class 7	0.180	5,521	863	1.6	2.5	3.0
Class 8	2.154	98,522	17,284	19.6	45.1	60.8
Total	**10.992**	**218,580**	**28,444**	**100**	**100**	**100**

SOURCE: Data for classes 3-8 from U.S. Department of Commerce, Bureau of the Census, 2002, Vehicle Inventory and Use Survey, 2002. Data for Class 2B from Davis, S.C. and L.F. Truett, Investigation of Class 2b Trucks (Vehicles of 8,500 to 10,000 lbs GVWR), ORNL/TM-2002/49, March 2002, Table 16. Classes 3-8, 2002 population; Class 2b, 2000 population. Totals are approximate due to rounding errors.

or diesel-fueled engines; their GVWs are from 10,000 to 26,000 lb.

- Classes 7 and 8 are heavy-duty vehicles, using primarily diesel engines.
- Class 8 combination trucks have a tractor and one or more trailers and a gross combined weight (GCW) of up to 80,000 lb, with higher weights allowed in specific circumstances.

Some vehicle classifications used by the EPA and the California Air Resources Board (CARB) for emissions regulations differ from those of the DOT and Figure 1-3. There is great variety among MHDVs. Examples of Class 7-8 vehicles include box trucks, refuse trucks, utility vehicles, buses, dump trucks, cement trucks, tractor trailers, and many others.

ANNUAL MILES AND FUEL CONSUMPTION BY VEHICLE WEIGHT CLASSES

As noted, there is a wide variety of medium- and heavy-duty vehicles. They are used in different applications—from refuse trucks that stop and go constantly and operate at low speeds, to long-haul tractor-trailers that spend much of their time at highway speeds. In addition, the numbers of trucks of different classes vary greatly. Consequently, the number of miles and fuel used vary greatly, depending on the application and the type of vehicle. Table 1-1 presents a summary for the United States of the approximate annual miles and fuel consumption for different vehicle classes, which was addressed in detail in NRC (2010a). The data presented in the table are based on the 2002 Vehicle Inventory and Use Survey (VIUS). According to Davis et al. (2010), the Census Bureau has not conducted a VIUS since 2002, so these are the latest survey data available. Note that Class 8, which includes tractor-trailers, represents about 20 percent of the fleet in total number of vehicles, but 61 percent of the fuel use in all heavy-duty vehicles. Note that Class 2B, Class 6, and Class 8 together account for more than 90 percent of the total fuel use for MHDVs (NRC, 2010a).

PROPOSED FUEL CONSUMPTION AND GREENHOUSE GAS REGULATIONS FOR MEDIUM- AND HEAVY-DUTY VEHICLES

As noted in this chapter, the EPA and NHTSA issued final standards on September 15, 2011, for GHG emissions and fuel efficiency standards for medium- and heavy-duty engines and vehicles. These standards will be phased in and apply to model years 2014 to 2018 and are tailored to each of three main regulatory categories of vehicles: (1) combination tractors, commonly known as semi-trucks that typically pull trailers (Classes 7 and 8), although the agencies are not regulating trailers; (2) heavy-duty pickup trucks and vans (Classes 2b and 3); and (3) vocational vehicles, which comprise a very wide variety of truck and bus types (Classes 2b through 8).[11]

The final rules for vocational vehicles and combination tractors, which are semi trucks that typically pull trailers, have separate fuel consumption standards for vehicles and engines. Standards for fuel consumption of tractors are expressed in gallons per 1,000 ton-miles and in gallons/100 bhp-hr for engines (EPA/NHTSA, 2011). The standards for

[11] The variety of vocational vehicles include delivery, refuse, utility dump, cement, shuttle bus, school bus, emergency vehicles, and others.

Class 7 and 8 tractor fuel consumption are voluntary for 2014 and 2015 model years and become mandatory in the 2016 model year. The new combination tractor standards for the 2017 model year reflect additional improvements in only the heavy-duty engines. The final standards will achieve from 9 to 23 percent reduction in fuel consumption from affected tractors compared to the 2010 baselines.

Vocational truck (Classes 2b through 8) standards are also expressed in gallons per 1,000 ton-miles and are set separately for light heavy-duty (Class 2b through 5), medium heavy-duty (Class 6-7), and heavy heavy-duty (Class 8). The agencies are regulating chassis manufacturers. Achieving standards for vocational vehicles is limited to tire technologies for reduced rolling resistance and engine improvements. The standards allow vocational truck manufacturers to quantify improvements due to hybrid powertrains as a means for compliance. The fuel consumption standards for vocational vehicles represent reductions from 6 to 9 percent compared to a 2010 baseline.

The fuel consumption standards for heavy-duty pickups and vans (Class 2b and 3) are expressed in gallons/100 miles with separate standards for gasoline-fueled and diesel-fueled vehicles. The EPA and NHTSA expect industry to apply similar technologies as the 2012-2016 light-duty vehicle program, but adapted to heavy-duty applications. The standards are fleetwide corporate average standards as in the case for light-duty vehicles. The fuel consumption standards are voluntary in 2014 and 2015. The final standards represent an average per-vehicle improvement in fuel consumption of 15 percent for diesel vehicles and 10 percent for gasoline vehicles, compared to a common baseline.

BUDGET TRENDS OF THE 21ST CENTURY TRUCK PARTNERSHIP

The 21CTP itself has only a small research budget at the DOE, and that had been diminishing during the past few years, although the FY 2011 level is about $37 million (see Table 1-2). Table 1-2 shows congressional appropriations to the heavy-vehicle R&D activities at the DOE from FY 1999 through FY 2010 and the DOE budget request for FY 2011. These appropriations have represented a declining proportion of the total of both the LDV and the heavy-duty vehicle funding from the Office of Vehicle Technologies. This trend was also noted in presentations to the NRC Phase 1 review.[12] In addition to the DOE, the other three agencies have their own, separate budgets that are associated with the Partnership. At the time of this review, the budget resolution in April 2011 for the FY 2011 appropriations indicated significant reductions for the DOE Office of Energy Efficiency and Renewable Energy from the FY 2011 budget request. However, how these reductions affected the FY 2011 funding for individual R&D areas for the DOE's part of 21CTP was unknown.

As noted in the NRC Phase 1 report, the challenge of analyzing multiagency partnerships is underscored by the fact that no one can tell the committee how much the various non-DOE parts of the 21CTP spend on their activities. Even the DOE parts are clouded by proprietary restrictions imposed by industrial partners. There are also other programs in the DOE in addition to the technical R&D areas listed in Table 1-2 that can be leveraged to promote advanced technologies for medium- and heavy-duty vehicles. For example, the Clean Cities Deployment Program, which provides funding for demonstration vehicles, also received more than $90 million in funding from the American Recovery and Reinvestment Act of 2009 (ARRA, or the Stimulus Program) for such activities.

The ARRA has injected a significant amount of funding into activities, including R&D, on vehicles. Although this funding is a "one-shot" infusion and is not included as part of the congressional appropriations of each of the agencies, it has allowed the initiation of a number of both LDV and MHDV activities that can help to promote technologies for reducing fuel consumption. For example, approximately $2.8 billion was provided to accelerate the manufacturing and deployment of the next generation of U.S. batteries ($1.5 billion), to manufacture electric-drive components ($500 million), and for transportation electrification ($400 million).[13] Such efforts, for example, can help to promote the more rapid development of battery technologies and to stimulate the demonstration and deployment of hybrid vehicles.

ARRA funding also allowed a solicitation to be announced and funded called Systems Level Technology Development, Integration, and Demonstration for Efficient Class 8 Trucks (SuperTruck) and Advanced Technology Powertrains for Light-Duty Vehicles (ATP-LD). The heavy-vehicle part of this solicitation has a goal "to develop and demonstrate a 50-percent improvement in overall freight efficiency on a heavy-duty Class 8 tractor-trailer measured in ton-miles per gallon."[14] Three contracts were awarded in response to this solicitation: the total funding for these contracts was about $115 million, with about $100 million associated with ARRA funding for two of the contracts (see Chapter 8).

In the part of the 21CTP program that is administered by the DOE/EERE, for example, the total appropriation each year is divided on the basis of the several "technical areas" of the DOE/EERE, which correspond to engines, lightweighting, idle reduction, and so on. In addition, the DOE/EERE must maintain funding to companies with multiyear cooperative agreements and with Cooperative Research and

[12] Ken Howden, Director, 21st Century Truck Partnership, "21st Century Truck Partnership," presentation to the Phase 1 review committee, Washington, D.C., February 8, 2007.

[13] P. Davis, DOE, "U.S. Department of Energy Vehicle Technologies Program Overview," presentation to the committee, September 8, 2010, Washington, D.C.

[14] See http://www07.grants.gov/search/search.do?&mode=VIEW&flag2006=false&oppId=47867.

TABLE 1-2 Department of Energy Budgets for Heavy-Duty Vehicle Technologies, 1999-2011 (millions of dollars)

FY		1999	2000	2001	2002	2003	2004	2005	2006	2007	2008	2009	2010	2011
						Appropriation ($ in Millions)								Request
Advanced Combustion Engine	***Subtotal:***	18.200	26.441	29.862	31.821	36.978	35.023	27.530	19.869	24.455	16.266	14.230	20.949	20.949
• Combustion & Emission Control		3.400	3.200	3.668	4.176	4.705	3.333	8.312	3.317	3.680	13.738	12.442	18.422	18.422
• Light Truck Engine		14.800	17.411	17.783	15.778	14.734	12.945	0.000	0.000	0.000	0.000	0.000	0.000	0.000
• Heavy Truck Engine		—	4.830	5.914	9.396	12.174	11.831	13.832	9.270	14.490	0.000	0.000	0.000	0.000
• WHR / Solid-State Energy Conversion		—	—	1.000	0.500	0.488	2.469	3.435	1.500	3.806	2.528	1.788	2.527	2.527
• Health Impacts		—	1.000	1.497	1.471	1.463	0.988	1.951	2.413	2.479	0.000	0.000	0.000	0.000
• Off-highway Engine R&D		—	—	—	0.500	3.414	3.457	0.000	3.369	0.000	0.000	0.000	0.000	0.000
Vehicle Systems (includes Hybrid Systems in 2010)	***Subtotal:***	1.500	2.915	4.730	9.869	10.548	10.582	8.863	8.553	5.922	5.870	2.916	4.605	2.800
Heavy Vehicle Systems R&D														
• Vehicle System Optimization		1.500	2.915	4.230	9.369	9.555	10.187	8.764	8.457	5.922	5.870	2.916	4.605	2.800
• Truck Safety Systems		—	—	0.500	0.400	0.397	0.395	0.099	0.096	0.000	0.000	0.000	0.000	0.000
• STICK Program		—	—	—	0.100	0.596	0.000	0.000	0.000	0.000	0.000	0.000	0.000	0.000
Hybrid and Electric Propulsion (see Vehicle Systems for '10)	***Subtotal:***	0.000	3.881	3.938	4.941	3.939	4.976	5.353	1.815	0.000	0.000	0.000	0.000	0.000
Subsys. Integ. & Dev. - Heavy Hybrid		—	3.881	3.938	4.941	3.939	4.976	5.353	1.815	0.000	0.000	0.000	0.000	0.000
Fuels Technology	***Subtotal:***	10.900	13.327	16.313	18.206	13.908	12.247	11.384	6.315	6.570	6.630	6.364	12.224	4.705
Advanced Petroleum-Based Fuels														
• Heavy Trucks		2.700	3.872	4.854	5.853	7.996	6.321	5.876	3.375	3.511	2.599	2.333	4.068	0.000
Non-Petroleum-Based Fuels & Lubes														
• Heavy Trucks		3.300	2.743	3.241	3.695	1.408	1.383	0.690	0.000	0.000	0.000	0.000	0.000	0.000
• Medium Trucks		4.700	2.712	3.266	3.903	1.316	1.284	1.282	0.000	0.000	0.000	0.000	0.000	0.000
• Fueling Infrastructure		0.200	2.000	1.979	1.966	0.906	0.889	1.183	0.000	0.000	0.000	0.000	0.000	0.000
• Renewable & Synthetic Fuels Util.		—	—	—	—	—	0.395	1.367	2.940	3.059	4.031	4.031	8.156	4.705
Environmental Impacts		—	2.000	2.973	2.789	2.282	1.975	0.986	0.000	0.000	0.000	0.000	0.000	0.000
Materials Technologies	***Subtotal:***	15.000	19.912	20.401	20.832	19.899	20.149	18.563	14.242	8.274	11.380	10.530	11.830	7.765
Propulsion Materials Technology														
• Heavy Vehicle Propulsion Matls.		5.300	5.871	6.009	5.756	5.705	5.778	4.858	4.259	3.900	4.816	4.860	5.668	5.645
Lightweight Materials Technology														
• High Strength Wt. Reduc'n Matls.		4.200	5.781	8.804	9.574	8.731	8.840	7.690	2.766	0.000	0.000	0.000	0.500	2.120
High Temp. Matls. Lab. (HTML)*		5.500	8.260	5.588	5.502	5.463	5.531	6.015	7.217	4.374	6.564	5.670	5.662	0.000
Technical Support Services / SBIR / Peer Review				*0.773*	*0.979*	*1.141*	*1.142*	*0.925*	*1.188*		*0.963*	*1.317*	*1.540*	*0.500*
TOTAL Heavy Vehicle Technologies		45.600	66.476	76.017	86.648	80.950	78.588	66.603	44.765	40.847	34.545	29.687	45.486	36.719

* *HTML was a separate line item from FY 2003 to 2010*

SOURCE: Submitted by Ken Howden, DOE, to the committee, October 28, 2010.

Development Agreements (CRADAs) in the DOE laboratories. Table 1-2 was provided by the DOE to indicate the level of funding in each of the main areas that are considered part of 21CTP and which are the main areas that the committee focused on during its review.

The President's FY 2012 budget request to Congress indicates a substantial increase in funding for the Office of Vehicle Technologies, from about $304 million in FY 2010 to a request of $588 million for FY 2012. It is unclear at the time of this review whether the Congress will appropriate this level of funding, and also what portion of this funding will be directed toward 21CTP activities.[15]

ORIGIN AND SCOPE OF THIS STUDY

In response to a request from the director of the DOE's Office of Vehicle Technologies, the National Research Council formed the Committee to Review the 21st Century Truck Partnership, Phase 2 (see Appendix A for biographical information on committee members). The committee was asked to fulfill the following statement of task:

> The committee will conduct an independent second review of the 21st Century Truck Partnership. In its review, the committee will critically examine and comment on the overall adequacy and balance of the 21st Century Truck Partnership to accomplish its goals, on progress in the program, and make recommendations, as appropriate, that the committee believes can improve the likelihood of the Partnership meeting its goals. In particular, the committee will:
>
> (1) Review the high-level technical goals, targets, and timetables for R&D efforts, which address such areas as heavy vehicle systems; hybrid electric propulsion; advanced internal combustion engines (ICEs); and materials technologies.
>
> (2) Review and evaluate progress and program directions since the inception of the Partnership towards meeting the Partnership's technical goals, and examine on-going research activities and their relevance to meeting the goals of the Partnership.
>
> (3) Examine and comment on the overall balance and adequacy of the 21st Century Partnership's research effort, and the rate of progress, in light of the technical objectives and schedules for each of the major technology areas.
>
> (4) Examine and comment, as necessary, on the appropriate role for federal involvement in the various technical areas under development.
>
> (5) Examine and comment on the Partnership's strategy for accomplishing its goals, which might include such issues as (a) program management and organization; (b) the process for setting milestones, research directions, and making Go/No Go decisions; (c) collaborative activities within DOE, other government agencies, the private sector, universities, and others; and (d) other topics that the committee finds important to comment on related to the success of the program to meet its technical goals.
>
> (6) Examine and comment on the response of the Partnership to the recommendations made in the Phase 1 report, "Review of the 21st Century Truck Partnership" issued by the NRC in 2008.
>
> After examining the 21st Century Truck Partnership activities and receiving presentations from federal government representatives and industry representatives, and outside experts, as appropriate, the committee will write a report documenting its review of the Partnership with recommendations for improvement, as necessary.

The statement of task as defined above contains a number of standard elements that the NRC has used for the review of a number of DOE R&D programs because it is general enough to allow a committee to make an assessment either narrowly, broadly, or both, as appropriate. In an ideal world, every technical area would have well-defined projects, budgets, milestones, and targets against which to assess progress. But in reality, given the multiagency-and-industry nature of the 21CTP, the identification of such well-defined projects that can fall under the 21CTP umbrella is not uniform across the various areas and agencies (see Chapter 2). However, as noted in this chapter, the Partnership has been focused around five technical areas and has white papers and goals for each of those areas, and a white paper for a new sixth area, efficient operations, has been drafted by the 21CTP. In some instances there are precise targets against which to measure progress; in others there are not. The assessments of the committee are contained in the respective technical chapters, which correspond to the areas addressed by the white papers. In some cases, such as in hybrid propulsion, the budgets have been zeroed out, but the Partnership has leveraged the work on various technical areas that are occurring at the DOE—in this example for light-duty hybrid vehicles. In following its statement of task to comment on the 21CTP strategy for accomplishing its goals (Item 5), the committee reviewed a 21CTP draft white paper for the new area on efficient operations; there are no goals and targets as yet but the committee has made suggestions to the Partnership on improving its white paper and on what might be addressed (see Chapter 9).

The present review has also been complicated by the fact that the white papers and some goals and targets have been undergoing revision during the committee's review. Further, the important new undertaking called SuperTruck is following on the NRC Phase 1 report recommendations for integrating technology into whole vehicle systems (see Chapter 8). Given that the SuperTruck progam is new, the committee has not been able during this review to comment on specific progress toward technical targets. The committee has done what is possible in assessing progress but with the understanding that in some areas there are not well-defined targets and committee judgment has been used. The situation is not dissimilar to that during the Phase 1 review, from which the committee's recommendations helped to focus some of the 21CTP efforts; the committee anticipates that the current

[15] Ken Howden, DOE, "FY2012 Budget Request," presentation to the committee, March 31, 2011, Washington, D.C.

report recommendations also will help the Partnership with its focus over the next few years.

ROLE OF THE FEDERAL GOVERNMENT

The role of the federal government in R&D varies depending on the administration and the Congress and the issues that they deem important for the nation to address. An extensive economics literature on the subject points to the importance of R&D to promote technical innovation, especially for research for which the private sector finds it difficult to capture the returns on its investment; this is especially true for basic research, the results of which can be broadly used. Such innovation, if successful, can foster economic growth and productivity, with improvements in the standard of living (Bernanke, 2011). Furthermore, in the energy area, the government generally has to confront issues of national security, environmental quality, or energy affordability. Many of these issues are addressed through policy initiatives or regulations, which place a burden on private firms to achieve. Thus there is a role for the federal government in supporting R&D not only to help the private sector achieve these policy goals but also to help U.S. firms remain competitive in the face of international competition.

The committee believes that the federal government plays an important role in the development of technologies that can help to address government policies and regulations aimed at reducing emissions and fuel consumption from medium- and heavy-duty vehicles. There are similar reasons for the government playing a role in R&D for light-duty vehicles as well. Such partnerships as the Partnership for a New Generation of Vehicles, the FreedomCAR and Fuel Partnership (which is now being replaced by U.S. DRIVE), and the 21CTP are examples of public-private efforts to support R&D and to develop advanced technologies for vehicles (NRC, 2001, 2010a,b). These partnerships generally include a variety of efforts (fundamental research, development, demonstration, and in some cases deployment). The federal government can support fundamental research through the national laboratories, and universities and industry can focus on development. The importance of having government-industry collaboration is that the private sector can help to transform improvements from research into cost-effective and marketable products. Generally, the contracting that is engaged in with the private sector is cost-shared, and those research contracts more closely associated with fundamental or basic research will have a majority of federal funding, whereas contracts with a strong development or product component will have significant support from the private sector. In its recommendations in each of the technical areas, the committee has considered what activities are most appropriate for the 21CTP to support. Implicit in all the recommendations that relate to the support of additional research, the committee believes that the federal government has a role in the R&D.

STUDY PROCESS AND ORGANIZATION OF THE REPORT

The committee held meetings to collect information through presentations on 21CTP activities by representatives of the four federal agencies involved in the Partnership, as well as individuals outside the program (see Appendix B for a list of the presenters and their topics). During the NRC Phase 1 review, the 21CTP had developed a roadmap and a series of white papers on the main technical areas that the Partnership had focused on (DOE, 2006); during the current, Phase 2 review the 21CTP was in the process of modifying these white papers. The white papers were very important to the committee in its information-gathering activities, because they provided the strategy, goals, and technical challenges from the viewpoint of the 21CTP for each of the technical areas under review. Drafts of the white papers were submitted to the committee in September 2010 (DOE, 2010c); updated versions were provided in March 2011 (DOE, 2011). A draft of a white paper for the new area, efficient operations, was submitted to the committee in March 2011. The committee provides feedback and suggestions to the 21CTP on this white paper in Chapter 9 of the present report.

To obtain clarifications on some aspects of the 21CTP, the committee sent written questions to 21CTP representatives and received very helpful answers in response. The committee also made site visits to Cummins Technical Center; Navistar Inc. Truck Development and Technology Center; Eaton Corporation's Eaton Innovation Center; EPA; Oak Ridge National Laboratory; and the U.S. Army Tank Automotive Research, Development and Engineering Center (TARDEC); each of the organizations visited are undertaking R&D under the 21CTP. The committee's findings and recommendations are based on the information gathered during the study and on the expertise and knowledge of committee members.

Following is an overview of the topics covered in the rest of this report. Chapter 2 addresses the overall management strategy and priority setting of the Partnership. Chapter 3 addresses various engine programs at the DOE, the EPA, and DOD, discusses fuels and aftertreatment research, health-related research, high-temperature propulsion materials, as well as the High Temperature Materials Laboratory, a user facility run by the Oak Ridge National Laboratory. Chapter 4 focuses on hybrid vehicles. Chapter 5 addresses vehicle power demands, which are referred to by many as parasitic losses; they represent the power needed to overcome such resistive forces as aerodynamics, rolling resistance, and friction losses in the drivetrain, or to power auxiliary systems on a vehicle. Chapter 6 addresses idle reduction technologies for reducing fuel consumption and emissions during truck idle time. Chapter 7 deals with safety, which is mostly under the initiatives in the DOT. Chapter 8 addresses the newly established SuperTruck efforts that are focused on three major project teams. Finally, Chapter 9 offers some guidance on a new area for the 21CTP, efficient operations.

Appendix A presents biographical sketches of the committee members. Appendix B lists all of the public presentations at the committee's four meetings. Appendix C contains the list of recommendations from the Phase 1 NRC report as well as the 21CTP responses to them. Appendixes D through I provide some background material for various chapters, including a list of abbreviations and acronyms used in the report.

REFERENCES

Bernanke, B.S. 2011. Promoting Research and Development: The Government's Role. Issues in Science and Technology. A publication of the National Academies and the University of Texas-Dallas. Summer. Available at http://www.issues.org/27.4/.

BP. 2011. BP Statistical Review of World Energy. June. Available at bp.com/statisticalreview. Accessed July 13, 2011.

Davis, S., S. Diegel, and R.G. Boundy. 2009. Transportation Energy Data Book, Edition 28. Report No. ORNL-6984. Knoxville, Tenn.: U.S. Department of Energy, Office of Energy Efficiency and Renewable Energy.

Davis, S., S. Diegel, and R.G. Boundy. 2010. Transportation Energy Data Book, Edition 29. Report No. ORNL-6985. Knoxville, Tenn.: U.S. Department of Energy, Office of Energy Efficiency and Renewable Energy.

DOE (U.S. Department of Energy). 2006. 21st Century Truck Partnership Roadmap and Technical White Papers. Doc. No. 21CTP-003. December. Washington, D.C.: Office of FreedomCAR and Vehicle Technologies.

DOE. 2007. FY 2008 Congressional Budget Request. Vol. 3, Office of Chief Financial Officer. Doc. No. DOE/CF-016. Washington, D.C. February.

DOE. 2010a. Office of Vehicle Technologies, Office of Energy Efficiency and Renewable Energy website. Available at http://www1.eere.energy.gov/vehiclesandfuels/about/partnerships/21centurytruck/21ct_goals.html. Accessed October 4, 2010.

DOE. 2010b. Office of Vehicle Technologies, Office of Energy Efficiency and Renewable Energy website. Available at http://www1.eere.energy.gov/vehiclesandfuels/about/partnerships/21centurytruck/index.html. Accessed October 4, 2010.

DOE. 2010c. Updated 21st Century Truck Partnership Roadmap and Technical White Papers. Working draft, September 1, 2010. Washington, D.C.: Office of Vehicle Technologies.

DOE. 2010d. Engine Systems White Paper. Working draft, September 1. Washington, D.C.: Office of Vehicle Technologies.

DOE. 2011. Updated 21st Century Truck Partnership Technical White Papers. Working draft, February 25, 2011. Washington, D.C.: Office of Vehicle Technologies.

DOE, FCVT (DOE, FreedomCAR and Vehicle Technologies). 2006. FreedomCAR and Vehicle Technologies Program, Multi-Year Program Plan: 2006-2011. Washington, D.C.: DOE Office of Energy Efficiency and Renewable Energy.

Ehlmann, J., and G. Wolff. 2005. Automobile Emissions—The Road Toward Zero. Pittsburgh, Pa.: Air and Waste Management Association. January.

EIA (Energy Information Administration). 2010. AEO [Annual Energy Outlook] 2011. Early Release Overview. December 16. Available at http://www.eia.doe.gov/forecasts/aeo/. Accessed January 15, 2011.

EPA (U.S. Environmental Protection Agency). 2010. Inventory of Greenhouse Gas Emissions and Sinks: 1990-2008. Report No. EPA 430-R-10-006. April 15. Available at http://www.epa.gov/climatechange/emissions/downloads10/US-GHG-Inventory-2010_Report.pdf.

EPA/NHTSA (U.S. Environmental Protection Agency/National Highway Traffic Safety Administration). 2010. Greenhouse Gas Emissions Standards and Fuel Efficiency Standards for Medium- and Heavy-Duty Engines and Vehicles. Dockets No. EPA-HQ-OAR-2010-0162 and No. NHTSA-2010-0079, October 25, 2010. Available at http://www.regulations.gov. Accessed November 15, 2010.

EPA/DOT NHTSA. 2011. Greenhouse Gas Emissions Standards and Fuel Efficiency Standards for Medium- and Heavy-Duty Engines and Vehicles, Final Rules, September 15, 2011. Available at http://www.nhtsa.gov/fuel-economy.

Johnson, J.H. 1988. Automotive Emissions. Pp. 39-46 in Air Pollution, the Automobile, and Public Health. Washington, D.C.: National Academy Press.

NAS-NAE-NRC (National Academy of Sciences, National Academy of Engineering, National Research Council). 2009a. America's Energy Future: Real Prospects for Energy Efficiency in the United States. Washington, D.C.: The National Academies Press.

NAS-NAE-NRC. 2009b. Liquid Transportation Fuels from Coal and Biomass: Technological Status, Costs, and Environmental Impacts. Washington, D.C.: The National Academies Press.

Newell, R. 2010. Annual Energy Outlook 2011 Reference Case. Presented at the Paul H. Nitze School of Advanced International Studies, December 16. Washington, D.C. Available at http://www.eia.gov/neic/speeches/newell_12162010.pdf. Accessed December 18, 2010.

NRC (National Research Council). 2000. Review of the U.S. Department of Energy's Heavy Vehicle Technologies Program. Washington, D.C.: National Academy Press.

NRC. 2001. Review of the Research Program of the Partnership for a New Generation of Vehicles. Washington, D.C.: National Academy Press.

NRC. 2008. Review of the 21st Century Truck Partnership. Washington, D.C.: The National Academies Press.

NRC. 2010a. Technologies and Approaches to Reducing the Fuel Consumption of Medium- and Heavy-Duty Vehicles. Washington, D.C.: The National Academies Press.

NRC. 2010b. Review of the Research Program of the FreedomCAR and Fuel Partnership, Third Report. Washington, D.C.: The National Academies Press.

2

Management Strategy and Priority Setting

INTRODUCTION

As part of its review of the 21st Century Truck Partnership (21CTP), the committee received presentations from the four participating agencies (Department of Energy [DOE], Department of Transportation [DOT], Department of Defense [DOD], and Environmental Protection Agency [EPA]) and the 21CTP industrial partners. These presentations included detailed responses to the concerns about the program's overall effectiveness, funding variations, priority setting, partnership performance, and other 21CTP issues raised in the National Research Council's Phase 1 report (NRC, 2008). The committee also collected information by formulating questions to which the 21CTP provided informative answers. In addition, the 21CTP provided responses to the recommendations from the NRC Phase 1 report (see Appendix C).

In this chapter the committee reviews each of these areas and reports its findings and recommendations. For background on how the Partnership functions, the chapter also includes and summarizes information from the NRC Phase 1 report (NRC, 2008).

PROGRAM MANAGEMENT

Overall management for the Partnership currently rests with the DOE's Office of Vehicle Technologies (the former name was the Office of FreedomCAR and Vehicle Technologies [FCVT]), in the Office of Energy Efficiency and Renewable Energy (EERE). DOE personnel lay out Partnership goals (DOE, 2006, 2010, 2011; further revisions are planned for 2011), lead the discussions for and preparation of the updated 21CTP roadmap and white papers, maintain the information-flow infrastructure (such as websites, e-mail lists), and organize meetings and conference calls. The management of individual projects under the 21CTP umbrella rests with the individual federal agencies that have funded the work. These agencies communicate among one another through the 21CTP information-sharing infrastructure to coordinate efforts and to ensure that valuable research results are communicated and that any overlap of activities among their respective efforts is reduced.

Figure 2-1 illustrates the interrelations among the key parties in setting 21CTP research programs. Government agencies request funding from Congress through the administration and work with the industrial partners and research organizations (including universities and government laboratories) to establish research programs that meet national priorities and the interests of industry partners. However, final funding levels are determined by congressional appropriations, with each agency overseen by different congressional committees. This makes prioritization of all of the 21CTP projects within the four agencies difficult, if not impossible.

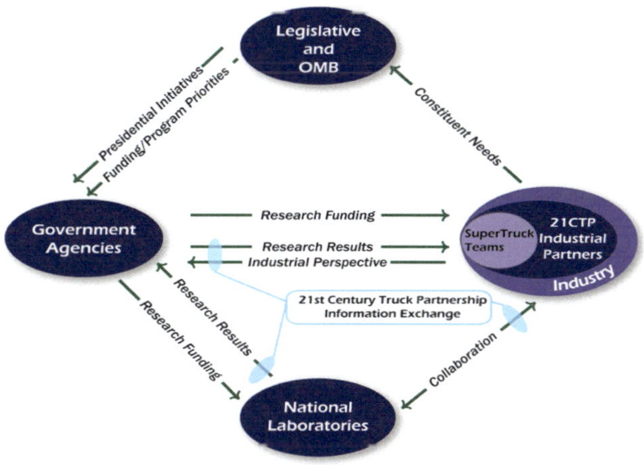

FIGURE 2-1 Interrelations among participants in the 21st Century Truck Partnership. OMB, Office of Management and Budget. SOURCE: Submitted to the committee by the DOE Office of Vehicle Technologies, January 29, 2010.

The committee requested that the budgets for the 21CTP projects of all four agencies be provided for its review, but only the DOE was able to provide a budget for 21CTP activities (see Chapter 1, Table 1-2). Because there are no specific 21CTP budget lines at the EPA, DOD, and DOT, the committee is not aware of a specific list of 21CTP projects and corresponding funding levels for these agencies; it never received a well-defined list of such projects and budgets. Even in the case of the DOE, light-duty vehicle and heavy-duty vehicle work overlaps in some cases, in areas such as combustion or lightweight materials, and so there is at times some difficulty in defining exactly what projects are considered part of the 21CTP, although leveraging the results of light-duty work for heavy-duty vehicles is appropriate. In addition, it was difficult for the committee to ascertain the level of resources that is being contributed by the private sector.

In the case of the DOE, technology programs are developed to meet a cascading series of goals that begin at the President's National Energy Policy and culminate (at the program level) with specific technology goals. Figure 2-2 illustrates that pattern schematically.

The DOE focuses its technology research and development (R&D) investments specifically in high-risk areas or on activities with uncertain or long-term outcomes that are of national interest but would most likely not be pursued by industry alone. Program activities include research, development, testing, technology validation, technology transfer, and education. These activities are aimed at developing technologies that could achieve significant reductions in vehicle fuel consumption and the displacement of oil by other fuels that ultimately can be produced domestically in a clean and cost-competitive manner.

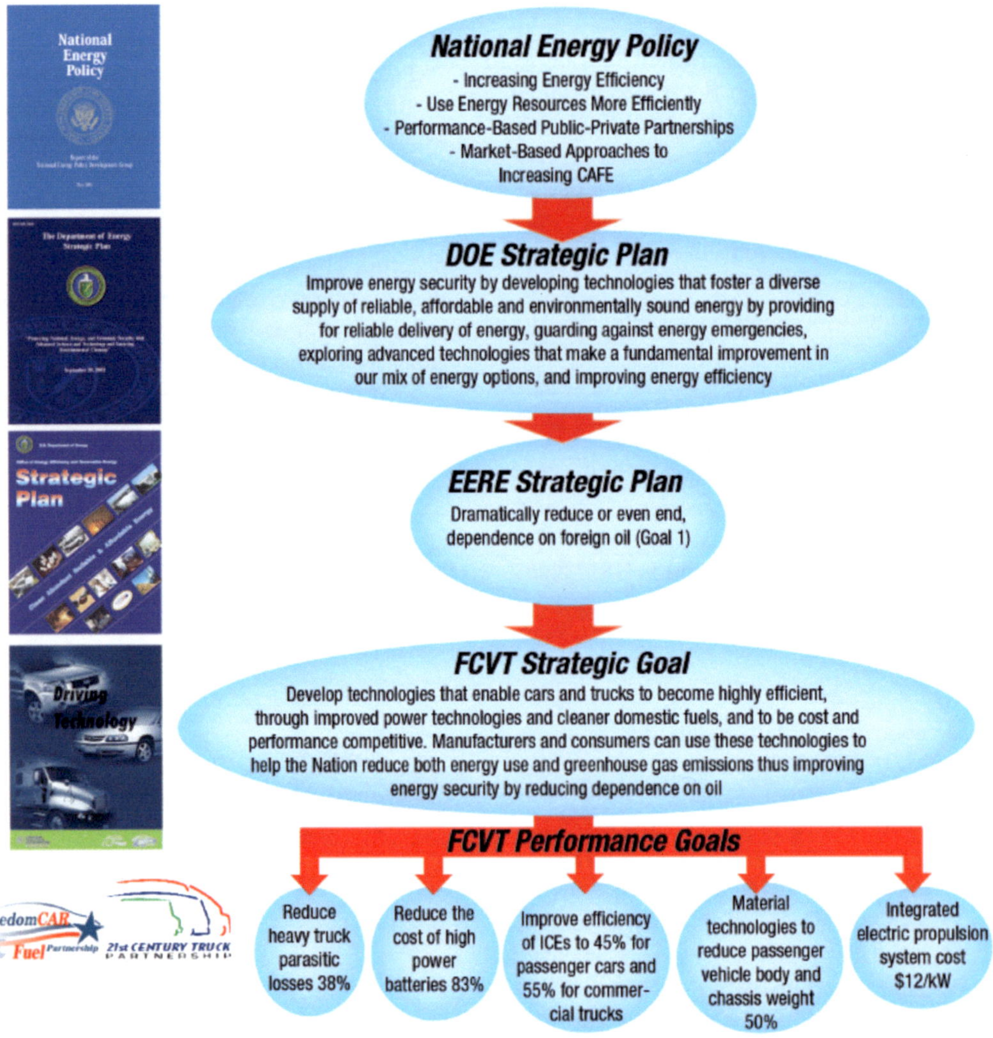

FIGURE 2-2 Department of Energy goal setting process for technology programs. SOURCE: DOE, Responses to Committee Queries on 21CTP, Management and Process Issues. Transmitted by e-mail from Ken Howden, DOE Office of Vehicle Technologies (formerly the Office of FreedomCAR and Vehicle Technologies [FCVT]).

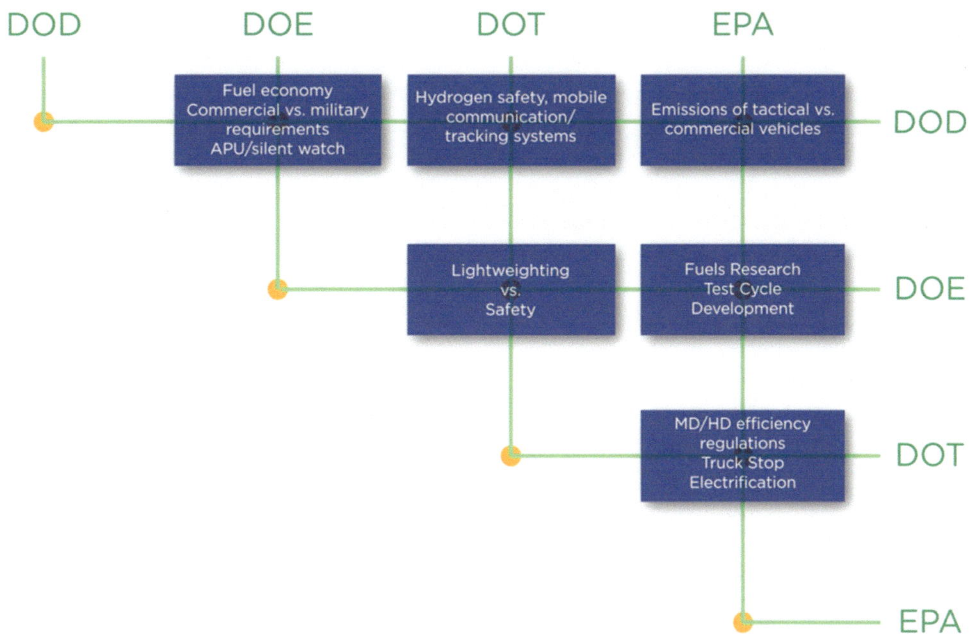

FIGURE 2-3 Some areas of common interest among the collaborative government agencies in the 21st Century Truck Partnership. Acronyms are defined in Appendix I. SOURCE: Submitted to the committee by the DOE Office of Vehicle Technologies, January 29, 2011.

In DOE vehicle research, which specifically addresses the national issue of energy security and the increasing pressures of the rising global consumption of oil, the Office of Vehicle Technologies has involved the affected industries in planning the research agenda and identifying technical goals that, if met, will provide the basis for commercialization decisions. The government's approach is intended to allow industry-wide collaboration in precompetitive research, which is then followed by competition in the marketplace.

The Partnership provides a forum for the exchange of technical information among the industry and government partners involved in heavy-duty transportation. At present, the coordination of initiatives takes place as part of this information exchange.

Specific areas in which the government partners have already coordinated initiatives include the following:

- *Diesel fuel sulfur standard development*—with coordination between the DOE and EPA on appropriate sulfur levels for low-sulfur diesel;
- *Idle reduction activities*—with cooperation between the EPA and DOT and their focus on deployment, and the DOE with its focus on technology R&D;
- *Development of heavy-duty truck fuel efficiency standards*—with coordination between the DOT and EPA to create the Notice of Proposed Rulemaking (NPRM) "Greenhouse Gas Emissions Standards and Fuel Efficiency Standards for Medium- and Heavy-Duty Engines and Vehicles," issued on October 25, 2010 (EPA/NHTSA, 2010), which led to a final rule issued on September 15, 2011 (EPA/NHTSA, 2011);
- *Truck aggressivity*—with the National Transportation Safety Board (NTSB) using the 21CTP as a forum for approaching all key government and industry participants involved with the issue; and
- *Hybrid powertrains*—with the DOE and EPA pursuing different technologies for hybridization, e.g., hydraulic hybrids at the EPA and electric hybrids at the DOE.

Figure 2-3 illustrates the general collaborative structure of the four government agencies and some areas of interest among them.

The full Partnership meets by conference call monthly, or at times biweekly, and meets face-to-face about four times per year. The Partnership's Executive Committee is made up of three industry members, one from each of three industrial sectors: truck original equipment manufacturers (OEMs), engine manufacturers, and hybrid/system component manufacturers (NRC, 2008).[1] Agendas for the conference calls typically include discussion of topics such as the following:

- Open funding opportunities (to bring these to the attention of members who may wish to apply),

[1] According to an e-mail from Michael Laughlin, 21CTP, to John H. Johnson, committee chair, dated May 17, 2011, Executive Committee conference calls are scheduled monthly to discuss issues related to 21CTP management and operations, and one full Partnership call per month is scheduled to discuss issues relevant to the entire group.

- Budget activities in the federal sector (where appropriate),
- Technical accomplishments or plans for individual areas of interest to the heavy-duty trucking industry,
- News articles of interest to the industry,
- Industry/government events (e.g., the Society of Automotive Engineers [SAE] Government-Industry Meeting, the SAE Commercial Vehicle Congress, and so forth) and any Partnership participation plans,
- Other Partnership activities (such as face-to-face meetings, special visits to laboratories or other facilities, and reviews such as the National Research Council review) and planning and participation in DEER (formerly "Diesel Engine-Efficiency and Emissions Research," now "Directions in Engine-Efficiency and Emissions Research") conferences.

These meetings typically last no more than 2 hours, with time reserved for industry partners to speak among themselves, for government personnel to speak among themselves, and for industry and government to speak together.

The foregoing description of the overall program management process, originally published in the NRC Phase 1 report (NRC, 2008), has been updated here to reflect the current Partnership practices. It reflects the Partnership's responses to questions from the committee during this Phase 2 review dated November 4, 2010.

The original Partnership structure—which has been characterized as a virtual network[2] of agencies and government laboratories, with agency personnel meeting frequently and industry partners meeting periodically for limited sharing and communication—was judged to be far from ideal. Accordingly, in the NRC Phase 1 report, the committee found that, in summary, the 21CTP effectiveness could be improved by:

- Adhering to the agreed program budget spanning the agencies,
- Appointing a full-time executive director to provide project management and set unified priorities,
- Setting realistic programmatic goals and objectives with stretch targets, and
- Empowering the 21CTP Executive Committee with authority to act collaboratively across agencies on program decisions and implementation, using a rigorous go/no-go process.

The formal Partnership response to this Phase 1 assessment was that "The Partnership continues to examine its organization and management structure as part of its ongoing self assessment efforts" (in Appendix C in this report, see Phase 1 Finding and Recommendation 2-1 and the 21CTP response).

On November 4, 2010, the Partnership elaborated further on the program management structure as follows: "By design, and de facto by statutory mandate, the 21CTP itself has no direct control over research activities, funding, or regulations in any of the participating agencies or by its industrial partners. Rather, each participating agency follows its own organizational structure and policies for both decision making and funding for research and development."

Overall, the committee found the Partnership's responses to the 51 NRC Phase 1 report recommendations to be very satisfactory, particularly with regard to the move toward funding the development of vehicle hardware and the demonstration of advanced concepts deemed to be of merit. Although the specific responses to two program management recommendations were somewhat disappointing (in Appendix C, see Recommendations 2-1 and 2-2 and the 21CTP responses), the committee understands that the Partnership was indeed formed as a virtual network, with each agency responsible for its own activities and budget, and that gives the DOE a very limited mandate. Within this limited mandate, the DOE has an effective process for reviewing and managing its own projects and for maintaining focus on the stated Partnership goals. Interagency collaboration is mixed, however: under the 21CTP umbrella, collaboration between the DOE and DOT appears strong, whereas that with and between the EPA and DOD is weak, as one would expect, because the two agencies have different objectives. The collaboration between the EPA and DOT on the NPRM on heavy-duty vehicle fuel consumption standards (EPA/NHTSA, 2010) was strong.

In addition to the committee's hearing a DOD presentation at its meeting on November 15, 2010, a committee subgroup visited the U.S. Army Tank Automotive Research, Development, and Engineering Center (TARDEC), in Warren, Michigan, on January 10, 2011, to review 21CTP-related projects. Although the DOD places a high priority on reduced energy consumption, it is necessarily focused on the needs of the soldier, with exemptions from emissions regulations and emphasis on high power density and JP-8 fuel. Consequently, there is little synergy with the needs of the commercial heavy-truck industry.

The national laboratories conduct many DOE programs as part of, or synergistically with the 21CTP. Examples of such programs include those on advanced combustion engine research, fuels research, aftertreatment, propulsion materials, lightweight materials, hybrid simulation, vehicle parasitic loss, and unregulated pollutants projects. In addition to developing new technologies in the respective areas, these programs foster ongoing technical interchange with industry at the working level, thereby facilitating collaboration

[2] The committee and others have referred to the organization of this program as a "virtual network" or "virtual structure" or "virtual management" because there are no clear lines of authority across the various agencies. As discussed in the NRC Phase 1 report (NRC, 2008) and in this chapter, there is no overall management structure with authority vested in a central manager whose direction is followed by other agency managers associated with the Partnership.

between the national laboratories, the government agencies, universities, and industry. This collaboration ensures that the national laboratories know industry's needs and have its input while ensuring that industry has knowledge of the developing technologies at the national laboratories.

Combustion modeling is an outstanding example of a technology developed collaboratively at the national laboratories and universities that has been adopted to fulfill the needs of industry for modeling to improve on and develop new combustion systems, identify promising engine operating configurations, and reduce hardware testing. By continuing to focus programs in the national laboratories on the fundamental aspects of the needs of industry and/or government agencies, timely transition of mobility technologies from the laboratory to practice can be facilitated.

PRIORITIZATION OF PROJECTS

The organizational structure of the 21CTP precludes any systematic prioritization of research projects for the total program. Each of the four agencies included in the 21CTP has its own separate budgets and priorities. The industrial partners also have their own needs, priorities, and resources. As a consequence, the program-wide prioritization that does occur is the result of a complex interaction (summarized in Figures 2-1 through 2-3) among government agencies, the industrial partners, the national laboratories, and the Congress and the Office of Management and Budget.

In summary, the primary intent of the 21CTP is to facilitate communication among the many partners to avoid duplication of effort, to communicate technical achievements, and to provide financial support to assist in moving new technology through development to commercialization.

In the NRC Phase 1 report, the committee recommended the creation of "a portfolio management process that sets priorities and aligns budgets among the agencies and industrial partners" (NRC, 2008, Recommendation 2-2). In response, the Partnership stated that although this recommendation "will be considered . . . the ability to directly align budgetary decisions across the agencies, however desirable, may be outside the scope of this voluntarily collaborative organization" (see Appendix C). For the reasons cited above, the committee concluded that, although indeed highly desirable, such a portfolio management process is simply not likely to happen with the decentralized nature of the Partnership.

Although prioritization across agencies is unlikely to happen in any meaningful way, the DOE has focused much of its 21CTP effort going forward on three SuperTruck projects, two funded with the American Recovery and Reinvestment Act of 2009 (ARRA, Public Law 111-5) funds and one receiving DOE internal funds. These projects, detailed in Chapters 3 and 8, are directed towards demonstrating feasibility, fuel efficiency, and emissions compliance with full vehicle hardware, as recommended in the NRC Phase 1 report. The committee applauds the prioritization of available ARRA and DOE funds on these projects.

In the process of moving a new concept from research idea to commercial product, DOE research organizations use the general process shown in Figure 2-4. The "Basic Research" steps are clearly dominated by DOE laboratories and "Commercial Research and Design" by industry. Research results and budget proposals are thoroughly reviewed. Those not approved or having marginal benefit go into the "Valley of Death" where they remain until circumstances change.

FINDINGS AND RECOMMENDATIONS

In summary, the 21CTP is operated as a virtual network of agencies, industry, and government laboratories, and it is difficult in many cases to identify individual Agency priorities and budgets. As in the Phase 1 review, the committee is con-

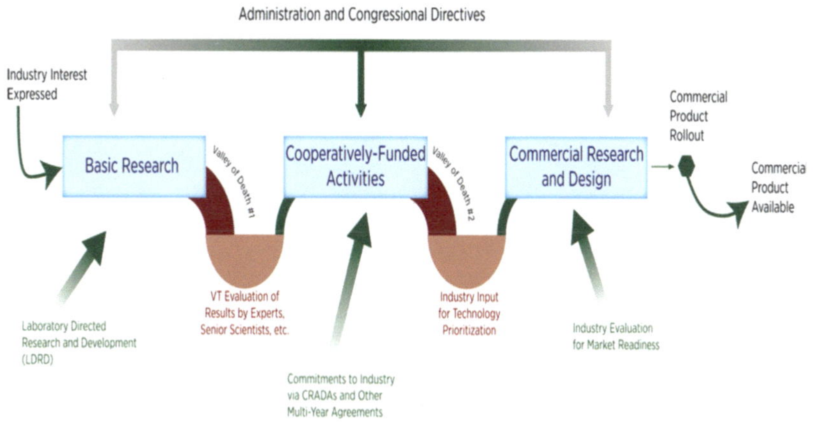

FIGURE 2-4 Department of Energy project management and innovation process. Acronyms are defined in Appendix I. SOURCE: Submitted to the committee by the DOE Office of Vehicle Technologies, January 29, 2011.

cerned about the lack of stable year-to-year funding. However, despite its unwieldy structure and budgetary process, it has made significant progress, and the outlook for continued success is bright, barring any major funding issues.

Following are the committee's findings and recommendations with respect to the management, strategy, and priority setting of the 21st Century Truck Partnership.

Finding 2-1. The 21CTP is a virtual organization facilitating communication among four agencies, government laboratories, and industry, but it has no direct control over research activities or funding across the agencies or by its industry partners. The committee continues to believe that the lack of single-point 21CTP authority is far from optimal, although it recognizes that this is necessary because of the various Congressional committees that the agencies report to and that provide their budgets.

Recommendation 2-1. The DOE is urged to continue to improve the functioning of the 21CTP "virtual" management structure in every way possible. Such improved functioning would include strengthening interagency collaboration (particularly that involving the EPA and DOD)[3] and documenting and publishing specific 21CTP activity within all four agencies.

Finding 2-2. The EPA, DOD, and DOT did not have a well-defined list of the projects and associated budgets that were included under the 21CTP umbrella. This stems in part from the virtual nature of the Partnership and partly, particularly within the DOE, from the natural overlap in activities on batteries, hybrids, materials, and other areas between the activities for light–duty vehicles and the 21CTP. While many of these activities are reviewed at the annual DOE Merit Review and at Directions in Engine-Efficiency and Emissions Research (DEER) conferences, and the new SuperTruck projects include an annual reporting requirement, there is no dedicated report for the 21CTP.

Recommendation 2-2. The DOE should issue a brief annual report documenting the specific projects within the 21CTP and the progress made. The annual report should provide references to published technical reports from the involved agencies. This would especially help outside groups, future review committees, the Congress, and others to understand the structure, activities, and progress of the Partnership.

[3] Subsequent to the committee's review of 21CTP programs, the DOE and the DOD entered into the Advanced Vehicle Power Technology Alliance (AVPTA) partnership on July 18, 2011. See, for example, "DOE, Army Alliance Underlines Achieving Energy Security" by Chris Williams, available at http://www.army.mil/article/62727/. Accessed October 18, 2011.

REFERENCES

DOE (U.S. Department of Energy). 2006. 21st Century Truck Partnership Roadmap and Technical White Papers. Document No. 21CTP-003. Washington, D.C.: Office of FreedomCAR and Vehicle Technologies.

DOE. 2010. Updated 21st Century Truck Partnership Roadmap and Technical White Papers. Working draft, September 1, 2010. Washington, D.C.: Office of Vehicle Technologies.

DOE. 2011. Updated 21st Century Truck Partnership Technical White Papers. Working draft, February 25, 2011. Washington, D.C.: Office of Vehicle Technologies.

EPA/NHTSA (Environmental Protection Agency and National Highway Traffic Safety Administration). 2010. Greenhouse Gas Emissions Standards and Fuel Efficiency Standards for Medium- and Heavy-Duty Engines and Vehicles. Dockets No. EPA-HQ-OAR-2010-0162 and No. NHTSA-2010-0079, October 25, 2010. Available at http://www.regulations.gov.

EPA/NHTSA. 2011.Greenhouse Gas Emissions Standards and Fuel Efficiency Standards for Medium- and Heavy-Duty Engines and Vehicles, Final Rules, September 25, 2011. Available at http://www.nhtsa.gov/fuel-economy.

NRC (National Research Council). 2008. Review of the 21st Century Truck Partnership. Washington, D.C.: The National Academies Press.

3

Engine Systems and Fuels

INTRODUCTION

Heavy-duty truck engines are central to all aspects of the 21st Century Truck Partnership (21CTP) vision: improved thermal efficiency, reduced oil dependency, low-exhaust emissions, low cost, and improved safety are aspects of that vision. Although diesel engines used in most trucks are the most efficient on-road transportation power plants available today, with diesel engines only approximately 42 percent of the fuel energy is converted to mechanical work, resulting in a loss of the energy input by means of the fuel of approximately 58 percent (DOE, 2010a). Additional improvements in the thermal efficiency of diesel and gasoline engines are still possible. However, there are thermodynamic limitations, which means that only a modest portion of the 58 percent of energy that is lost today can ever be recovered. In addition to petroleum-based fuels, these engines can be powered by nonpetroleum fuels from a number of feedstocks. The engine, together with the fuel characteristics and exhaust emission control devices, governs the level of exhaust emissions so critical for compliance with regulations, environmental impact, and public perception. The engine is critical to the safety of the heavy vehicle because it provides braking power, as well as adequate power for the vehicle to blend with traffic. This chapter covers engine programs, fuel programs, aftertreatment systems, high-temperature materials, and health concerns related to emissions from diesel engines.

Several members of the 21CTP are global companies, and they bring to the Partnership the perspective of technologies used in markets around the world. The committee also was given a presentation on a joint Department of Energy (DOE)/ Swedish Ministry of Energy research project that combined U.S. and European technology.

ENGINE PROGRAMS: STATE OF TECHNOLOGY

Diesel engines derive their efficiency from high-efficiency thermodynamic processes, fuel with a high heating value, and minimal mechanical losses. These engines achieve their efficiency by means of a high compression (expansion) ratio, high rates of combustion under overall lean fuel conditions, and the use of air-to-fuel ratio (instead of throttling) for managing load control, thus avoiding part-load pumping losses. Turbocharging increases engine power density and recovers some of the exhaust energy. Diesel engines operate at relatively low speeds, which reduce mechanical friction losses, and high power density is achieved primarily through high brake mean effective pressure (BMEP).[1]

Owing to its low fuel consumption, reliability, and low life-cycle cost, the diesel engine has continued to be the preferred power source for commercial vehicles, urban buses, and military vehicles worldwide. The cost of complying with emissions regulations for traditional diesel combustion has given rise to reconsideration of alternative power plants such as heavy-duty spark-ignition engines, and gasoline engines have even regained market share in Class 6 trucks (DOE, 2011a; NRC, 2010). High worldwide demand for diesel fuels has driven their price above gasoline in the United States, furthering this trend. In addition, U.S. national average diesel fuel taxes were 5 cents per gallon higher than for gasoline in January 2011, as reported by the American Petroleum Institute.

Modern highway truck diesel engine brake thermal efficiency (BTE) peaks at about 42 percent, compared to 33 percent for commercial gasoline, spark-ignition engines. This 42 percent peak efficiency represents significant improvement since the 1970s when highway diesel BTE peaked around 35 percent. Key elements for achieving a BTE of 50 percent have already been demonstrated in test laboratories, and demonstration is expected within the next few years in research vehicles meeting emissions standards.

[1] The brake mean effective pressure is the ratio of the shaft work leaving the engine to the displacement of the engine. This ratio is expressed in units of pressure, hence the name. BMEP is a useful metric in that it assesses the work output per unit of engine displacement, so it can be used to compare the performance of engines of different displacements.

Most advances in thermal efficiency will be achieved through continued improvements in combustion, air handling, fuel injection equipment, and other subsystems. In addition, an effective exhaust heat recovery system may be necessary for achieving 50 percent BTE. However, the design of a waste heat recovery (WHR) system must take into account the temperature requirements of exhaust emission control devices as well as considerations such as weight, space, cost, reliability, and durability. In order to be commercially viable, the WHR system needs to last for the life of the engine, which carries an emissions warranty of 435,000 miles, but typically has a design life of 600,000 to 1 million miles. The 55 percent BTE stretch goal will require the research and development (R&D) of technologies discussed below in this chapter and should include comparable BTE improvements over the entire engine operating map, especially for those conditions used in a duty-cycle-weighted BTE.

Exhaust Emissions

Exhaust emissions of diesel engines have been regulated since 1973 by the California Air Resources Board (CARB) and since 1974 by the U.S. Environmental Protection Agency (EPA). After 1974, diesel engine manufacturers achieved remarkable reductions in oxides of nitrogen (NO_x) (~99 percent) and particulate matter (PM) (99 percent) emissions by modifying their engines and adding aftertreatment devices. Through 2006 heavy-duty diesel engines were certified at 2.5 g/bhp-h of NO_x + HC and 0.10 g/bhp-h PM (<0.05 g/bhp-h for transit buses). In 2007 the regulations allowed a phase-in of sales-averaged NO_x at approximately 1.2 g/bhp-h[2] and PM at 0.01 g/bhp-h (DOE, 2006).

Compliance with the 2007-2010 federal emissions standards is perhaps the strongest example of progress by diesel engine manufacturers since the National Research Council (NRC) Phase 1 review of the 21CTP in 2007 (NRC, 2008). Until 2007, exhaust aftertreatment had not been required or utilized to meet emissions standards for heavy-duty diesels (except for limited use of oxidation catalysts on buses and medium-sized trucks). The 2007-2010 regulations were intended by the EPA to be "aftertreatment-forcing." Aftertreatment technologies for PM were necessary in 2007, and all new truck heavy-duty diesel engines were equipped with diesel particulate filters (DPFs). Catalyst-based DPFs used with ultra-low-sulfur diesel fuel (<15 parts per million [ppm]) achieve PM reductions in excess of 90 percent from 2006 levels. In October 2006, ultra-low sulfur diesel fuel became the mandatory on-highway fuel, thus enabling the use of DPFs and other types of exhaust aftertreatment (NRC, 2008).

For 2010, NO_x emissions standards were lowered another 83 percent to 0.20 g/bhp-h NO_x + HC, along with 0.01 g/bhp-h PM. These standards have been met by most engine original equipment manufacturers (OEMs) by a combination of cooled exhaust gas recirculation (EGR) and selective catalytic reduction (SCR) for NO_x control and an actively regenerated DPF for particulate control. Meeting the requirements of 2010 exhaust emissions regulations required significant in-cylinder control, high-performing aftertreatment for NO_x and PM systems, and engine thermal management that includes a degree of control over exhaust mass flow rate, exhaust temperature, and exhaust oxygen. Thermal management, an essential element in the integration of engine and aftertreatment, allows both the DPF and the SCR to operate at peak efficiency over a wide range of duty cycles. In contrast to most manufacturers, which use SCR for NO_x control, Navistar is planning to achieve the NO_x standard with an EGR-only system and the PM standard with a DPF (Navistar, 2010).

Substantial effort across the industry went into the design of systems for storing and metering urea on the vehicle; these systems are required to support SCR systems. Considerations of freeze protection, contamination, labeling, and stability had to be accounted for. In addition, the infrastructure for distributing and dispensing urea at refueling outlets had to be developed. The industry adopted the name of Diesel Exhaust Fluid (DEF) for the aqueous urea solution. It was found that there is an optimum balance between in-cylinder control of NO_x and PM and aftertreatment control of NO_x and PM. The primary parameter determining this optimum balance is the operating cost, driven by both fuel consumption and DEF consumption. Fuel consumption has been affected by some of these emission control strategies, such as fuel used to regenerate particulate filters and DEF usage for the SCR system.

Another key enabling system technology is high-pressure common rail fuel systems with high-pressure capabilities exceeding 2,400 bar and allowing multiple injections per cycle. In addition, advancements in turbomachinery have resulted in variable-geometry turbochargers, the use of multiple turbochargers (in series and parallel) with aftercooling and intercooling, and turbocompounding. The turbomachinery serves several purposes in engine performance and emissions control, including airflow for high BMEP and transient response, EGR delivery and control, enhanced engine braking, and exhaust thermal management. EGR systems were introduced in 2002 and have mainly been high-pressure loop with cooling by means of the engine coolant. Some low-pressure loop EGR systems have also been introduced to the market. The first stage of implementation of on-board diagnostics (OBD) was completed on heavy-duty 2010 engines with a second stage in 2013.

As discussed in DOE (2006), engine controls deserve mention here, because historically, controls requirements for diesel engines have lagged those for gasoline engines in passenger cars. For the truck diesel engine, controls were primarily limited to one or two degrees of freedom (i.e.,

[2] The NO_x and nonmethane hydrocarbon (NMHC) standards were phased in for diesel engines between 2007 and 2010. The phase-in was on a percent-of-sales basis: 50 percent from 2007 to 2009 and 100 percent in 2010. In 2007, most manufacturers opted to meet a Family Emission Limit (FEL) around 1.2 to 1.5 g/bhp-hr NO_x for most of their engines (average of 0.2 g/bhp-h NO_x standard for 2010 and about 2.2 g/bhp-h NO_x portion of the 2.5 g/bhp-h NMHC + NO_x standard for 2006).

fuel injection delivery and timing). The future controls requirements in the heavy-duty diesel engine environment was realized with the introduction of EGR and the ongoing implementation of more sophisticated multipulse fuel injection systems and strategies. With the introduction of single- and multistage exhaust aftertreatment systems in 2007 and 2010 and the continuing progress of multimode combustion toward production feasibility, coupled with legislated or customer-demanded expansion of onboard sensing and diagnostic features, the minimum required capability of heavy-duty control systems hardware and software has increased as much as several orders of magnitude.

At present there is no expectation that new regulations will be promulgated to further reduce criteria emissions from new engines. Regulation for PM on a particle number basis has been introduced for vehicles and engines in Europe, and California has studies under way on this subject. The EPA and the Department of Transportation/National Highway Traffic Safety Administration (DOT/NHTSA) announced proposed standards for fuel consumption and greenhouse gas (GHG) emissions for medium- and heavy-duty vehicles and engines on October 25, 2010 (EPA/NHTSA, 2010) and issued final standards on September 15, 2011 (EPA/NHTSA, 2011).

DEPARTMENT OF ENERGY ENGINE PROGRAMS

DOE funding for FY 2007 to FY 2010 was provided for engine R&D in the following areas as shown in Table 1-2 (see Chapter 1 in this report):

- Advanced Combustion Engine
  —Combustion and Emission Control (shared between light- and heavy-truck engines)
  —Heavy Truck Advanced Combustion Engine
  —Waste Heat Recovery/Solid-State Energy Conversion
  —Health Impacts
- Fuels Technology
  —Advanced Petroleum-Based Fuels–Heavy Trucks
  —Non-Petroleum-Based Fuels and Lubes–Heavy Trucks
- Materials Technologies
  —Propulsion Materials Technology–Heavy Trucks
  —High Temperature Materials Laboratory
  Progress in these areas since 2007 is discussed below in this chapter.

Brake Thermal Efficiency Improvements

The DOE's targets for the improvement of engine brake thermal efficiency have evolved over the years, sometimes in ways that can be confusing, because different statements of a given goal can lead to varying interpretations. At the time of the NRC Phase 1 report published in 2008, the goals were set at 50 percent BTE in 2010 and 55 percent in 2013. These goals were in terms of the peak efficiency demonstrated by an engine in a test cell, where "peak efficiency" means the efficiency achieved by the engine at its best operating point. In general, peak efficiency occurs at relatively low speed and high load. As shown later in this section, the goal of a 50 percent peak BTE demonstration by 2010 was not met, but the technologies required to achieve this goal have been demonstrated successfully.

The DOE has now revised the goal to require 50 percent BTE at a load representative of over-the-road vehicle cruise conditions by 2015. This change was recommended in the NRC Phase 1 report (NRC, 2008). The actual speed and load point are not defined by the DOE in its goal statements, but in the SuperTruck projects (see Chapter 8 in this report), 65 miles per hour (mph) level road cruise is established as the operating point for 50 percent BTE. The committee believes that this is a reasonable operating point to target for high BTE, but it should be understood that this is a more difficult target than the old peak point target. At cruise, the engine must run at a load that is likely to be lower than the peak efficiency operating load, and the engine speed may also be higher than the peak efficiency speed. The goal of 55 percent peak BTE has now been delayed from 2013 to 2018.[3] This stretch thermal efficiency goal of 55 percent in prototype engine systems would lead to a corresponding 10 percent gain in over-the-road fuel economy relative to the earlier 50 percent BTE goal at a corresponding condition representing 65 mph level road load cruise. The committee believes that this will be a very difficult goal to achieve.

Research and Development Programs

The DOE and the heavy-duty engine industry have been working in public-private partnerships to develop and demonstrate advanced diesel engine technologies and concepts that improve engine thermal efficiency while meeting the EPA's 2010 emissions standards. The two technology goals established by the 21CTP for improving brake thermal efficiency of heavy-duty engines are discussed in this section (DOE, 2011a).

Programs and Projects Directed Toward Achieving 21CTP Goal 1

21CTP Goal 1: Develop and demonstrate an emission compliant engine for Class 7-8 highway trucks that achieves 50% brake thermal efficiency in an over-the-road cruise condition, improving the engine fuel efficiency by about 20% (from approximately 42% thermal efficiency today) by 2015 (DOE, 2011a).

The goal for 50 percent BTE discussed in the NRC Phase 1 report (NRC, 2008) was for the peak efficiency condition.

[3] As noted in this chapter in the discussion on the 21CTP Goal 2, the 55 percent BTE goal in the 21CTP updated white paper, "Engines" (DOE, 2011a) is for a prototype engine system in the laboratory by 2015. In the DOE Multi-Year Program Plan, the goal is for 2018 in a prototype engine (DOE, 2010d).

TABLE 3-1 Accomplishments of High Efficiency Clean Combustion Projects

Contract (Company, Project Leader, Number)	Funding	Dates	Accomplishments
Caterpillar, Chris Gehrke, DE-FC26-05-NT42412	DOE: \$10,309,000 CAT: \$10,309,000	Start: August 2005 Finish: July 2010	4% BSFC[a] improvement below 750 kPa BMEP
Cummins, Donald Stanton, DE-FC26-05-NT42418	DOE: \$13,420,136 Cummins: \$13,629,115	Start: October 2005 Finish: March 2010	15 L engine: 10.2% improvement in brake thermal efficiency (2010 in-cylinder NO_x control) 15 L engine: 16.4% improvement in brake thermal efficiency (2010 emissions with SCR NO_x technology) 6.7 L engine: 14% improvement in fuel economy (Tier 2 Bin 8 emissions met without NO_x aftertreatment)

NOTE: Acronyms are defined in Appendix I.

[a] Brake specific fuel consumption (BSFC) is a common and convenient measure of the thermal efficiency of an engine. It is the ratio of the mass of fuel consumed to the work produced by the engine; lower BSFC means higher thermal efficiency. It is different from efficiency in that it measures fuel consumption, whereas the efficiency is a dimensionless ratio that measures the portion of the fuel energy input that gets converted into work output. The two are related to one another through the energy content of the fuel.

SOURCE: DOE (2010b) Annual Merit Review.

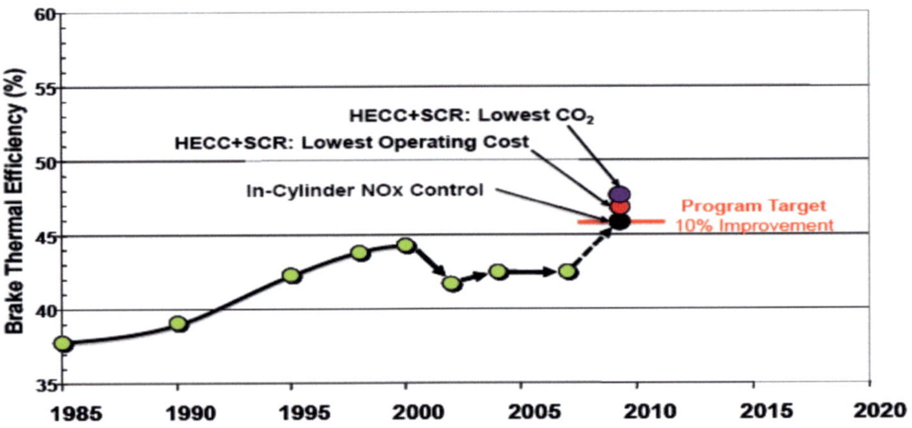

FIGURE 3-1 High Efficiency Clean Combustion (HECC) engine efficiency improvements. SCR, selective catalytic reduction. SOURCE: Stanton (2010).

However, as noted above, the DOE revised the goal for 50 percent BTE so that it is now for an over-the-road cruise condition. Programs and projects directed toward the development of technology required to achieve Goal 1 include High Efficiency Clean Combustion (HECC), Waste Heat Recovery (WHR), the "NZ50" project at Detroit Diesel Corporation (DDC), and others (DOE, 2011a). These programs and projects are reviewed in this section. Cost-effectiveness was not part of these R&D projects, but cost-effectiveness evaluations will be a required outcome of the SuperTruck projects, as discussed in Chapter 8.

High Efficiency Clean Combustion

A key objective of the HECC program has been to design and develop advanced engine architectures that improved BTE by 10 percent compared to the 2006 product, according to the presentation to the committee.[4] The essence of the work was the development of clean combustion in the form of low-temperature, highly premixed combustion combined with lifted flame diffusion controlled combustion. Using these technologies, 10 percent engine BTE improvement targets have been achieved using no NO_x aftertreatment. When integrating HECC-developed technologies with SCR NO_x aftertreatment system, further engine efficiency enhancements were demonstrated. The results from the HECC projects are listed in Table 3-1 and shown in Figure 3-1.

[4] Donald Stanton, Cummins, Inc., "Cummins-Peterbilt Super Truck Program," presentation to the 21CTP committee, September 9, 2010, Washington, D.C.

TABLE 3-2 Accomplishments of Waste Heat Recovery (WHR) Projects

Contract Company, Project Leader, Number	Funding	Dates	Accomplishments
Caterpillar, Richard W. Kruiswyk, DE-FC26-05-NT42423	DOE: \$2,188,000 CAT: \$2,188,000	Start: October 2005 Finish: June 2010	10% improvement in BTE was not achieved. 9% improvement target: 7 percent "virtual" demonstration. Demonstrated turbocompounding vs. Brayton cycle for bottoming. Demonstrated improvements in turbocharger compressor and turbine efficiencies.
Cummins, Chris Nelson, DE-FC26-05-NT42419	DOE: \$4,245,906 Cummins: \$4,488,836	Start: June 2005 Finish: March 2010	10% improvement in BTE (8% from WHR). (Generation 1, No NO_x AT, electric accessories) 10% improvement in BTE (6% from WHR). (Generation 2, SCR NO_x AT, no electric accessories) Both generations included Organic Rankine Cycle.

NOTE: Acronyms are defined in Appendix I.
SOURCE: DOE (2010b).

The historical thermal efficiency data shown in Figure 3-1 indicate a drop in 2002, with only minor improvements until 2010. This drop was caused by measures applied by engine manufacturers to meet requirements for lower NO_x emissions (NRC, 2010, Figure 4-2). With NO_x requirements now stabilized, engine manufacturers will be able to refocus efforts on thermal efficiency improvements.

Waste Heat Recovery

The objective of the WHR program is to improve engine BTE by 10 percent (i.e., from 42 percent to 46 percent BTE) by capturing and converting wasted heat energy to useful work. A bottoming cycle or turbocompounding captures heat from engine exhaust gas recirculation, charge air, and exhaust streams. WHR systems were designed and developed so that a bottoming device could be coupled to the engine either mechanically or electrically through a high-speed generator. The results from these projects are listed in Table 3-2. The Cummins WHR system is shown in Figure 3-2 (Nelson, 2010).

The highest-quality source of heat for WHR comes from the EGR stream, as illustrated in Figure 3-2. As the effectiveness of SCR is further improved, the use of EGR is likely to

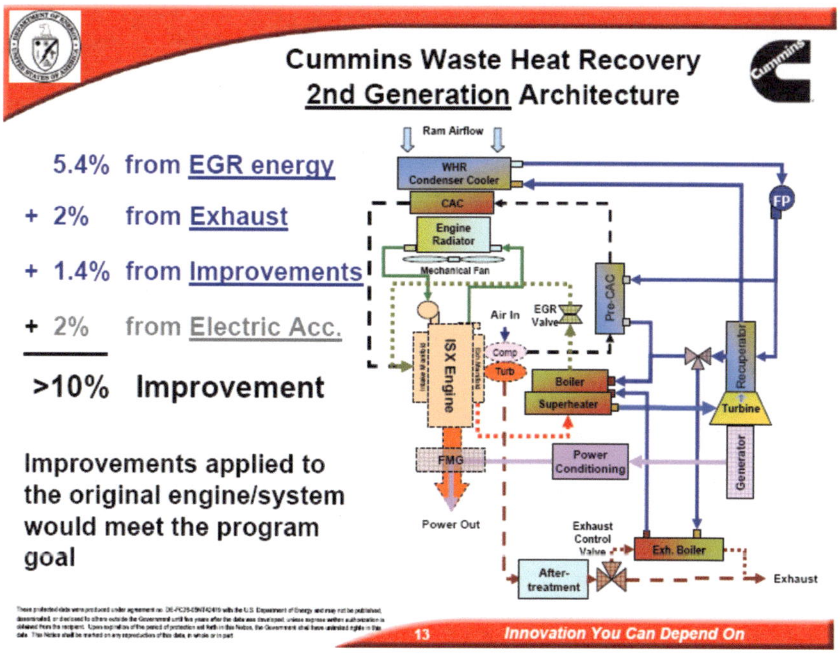

FIGURE 3-2 Cummins Waste Heat Recovery (WHR) system, second-generation architecture. Acronyms are defined in Appendix I. SOURCE: Nelson (2010).

TABLE 3-3 Accomplishments of Other Engine Projects Funded in Part by the Department of Energy

Contract	Funding	Dates	Accomplishments
DDC/Daimler NZ50, Kevin Sisken and Houshan Zhang, DE-FC05-00-OR22805	DOE: \$5,000,000 DDC: \$5,000,000	Start: February 2007 Finish: 2010	Goal: 10% improvement in BTE. Results: Combustion, 4% (2% more expected); controls, 3%; turbocharger, 1%.
Navistar—Low Temperature Combustion Demonstration for High Efficiency, William de Ojeda, DE-FC26-05-NT42413	DOE: \$4,021,234 Navistar: \$5,153,881	Start: October 2005 Finish: May 2010	Goal: 5% BSFC at 2010 emissions, no AT Fuel economy improved 4% to 5.5% (with PCCI, VVA, and combustion feedback). Load range for LTC extended to 16.5 bar BMEP
General Motors, Kenneth Patton,	DOE: \$5,000,000 GM: \$5,000,000	Start: 2005 Finish: 3rd Quarter, 2009	Achieved 5% to 8% lower BSFC with internal EGR and VVA.
Ford Motor Company, Harold Sun, DE-FC26-07-NT43280	DOE: \$1,500,000 Ford: \$1,500,000	Start: October 2007 Finish: June 2010	Goal: Demonstrate a turbocharger with 3% to 5% fuel economy improvement at Tier II Bin 5 emissions. 2010 DEER presentation (Sun et al., 2010): Not obvious if the goals were met.

NOTE: Acronyms are defined in Appendix I.
SOURCE: DOE (2010b).

decline, which would reduce the potential benefit of WHR. If only the lower-quality heat sources (post-aftertreatment exhaust gas, charge air, and coolant) were available to be captured, the efficiency of the WHR system would be expected to decrease. Therefore, the trade-off between the use of higher-efficiency SCR systems with the decrease in effectiveness of the WHR system will need to be balanced.

Other Projects

Detroit Diesel/Daimler, Navistar, General Motors, and Ford Motor Company also had projects summarized in Table 3-3.

Volvo had a project that started in October 2007 and finished in September 2009. It was funded at \$9 million, with the DOE contributing \$3 million and Volvo \$6 million (subtopic 6B of DOE-FOA-0000239[5]). The results showing lower brake specific fuel consumption (BSFC) by 3 percent were presented at the DOE 2009 Merit Review. It appears that this project is also reported as the Bilateral Program being conducted by Volvo Powertrain North America and Volvo AB of Sweden.[6] It is jointly funded by the DOE and the Swedish Energy Agency. The objectives are to reduce CO_2 by 10 percent while meeting 2010 EPA criteria emissions, develop an engine platform capable of biodiesel fuel, and develop multifuel vehicles and drivelines. The FY 2010 accomplishments were as follows (DOE, 2009):

- A turbocompound engine showed 5 percent lower fuel consumption. Computational fluid dynamics (CFD) modeling shows that a high-performance piston can improve BTE by 1 percent.
- A parallel Rankine cycle WHR system can improve BTE by approximately 7 percent at B50[7] and 8.5 percent at C100.[8]
- Analysis of mechanical, electrical, and electromechanical transmission alternatives to connect the turbocompound power turbine to the crankshaft showed the mechanical transmission best for fuel consumption.
- Hybridization simulation showed 2 percent to 8 percent fuel consumption reduction in European applications and zero percent to 2.2 percent in typical U.S. applications.
- A biofuel B20[9] endurance truck test was initiated (DOE, 2009).

The nine diesel engine programs and projects reviewed above are examples of DOE public-private partnerships. As indicated previously, several programs exceeded their goals of 5 percent or 10 percent improvement in BTE. Not all programs met their goals, but some programs sorted out technologies and identified them for further development.

[5] Funding Opportunity Announcement (DOE, 2010b) available at http://www1.eere.energy.gov/vehiclesandfuels/pdfs/de-foa-0000239.pdf .

[6] E-mail, Rich Bechtold to committee member, David Merrion, November 18, 2010, report II.A.20 "Very High Fuel Economy, Heavy-Duty, Narrow-Speed Truck Engine utilizing Biofuels and Hybrid Vehicle Technologies."

[7] B50, a point on the EPA supplemental engine test (SET), 50 percent load, mid-speed.

[8] C100, a point on the EPA SET, 100 percent load, high speed.

[9] B20, 20 percent biofuel mixed with diesel fuel.

The demonstrations of BTE improvements were accomplished mainly at peak BTE operating conditions, with 42 percent as the baseline. As discussed in the NRC Phase 1 report (NRC, 2008), the BTE goal should be demonstrated at a more representative road load cruise condition, which is now incorporated in 21CTP Goal 1. The Cummins demonstrations, such as those shown in Figure 3-1, were at a highway cruise condition.

The current SuperTruck projects call for the contractors to demonstrate the 50 percent BTE target at cruise load on engines installed in vehicles by the end of 2014. If the contractors are able to meet the target, this would enable the 21CTP Goal 1 to be met by 2015. It should be noted that the SuperTruck projects include efforts to reduce the load that the vehicle places on the engine as well as to improve the efficiency of the engine. By lowering the load on the engine, the SuperTruck projects will actually make reaching the 50 percent efficiency goal at cruise conditions more difficult (see Chapter 8 for more detail). There is some risk that the engine thermal efficiency goal will not be met, or that meeting the goal will require complex and expensive technology that would be difficult to implement in production and which may not be cost-effective.

Finding 3-1. The committee reviewed nine diesel engine programs that were funded at a total of more than $100 million by the DOE and industry and that included the HECC program, the WHR program, and others. Some programs met or exceeded their goals, for example achieving a 10.2 percent improvement in BTE versus a 10 percent goal, whereas others did not quite meet the goals of 5 percent or 10 percent improvement in BTE. By combining HECC and WHR, each demonstrating greater than 10 percent improvement in BTE, together with other technologies, it should be possible to improve BTE by 20 percent to achieve the original DOE target of 50 percent peak BTE. However, the DOE target of 50 percent peak BTE was not met by the original goal of 2010.

Finding 3-1a. The DOE has shifted the original target of 50 percent peak BTE by 2010 to a new target of 50 percent BTE at an operating point representative of vehicle load during highway cruise operation. This makes the efficiency target more difficult to meet and may require complex and expensive technology that extends beyond the technologies demonstrated on engines to date. These technologies will not necessarily be production-feasible or cost-effective.

Programs and Projects Directed Toward Achieving 21CTP Goal 2

21CTP Goal 2: Research and develop technologies which achieve a stretch thermal efficiency goal of 55% in prototype engine systems in the laboratory by 2015 (DOE, 2011a).

In the public-private DOE partnerships discussed, industry participants have provided technology roadmaps of their strategies for achieving brake thermal efficiencies of 50 and ultimately 55 percent. Improving the efficiency of the in-cylinder conversion of fuel energy into shaft work is an important component of each participant's roadmap. To do this, all of the participants project incorporating new modes of combustion into their engine operation. Figure 3-3 is an example of the integration of new combustion technologies, like HECC and low-temperature combustion (LTC), that will be needed to achieve the efficiency targets of the program (NRC, 2008).

Gaining the requisite fundamental understanding of the phenomena governing advanced combustion modes, such as those shown in Figure 3-3 (early pre-mixed charge compression ignition [PCCI, LTC] and lifted flame diffusion

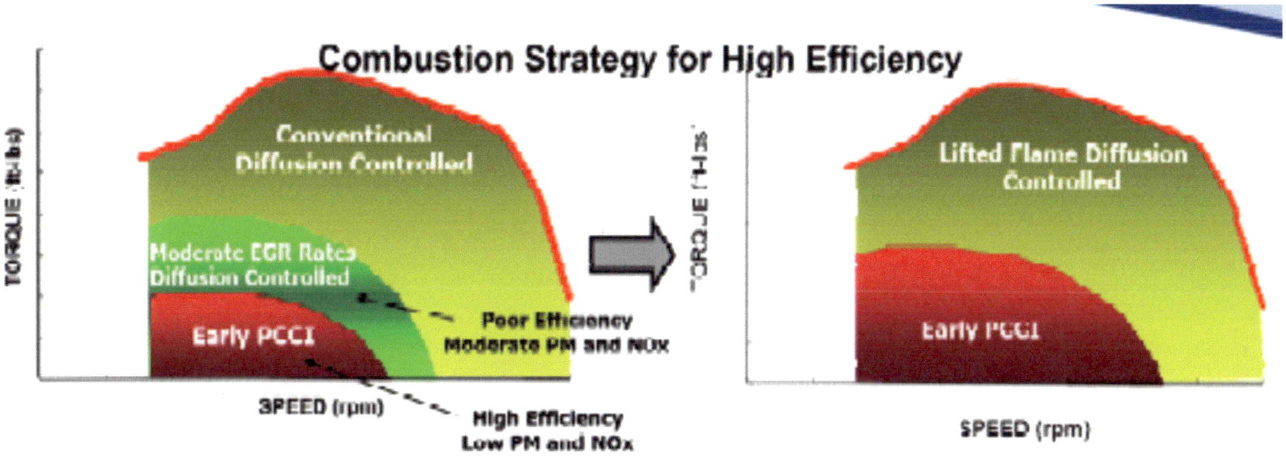

FIGURE 3-3 Schematic representation of the evolution of combustion processes to be used at different engine loads and speeds. Acronyms are defined in Appendix I. SOURCE: Gurpreet Singh, DOE, "Overview of the DOE Advanced Combustion Engine R&D Subprogram," presentation to the committee, November 15, 2010, Washington, D.C.

controlled combustion), is the focus of the DOE combustion activities within the national laboratories, industry, and academia. These advanced modes of combustion are achieved through synergistic interactions between the auto-ignition and reaction kinetics of the fuel, the thermodynamic states within the cylinder—the pressure, and distributions of temperature, fuel, oxygen, nitrogen, and the recirculated exhaust gases—and the rate of the mixing within the cylinder. The mixing processes are manipulated through fuel injection characteristics and in-cylinder fluid mechanics, which in turn are influenced by the shape of the combustion chamber and subject to manipulation through the intake and exhaust processes, which can be controlled by engine valve actuation. To maintain robust combustion throughout the engine operating range with multimode combustion requires the dynamic manipulation of a wide range of engine control variables. The goals of the DOE laboratory and university research programs are to develop the fundamental understanding and tools necessary to facilitate such a high level of engine combustion control.

This fundamental research is conducted under the Advanced Engine Combustion Memorandum of Understanding (AEC MOU) between industry and national laboratories that was initiated in 2003. The partners involved in the AEC MOU include 10 engine producers (Caterpillar, Cummins, Detroit Diesel, Navistar/International, John Deere, Mack/Volvo, General Electric, General Motors, Ford, and Chrysler), 5 energy companies (Chevron, ConocoPhillips, Shell, ExxonMobil, and BP), and 6 national laboratories (Argonne National Laboratory [ANL], Lawrence Livermore National Laboratory [LLNL], Los Alamos National Laboratory [LANL], Oak Ridge National Laboratory [ORNL], Sandia National Laboratory [SNL], and the National Renewable Energy Laboratory [NREL]). The energy companies joined the AEC MOU in late 2006 and brought a research focus involving fuel effects on advanced combustion strategies. The MOU was recently unanimously renewed by the partners through 2013.

As part of the 21 CTP Phase 2 review process, a list of nine DOE-funded projects focused on advanced combustion that were directly attributed to the 21CTP was supplied by the DOE to the committee. These projects are listed in Table 3-4 and were budgeted in FY 2010 for $5.66 million.

There is interaction and leverage between these projects and other DOE-supported combustion projects, but these projects were specifically designated by the DOE as being under the 21CTP umbrella, so the discussion below is limited to these projects.

Researchers participating within each of the nine projects have developed collaborative teams consisting of industry and academic partners. Academic participants are required to establish a 20 percent cost share. Everyone participates in two group research meetings per year and in an annual merit review as part of the DOE's Energy Efficiency and Renewable Energy (EERE) program. Project directions and continuation are based on the scores received in the merit review. The presentations given at the merit review are available to the public (DOE, 2010b).

TABLE 3-4 Major 21CTP-Related Projects Addressing Advanced Combustion Fundamentals

Title	Organization	DOE Funding: FY 2009 Budget	DOE Funding: FY 2010 Budget
KIVA 4 Development (advanced computer program)	LANL	$290,000	$290,000
Computationally Efficient Modeling of High-Efficiency Clean Combustion	LLNL	$1.0 million	$1.0 million
Chemical Kinetic Research on HCCI and Diesel Fuels	LLNL	$400,000	$400,000
Stretch Efficiency for Combustion Engines: Exploiting New Combustion Regimes	ORNL	$250,000	$250,000
HCCI and Stratified-Charge Compression Ignition Engine Combustion Research	SNL	$750,000	$750,000
Heavy-Duty Low-Temperature and Diesel Combustion Modeling	SNL	$580,000 with $115,000 subcontract to UW-Madison	$660,000 with $115,000 subcontract to UW-Madison
Combustion Modeling Large Eddy Simulation Applied to LTC/Diesel/Hydrogen Engine Combustion Research	SNL	$450,000	$450,000
Low-Temperature Diesel Combustion Cross-Cut Research	SNL	$570,000	$660,000
Optimization of Advanced Diesel Engine Combustion Strategies	University of Wisconsin–Madison	$360,000	$1.2 million

NOTE: Acronyms are defined in Appendix I.
SOURCE: DOE (2010b). Available at http://www1.eere.energy.gov/vehiclesandfuels/resources/proceedings/2010_merit_review.html.

The focus of the research projects listed in Table 3-4 is the development of a greater fundamental understanding of the processes that need to be sensed, and controlled, for the integration of advanced, high-efficiency clean combustion processes, into the engine operating map. The work includes the continued development of the advanced computer program KIVA, which is the framework that researchers use for model development, and the development of higher-resolution turbulence models for better spray and fluid-mixing simulation. Comprehensive kinetic routines for diesel-like, gasoline-like, and representative biofuels are an important aspect of the research. These routines identify the critical chemical kinetic pathways that must be included in the simulation for accurate results, which in turn allows for the development of reduced kinetic schemes that are critical for computational efficiency. Reduced kinetic schemes are only one aspect of developing computationally efficient simulations. Research is being supported that addresses computationally efficient methods of solving thermodynamic, fluid mechanic, and chemical kinetic coupled systems.

An integral part of this fundamental research consists of the detailed experimental measurements that identify the important phenomena occurring within the reacting systems, which aids in the model development and serves as a basis for the comparison of model predictions. Spatially and temporally resolved data are being obtained with advanced optical diagnostics techniques. Such measurements have elucidated the effects of in-cylinder temperature and mixture stratification on LTC processes, differences in particulate formation processes for varying combustion regimes, and details of the fuel distribution and wall interactions associated with pulsed injection, an important component of controlling LTC processes. Finally, work is being supported that endeavors to combine all of the advanced understanding and perform optimization studies and to push the efficiency boundaries by pursuing advanced combustion concepts to reach the stretch efficiency goals.

The fundamental combustion and emissions research to date under the AEC MOU has led to significant advances in the understanding of various strategies for achieving LTC. Critical aspects of how HCCI and diesel LTC progress, how their heat release rate and combustion phasing can be controlled, the sources of hydrocarbon (HC) and carbon monoxide (CO) emissions when the LTC approaches are pushed to limits of operation, and fuel effects on LTC are being unraveled.

Higher efficiencies in heavy-duty truck engines have also been shown in the laboratory. Implementation of diesel LTC approaches has begun in heavy-duty diesels for a portion of the fuel burned during moderate- to light-load parts of the engine operating range, providing significant engine-out emissions reduction. In general, higher injection pressure, multipulse injection, and EGR use have allowed a greater fraction of the reactive mixtures during diesel combustion to be pushed toward LTC conditions, contributing to the lower engine-out emissions that have been achieved. Highlights of two of these research projects are summarized here.

- *Gasoline HCCI:* HCCI—or, more generally speaking, low-temperature combustion strategies—applied in the laboratory environment, using conventional gasoline, have achieved light-load operation down to engine idle conditions and loads as high as 16 bar brake mean effective pressure (limited by the laboratory engine head design) offering the possibility of an engine operating with LTC over a large portion of the engine map. Peak indicated efficiencies in a light-truck-size engine of 48 percent were achieved with NO_x levels less than the 2010 standard, near-zero soot levels, and controlled heat release to provide low noise and preclude engine knock (Ra et al., 2010).
- *Dual-Fuel LTC:* Recent research in heavy-duty engines with dual-fueled (gasoline- and diesel-fueled) HCCI/LTC approaches are indicating potential for 50+ percent brake thermal efficiencies, controlled heat release rates, and 2010 emissions levels. Dual-fuel operation has been achieved from 4 to 17 bar BMEP in a heavy-duty laboratory engine, while achieving NO_x levels below the 2010 standard and near-zero soot levels. The start of combustion timing is controlled by the diesel fuel injection, and the heat release rate is controlled by the gasoline fraction.[10,11] This work has recently been extended, with encouraging results to bio-based fuels and single fuels with the addition of a cetane improver (Splitter et al., 2010).

The improved fundamental understanding has also advanced computational tools for engine design. Most engine designers are increasingly and aggressively using computational tools developed through support by DOE's Vehicle Technologies Program for experimental research and engine CFD development efforts. The growing use of computational tools for engine design is exemplified by Cummins introduction of the ISB 6.7-liter (L) light-truck (Class 2b) diesel in 2007. This diesel engine was computationally designed with much reduced testing to confirm performance. The design process led to reduced design time and a more robust design with reduced fuel consumption while meeting 2010 emissions standards (for trucks over 8,500 lb using chassis dynamometer certification). The future introduction of more robust computational design tools able to simulate the full range of engine combustion approaches (conventional mixing-controlled diesel combustion premixed and stratified flame propagation, the LTC bulk ignition and combustion

[10] Gurpreet Singh, DOE, "Overview of the DOE Advanced Combustion Engine R&D Subprogram," presentation to the committee, November 15, 2010, Washington, D.C.

[11] Kevin Stork, DOE, "DOE Fuel and Lubricant Technologies R&D," presentation to the committee, November 15, 2010, Washington, D.C.

processes) has very strong potential to lead to even faster evolution and improvement of cost-effective engines.

Finding 3-2. The DOE-funded research in advanced engine combustion at the national laboratories, in industry, and at universities is well managed and addresses important aspects for achieving an integration of advanced combustion processes that should be important enablers for achieving the 55 percent BTE goal as well as providing ongoing improvements. There also appears to be good interaction between the researchers performing the work and the industry stakeholders. Efforts to achieve 55 percent BTE are going to require complex and expensive technologies. It will be some time before it becomes clear whether there is a production-feasible and cost-effective way to achieve the 55 percent BTE target. The committee believes that this target carries considerable risk, even at the test cell demonstration stage.

Recommendation 3-1. The 21CTP fundamental research program should continue to provide important enablers for the 55 percent BTE goal, and the DOE should continue to look for leverage opportunities with other government- and industry-funded projects.

Future Engine R&D in the SuperTruck Program

The DOE, in 2010, announced three SuperTruck project awards to demonstrate engine and truck efficiency enhancements for Class 8 vehicles in real-world conditions (see Chapter 8). The aim of these developments is to foster quicker introduction of new technologies into the marketplace, thereby achieving energy savings later in this decade. The DOE made the decision to carry out future engine R&D under the SuperTruck program, but it retained Goals 1 and 2 discussed above. Key engine technology demonstrations under the 5-year SuperTruck program include the following:

- *Engine Goal 1:* Develop and demonstrate a 2010 emissions compliant engine system for Class 8 trucks that achieves 50 percent BTE at an over-the-road cruise condition by improving engine efficiency 20 percent from 42 percent BTE today, by 2015 (DOE, 2011a). This improvement in engine BTE will provide at least 20 percentage points of the 50 percent improvement in vehicle freight efficiency (ton-miles per gallon) (equivalent to 33 percent improvement in fuel consumption [gallons per 1,000-ton-miles]), which is the overall goal of the SuperTruck program.
- *Engine Goal 2:* Research and develop technology pathways using modeling and analysis to achieve a stretch goal of 55 percent BTE in a 2010 emissions compliant engine system in the laboratory, by 2015 (DOE, 2011a).[12] This efficiency gain would be equivalent to an additional 10 percent gain in over-the-road fuel economy when prototype concepts are fully developed for the market.

The SuperTruck engine programs are discussed in this section; the vehicle programs are discussed in Chapter 8.[13,14,15]

- *Daimler Trucks/Detroit Diesel:* Technologies listed in the September 9, 2010, presentation to the committee are as follows
  —Advanced fuel injection,
  —Optimized combustion,
  —Air/EGR refinement,
  —Friction reduction,
  —Accessories,
  —More efficient operating point,
  —Waste heat recovery,
  —Next generation controller,
  —Higher engine out NO_x,
  —Engine downsizing,
  —Optimized aftertreatment.
- *Cummins:* Technologies listed in Cummins's September 9, 2010, presentation to the committee, the Cummins's presentation to the DOE DEER Conference September 29, 2010, and the company's presentation during the committee's site visit to the Cummins on November 8, 2010 (see Appendix B) are as follows:
  —Fuel system,
  —Advanced LTC,
  —Controls: electronic,
  —Electrically driven components,
  —Waste heat recovery: EGR, charge air, exhaust heat, mechanical coupling,
  —Aftertreatment,
  —Turbo technology,
  —EGR loop,
  —Variable valve actuation (VVA),
  —Base engine: peak cylinder pressure, friction, parasitic.
- *Navistar:* Technologies listed in Navistar's September 9, 2010, presentation to the committee and its presentation during the committee's site visit to the Navistar,

[12] The DOE uses the term "stretch" for goals that cannot be achieved by incremental or small improvements and are thus very difficult to achieve, which seems appropriate for federal programs. The committee supports the use of this terminology because this thermal efficiency goal will be difficult to achieve.

[13] Derek Rotz, Daimler, "Daimler's SuperTruck Program," presentation to the committee, September 9, 2010, Washington, D.C.

[14] Donald Stanton, Cummins, "Cummins-Peterbilt SuperTruck Program," presentation to the committee, September 9, 2010, Washington, D.C.

[15] Anthony Cook, Navistar, "Navistar's Super Truck Program," presentation to the committee, September 9, 2010, Washington, D.C.

Inc., Truck Development and Technology Center on January 13, 2011, are as follows:
—Combustion efficiency improvement:
Injection pressure, nozzle, bowl optimization, combustion feedback control, cylinder head and port;
—Air system enhancements:
Turbocharger efficiency, hybrid EGR system;
—WHR (turbocompounding and Rankine):
Rankine cycle—Common cycle used to generate electricity, fluid development for typical truck engine heat range;
Turbocompounding—Dual turbines, visco-mechanical drive to the crankshaft, microturbine;
—Aftertreatment:
Minimize regeneration, opportunistic regeneration, PM-NO_x balance;
—Friction reduction/insulation:
Low friction, weight reduction, electrification of engine accessories, drive mechanism improvements, heat rejection reduction, insulated exhaust ports and manifolds;
—VVA:
Effective compression ratio control, cylinder deactivation;
—Dual fuel (gasoline and diesel).

The SuperTruck engine programs are a continuation of the previously discussed DOE diesel engine programs and have the same goals as those previously discussed. It is unlikely that other heavy-duty diesel engine programs will be funded by the DOE. (Also see the Findings and Recommendations in Chapter 8 regarding the SuperTruck projects.)

In the 21CTP projects to date, as described earlier in this chapter, 50 percent BTE has not been demonstrated in an engine in a vehicle. Significant advancements have been made for individual technologies, which, if combined in an engine, are expected to provide the key elements required to improve BTE by 20 percent to achieve the 50 percent BTE goal. The requirement to demonstrate 50 percent BTE at an over-the-road cruise condition poses an additional task, because the best point for BTE is typically at a higher load. Adding to this task will be the significantly reduced power demand at the over-the-road cruise condition resulting from the SuperTruck vehicle improvements.

The SuperTruck project teams revealed few, if any, plans concerning the research and development of technology pathways to achieve a stretch goal of 55 percent BTE. Moving toward this goal is expected to build on the future achievement of the 50 percent BTE goal and to rely heavily on the DOE-funded research programs in advanced engine combustion discussed in the previous section. Without having specific plans to review for this goal, the committee considers the 55 percent BTE very high risk, although it might be achievable.

Finding 3-3. Future engine R&D for Goal 1, develop and demonstrate 50 percent BTE at over-the-road cruise conditions by 2015, and for Goal 2, research and develop technology pathways to achieve a stretch goal of 55 percent BTE in a 2010 emissions-compliant engine system in the laboratory by 2015, will be carried out under the SuperTruck program. The engine programs outlined by the three SuperTruck project teams appear to be comprehensive and are expected to achieve the 50 percent BTE goal, although there is risk in being able to achieve the goal at a cruise condition with the significantly reduced power demand level of the SuperTruck. Developing engine technology pathways to achieve the stretch goal of 55 percent BTE in an engine in a laboratory by 2015 is considered very high risk, but might be achievable.

Recommendation 3-2. The DOE should ensure that the engine R&D for the goal of 50 percent BTE at over-the-road cruise conditions and the stretch goal of 55 percent BTE in an engine in a laboratory that will now be carried out under the SuperTruck program receive the appropriate share of the SuperTruck funding and benefit extensively from the DOE-funded research programs in advanced engine combustion.

ENGINE PROGRAMS OF THE ENVIRONMENTAL PROTECTION AGENCY

The EPA is developing several candidate types of engines specifically for application to its series hydraulic hybrid truck programs.[16] Although the DOE is not funding the EPA, the EPA's work fits under the 21CTP. The engines that are under development by the EPA are (1) optimized alcohol engines and (2) homogeneous-charge compression ignition engines. The BTE values of these engines are approaching the 42 percent BTE of current diesel engines, as shown in Table 3-5.

High-Efficiency Alcohol-Fuel Engines

The EPA has been developing optimized alcohol-fueled engines for the high-octane E85 fuels that can potentially provide a cost-effective engine technology suitable for both conventional and hybrid vehicles for the medium-duty fleet truck market. The high-octane number of alcohol, together with its latent heat of vaporization, enable the use of high compression ratios and boosted applications. Alcohol's high laminar flame speed relative to gasoline permits greater charge dilution with EGR, reducing the need for throttling at light to moderate loads while still allowing stoichiometric operation that facilitates the use of a three-way catalyst (Brusstar and Gray, 2007). The EPA has achieved current diesel levels of BTE with E85 fuel using conventional three-way catalysts to meet the 2010 emissions standards.

[16] John Kargul, EPA, "Clean Automotive Technology, Cost Effective Solutions for a Petroleum and Carbon Constrained World," presentation to committee subgroup, October 26, 2010, Ann Arbor, Michigan.

TABLE 3-5 Engines Under Development by the Environmental Protection Agency for Series Hydraulic Hybrid Trucks

Engine Type	Exhaust Gas Recirculation Rate (%)	Brake Thermal Efficiency (%)
Alcohol fueled engine: E85	20	41
Homgenous-charge compression ignition engine: Gasoline port injection	50-60	39

SOURCE: John Kargul, EPA, "Clean Automotive Technology, Cost Effective Solutions for a Petroleum and Carbon Constrained World," presentation to the committee subgroup, October 26, 2010, Ann Arbor, Michigan.

Finding 3-4. The EPA has demonstrated that optimized E85 alcohol-fueled engines using conventional three-way catalysts for meeting 2010 emissions standards can achieve current diesel levels of BTE that can potentially provide engine technology suitable for both conventional and hybrid vehicles for the medium-duty fleet truck market.

Homogenous-Charge Compression Ignition Gasoline Fuel Engine

The EPA's early development of an HCCI engine was focused on potential applications in a series hydraulic hybrid vehicle (Sun et al., 2004). This HCCI engine operated on commercial 87 octane gasoline. In 2008, the California South Coast Air Quality Management District, CARB, and IC Bus (a Navistar company) became interested in demonstrating gasoline HCCI engine technology in a shuttle bus application. A Cooperative Research and Development Agreement (CRADA) project was formed, and the EPA applied its gasoline HCCI technology to a Navistar 6.4 L diesel engine (provided by Navistar) for use with the series hydraulic hybrid technology in a shuttle bus (provided by Navistar). Features of this engine are shown in Table 3-6. The engine was mapped to meet the following control strategy targets: best efficiency, stable operation with the coefficient of variation of indicated mean effective pressure (IMEP) less than 3 percent, maximum rate of pressure rise approximately equal to 6 bar/deg, and NO_x emissions less than 0.2 g/kW-h on the unique test cycle (discussed later in this section).

The EPA emphasized that the HCCI engine can operate successfully because the series hydraulic hybrid engine operates over a narrow operating region of the engine map and has only slow transients. The engine is controlled to operate through a narrow region that encompasses the best BTE at each engine speed encountered. The response time for a transient ramp-up of power (from idle to a demanded power level) was controlled to 3 seconds to maintain stable combustion, whereas a typical response time for an engine directly driving the wheels is in the 50 to 100 millisecond (msec) range. Likewise, the response time for a down-power command was controlled to 1 second. The series hydraulic hybrid application reduces the need for rapid transients. However, the EPA is continuing its research to maintain stable combustion through the entire engine operating range to broaden the potential applicability of the HCCI engine concept beyond the series hydraulic hybrid application. EPA tests on the HCCI engine, using a unique test cycle that reproduces the engine operating conditions for the series hydraulic hybrid heavy-duty vehicle, have shown that NO_x and PM emissions are below the levels required by the 2010 emissions standards without aftertreatment.[17] HC and CO emissions are controlled with oxidation catalysts.

The EPA is continuing to address the following challenges for the HCCI combustion concept: controlling ignition and

TABLE 3-6 EPA's Homogenous-Charge Compression Ignition (HCCI) Engine Features for a Series Hydraulic Hybrid Demonstrator Shuttle Bus

Feature	Description
Base engine	6.4 L base engine (base was a diesel engine)
Compression Ratio	16.5:1 CR
Turbocharger	Variable geometry turbocharger with intercooler
Cooling	Coolant and air-cooled EGR
Fuel Injection	Port fuel injection
Throttle	Unthrottled except for starting
Ignition	Spark Plugs in place of diesel fuel injectors (used for 10-20 seconds during starting)
Sensors	Combination combustion pressure sensors and glow plugs (in place of glow plugs)
	Four knock sensors (each sensor serves a pair of adjacent cylinders)
Aftertreatment	No SCR or DPF
Power	Engine performance 120 kW versus 130 kW for base diesel engine
Fuel	Can operate on gasoline, diesel, M25, and M50

SOURCE: John Kargul, EPA, "Clean Automotive Technology, Cost Effective Solutions for a Petroleum and Carbon Constrained World," presentation to the committee subgroup, October 26, 2010, Ann Arbor, Michigan.

[17] Personal communication from John Kargul, EPA, to the committee, February 1, 2011.

combustion, expanding the useful operating range to lower BMEP levels and idle, operating without pressure transducers, managing transient operation, and reducing HC and CO emissions.

Finding 3-5. The EPA has developed an HCCI engine that operates in the HCCI mode at all times using low-pressure, port fuel injectors suited to the unique operating conditions of a series hydraulic hybrid vehicle. The unique operating conditions include a narrow range of operation at the best BTE condition for each engine speed, with only slow transient response times for changes in power demands. At these unique operating conditions, NO_x and PM are below the levels required by the 2010 emissions standards without aftertreatment; HC and CO emissions are controlled with oxidation catalysts.

DEPARTMENT OF DEFENSE ENGINE PROGRAMS

Introduction

The U.S. Army Research, Development and Engineering Command's Tank-Automotive Research, Development and Engineering Center (TARDEC) provided input on engines, fuels, and hybrid vehicles.[18,19] The presentation on U.S. Army engine programs, at the November 15, 2010, committee meeting, prepared by Paul Skalny, mainly covered vehicle programs under the National Automotive Center (NAC). The NAC serves as the Army focal point for the development of dual-use automotive technologies and their application to military ground vehicles. The engine program presented was part of a program on the M1114 High Mobility Multipurpose Wheeled Vehicle (HMMWV), with a goal of 70 percent fuel economy improvement over a blended cycle, and included a "high efficiency engine." Other vehicle improvements contributing to this goal included an integrated starter/generator (ISG) with new technology batteries and a solar panel, reduced rolling resistance, electrification of accessories, driveline improvements including a 6-speed high-efficiency automatic transmission, and weight reductions through use of lightweight materials (aluminum, titanium, and carbon-fiber composites) coupled with component design modifications.

The meeting at TARDEC of a committee subgroup on January 10, 2011 (see Appendix B), included discussions of engine programs for both combat and tactical vehicles. The Army has National Security Exemptions from the EPA for the following: (1) JP-8 fuel exclusion from on-road 2006 and off-road 2007 diesel fuel regulations, (2) 2007+ heavy-duty on-road emissions standards, (3) 2004 and later EPA on-road emissions standards, and (4) Tier IV EPA nonroad emission standards. Thus, in the near term the Army buys older-emissions standard engines that meet Tier II or Tier III emissions standards, export engines, and in some cases older, remanufactured engines. In the midterm, in addition to the foregoing, the Army plans to buy modified on-road commercial off-the-shelf (COTS) and Tier IV engines minus cooled EGR and exhaust aftertreatment. The Army's primary bulk fuel is JP-8; diesel is used for specialized engines and aviation gasoline is used in light aircraft. JP-8 requires the addition of an oil lubricant for some fuel systems, including common rail fuel systems. Long-term plans are under development.

[18] Bill Harris, U.S. Army, "Review of the TARDEC Programs," presentation to the committee, November 15, 2010, Washington, D.C.

[19] Peter Schihl and John Rzepecki, "TARDEC Programs," presentation to a committee subgroup, January 10, 2011, Detroit, Michigan.

Engine Progams

TARDEC conducts tests of high power density propulsion systems in which additional requirements of operating at high ambient temperatures (125°F) and low heat rejection are required. The Army defines power density as follows:

Power density = sprocket wheel power/ total propulsion system volume.

The Army's goals for power density are as follows:

- Bradley Vehicle (baseline): Power density = 3 bhp/ft^3
- Future Combat Systems: Power density = 4.6 (goal = 6) bhp/ft^3
- Research Target: Power density = 8 to 10 bhp/ft3

Low heat rejection helps to improve power density by reducing excess fan power. Cooling, fuel effects, and air filtration are continuing challenges for the Army's ground vehicle propulsion systems. Examples of some of these high power density programs are listed in Table 3-7.

TABLE 3-7 High Power Density–Low Heat Rejection Program Targets of the Tank-Automotive Research, Development and Engineering Center

Engine	Specific Power	Output	Revolutions per Minute	BSHR[a]	Power Density
I-4	125 bhp/L	440 kW (590 bhp)	4,250	0.6 kW/kW	NA
V-6	125 bhp/L	750 bhp	4,250	0.6 kW/kW	NA
OPOC[b]	Greater than 125 bhp/L	220 kW (295 bhp)	3,800	0.45 kW/kW	8 bhp/ft^3
4.7 L I-6	200 bhp/L	940 bhp	5,400	0.6 kW/kW	8 bhp/ft^3

NOTE: NA, not available.
[a] Brake specific heat rejection.
[b] Opposed piston opposed cylinder.

Auxiliary Power Unit for M1 Main Battle Tank

At tactical idle, the M1 Main Battle Tank uses 17 gal/h of fuel. A renewed investigation of auxiliary power units (APUs) that can be packaged in the available 8 ft^3 is under way, with the goal of significant reductions in tactical idle fuel consumption. The following two candidates are being investigated: Patrick Power Rotary Diesel 1.5 gal/h at 9 kW power output; the next-generation target power output is 17 kW; and Altec Tech. Corp. Fuel Cell (1.1 gal/h at 10 kW power output).

Adaptation of Commercial Engines for Military Use

The purpose of the programs on the adaptation of commercial engines for military use is to assess the minimum modifications necessary to adapt 2007- or 2010-compliant engines for use in Army ground vehicles. Targets for these programs are as follows: 48 percent BTE, brake specific heat rejection (BSHR) = 0.6 kW/kW, removal of both EGR and diesel particulate filter, and meeting of 1998 EPA heavy-duty emissions standards (for unarmored wheeled vehicles). Meeting the targets for these programs will require the peak BTE to be improved by approximately 20 percent and the heat rejection to be reduced by approximately 20 percent in order to reduce cooling fan parasitic losses. These programs, together with results to date, are listed in Table 3-8. As shown in the table, none of the engines tested to date has met the aggressive BTE goal of 48 percent, although all of them reached the heat rejection target of 20 percent reduction.

University Programs

The Automotive Research Center (ARC) is a university-based U.S. Army Center for Excellence for advancing the technology of high-fidelity simulation of military and civilian ground vehicles. The ARC was established at the University of Michigan in 1994 and now includes Wayne State University, Oakland University, the University of Iowa, Clemson University, Virginia Polytechnic Institute and State University, and the University of Alaska, Fairbanks. The center has increased emphasis on energy-efficient propulsion systems for ground vehicles (Genzale et al., 2010). The research topics of the ARC are directly relevant to the military—for example, lightweight blast-resistant materials and issues of track vehicle dynamics. However, there is overlap in the areas of propulsion. The military is actively researching the potential of hybrid powertrains, both hydraulic and electric. There is active research on the fundamentals of batteries and battery performance in hybrid vehicles, and there is work in understanding engine operation with fuels other than diesel, mainly JP-8.

Three Broad Agency Announcements (2010-2013)

The U.S. Army is launching a 4-year program to advance the state of the art by developing new powertrain technologies that will improve overall efficiency by reducing fuel consumption, providing exportable electrical power, reducing noise, and developing powertrains that consume a wide range of fuels. The three programs are summarized in Table 3-9.

The DOD engine programs have completely different objectives from those of the DOE and EPA programs. Those of the DOD programs include, for example, high power density, low heat rejection, 1998 emissions requirements (except 20- to 40-ton vehicles), and primarily JP-8 fuel. The DOD fuel programs are discussed later in this chapter, and the hybrid programs are addressed in Chapter 4.

Finding 3-6. The DOD has engine programs that are cooperative between industry and universities and have goals of improved BTE and other goals more specific to the Army.

Recommendation 3-3. The DOD and the DOE should increase their awareness of one another's programs and look for opportunities to share technologies on areas of joint interest, such as thermal efficiency. One way to encourage interaction is for the DOE to invite DOD program participants to

TABLE 3-8 Adaptation of Commercial Engines for Military Use

Awards	Contractor and Engine	Brake Thermal Efficiency Achievement and Baseline	Amount of Contract
FY 2009	Cummins: 2007 ISL 8.9 L I-6	43.5 to 45.2% (Baseline: ~42%)	\$2.5 million to \$3.0 million
	Mack: Euro IV MP8 13.1 L I-6	~44% (Baseline: ~39%)	\$1.0 million
FY 2010	AVL: 2010 Ford 6.7 L V-8	43 to 43.5% (Baseline: ~40%)	\$1.0 million
	AVL: Euro Ford Euro V Lyon 4.4 L V-8	NA	NA
	Ricardo: 2007 Navistar 6.4 L V-8	NA	\$2.5 million

NOTE: NA, not available.

TABLE 3-9 Army Broad Agency Announcements (2010-2013): Summary of Three Programs for Powertrain Technology Development

Vehicle	Contractor	Engine/Transmission	Horsepower	Amount of Contract
7- to 9-ton light vehicle	Cummins	Cummins 6.7 L ISB Eaton Ultra-Shift	150 to 300	$7.0 million
15- to 19-ton medium vehicle	Sapa Group.	DDC 15 L Binary Logic Longitudinal	350 to 500	$7.8 million
20- to 40-ton heavy vehicle	SWRI	Up-Powered Cummins 15 L ISX Kertrain 32-speed cross drive	750 to 1,000	$9.67 million

NOTE: All with 44 percent BTE or greater and 0.6 kW/kW BSHR or less. Acronyms are defined in Appendix I.

present their findings at the DEER (Diesel Engine-Efficiency and Emissions Research) Conference.[20]

RESPONSES TO RECOMMENDATIONS ON ENGINES FROM NRC PHASE 1 REVIEW

The NRC (2008) Phase 1 includes 12 findings and 12 recommendations regarding engines (Findings 3-1 through 3-12 and Recommendations 3-1 to 3-12). The 21CTP concurred with many of the recommendations and incorporated several of them in the SuperTruck contracts (see Appendix C). The 21CTP responses to Findings 3-1 and 3-8 and Recommendations 3-1 and 3-8 did not concur with the NRC (2008) comments about the 50 percent BTE goal for 2010 and the 55 percent BTE goal for 2013. However, the committee observes that the 21CTP has now changed the year for meeting the 50 percent BTE goal to 2015 and that the 55 percent BTE is now a stretch goal for a prototype engine system in the laboratory for 2015.

FUEL PROGRAMS

Introduction

Fuels are important in attaining the vision of the 21CTP in three ways:

1. Fuel formulation impacts the conditions that must be established within the cylinder to achieve advanced combustion regimes;
2. Nonpetroleum fuels are a direct route to displacing petroleum-based liquid energy carriers, with biofuels offering the potential for reducing CO_2; and
3. Improved properties of petroleum-based fuels can improve engine operation and reduce emissions.

In December 2000, regulations were finalized that required, by 2006, much lower sulfur content in diesel fuel (a maximum of 15 ppm). Ultra-low sulfur diesel (ULSD) fuel was deemed necessary to enable catalyzed diesel particulate filters, including the use of a broader range of catalytic NO_x aftertreatment devices, and mitigate engine damage from potential sulfuric acid formation in recirculated exhaust gases. Other fuel properties have been shown to impact engine-out emissions (DOE, 2006). For example, oxygen-containing fuels and additives have been found to reduce PM emissions. However, the understanding of the effects of fuel properties on emissions is still highly empirical. Similarly, the relation between fuel properties and low-temperature combustion modes is far from well understood, although considerable progress has been made in the past few years. Modified fuel specifications and new fuel formulations may facilitate expanding the operating range of new combustion regimes like homogeneous-charge compression ignition as well as improving the operation of conventional diesel engines. The changing of fuel specifications to achieve specific goals needs to be balanced with the cost to refineries for making the needed changes. It is unlikely that changes in diesel fuel properties will occur without EPA regulations to require them, and this can only occur if the changes facilitate reduced emissions.

Nonpetroleum diesel fuels can be produced from renewable resources such as seed oils and animal fat, as well as synthesized from natural gas, oil sands, coal, biomass, and other resources. Processes for the production of diesel fuel from these sources continue to be explored, but commercial production has been limited. The production of biodiesel is growing slowly in the United States. The use of syncrudes from oil sands in Canada has grown considerably. Fischer-Tropsch (F-T) diesel fuels, synthesized from natural gas, have been studied in numerous engine tests to determine

[20] Subsequent to the committee's review of 21CTP programs, the DOE and the DOD entered into the Advanced Vehicle Power Technology Alliance (AVPTA) partnership on July 18, 2011. See, for example, "DOE, Army Alliance Underlines Achieving Energy Security" by Chris Williams, available at http://www.army.mil/article/62727/. Accessed October 18, 2011.

TABLE 3-10 Major 21CTP-Related Projects Funded in FY 2010 Addressing Advanced Petroleum-Based Fuels and Non-Petroleum-Based Fuels

Project Title	Organization	FY 2009 Budget	FY 2010 Budget
Advanced Petroleum Based Fuel Effects on Combustion	ORNL	$750,000	$950,000
Fuels for Advanced Combustion Engines	NREL	$550,000	$550,000
Advanced Petroleum Based Fuels	NREL	$1.8 million	$1.5 million
Quality, Performance and Emission Impacts of Biofuels and Biofuel Blends	NREL	$3.2 million	$1.8 million
Kinetic Modeling of Fuels	LLNL	$325,000	$500,000
Non Petroleum-Based Fuels: Effects on Emission Control Technologies	ORNL	$845,000	$1.1 million
Non Petroleum-Based Fuels: Effects on Advanced Combustion	ORNL	$895,000	$1.47 million
Fuel Effects on Advanced Combustion: Optical Heavy-Duty Engine Research	SNL	$600,000	$730,000
Advanced Lean Burn Direct Injection Spark Ignition Fuel Research	SNL	$600,000	$630,000

NOTE: Acronyms are defined in Appendix I. Some of the projects included in Table 1-2 for the fuels budget are applicable to both light- and heavy-duty vehicles so that the total of $12.244 million is not the same as noted in this table, Table 3-10.

SOURCE: U.S. Department of Energy, Office of Energy Efficiency and Renewable Energy. Available at http://www1.eere.energy.gov/vehiclesandfuels/resources/proceedings/2010_merit_review.html.

their impact on emissions, because higher cetane number[21] and reduced PM are the primary benefits. Imported F-T liquids have been used as blending material in California diesel fuels since 1993. F-T diesel and biodiesel have a lower energy density than that of conventional diesel fuel, so the uniformity and quality of these new fuels need to be defined and improved to allow for the engine and emission control system to take advantage of them. Furthermore, if alternative fuels are to be used interchangeably with conventional diesel, it may become necessary to recognize the fuel and to compensate within the engine in order to achieve optimal efficiency and to maintain emissions control. Sensors (real or virtual) may be needed to accomplish this, along with control hardware and software.

Lubricant properties can also have a profound effect on emissions, by impacting the durability of exhaust aftertreatment devices. The sulfur and "ash" content of lubricants need to be minimized to prevent degradation of NO_x adsorber catalyst performance and to optimize cleaning intervals and regeneration phenomena in DPFs. Research continues on the fuel-savings potential of low-friction and low-viscosity lubricants while maintaining engine and transmission durability and reliability.

The remainder of this section addresses DOE fuels programs, DOD fuels programs, advanced petroleum fuels, biofuels, alternative fuels, and lubricants.

DOE Fuel Programs

To address the challenges described above, the 21CTP has maintained a portfolio of research programs investigating advanced petroleum-based fuels (budget for this was eliminated for 2011, which is discussed later in this section) and non-petroleum-based fuels. The DOE supplied to the committee a list of nine DOE-funded projects focused on advanced petroleum based fuels (APBF) and non-petroleum-based fuels (NPBF) that were directly attributed to the 21CTP. These projects are listed in Table 3-10; the total budget for FY 2010 is $9.23 million.

As with the advanced combustion research programs, researchers participating within each project on APBF and NPBF have developed collaborative teams consisting of industry and academic partners. Participants are involved with group research meetings and an annual merit review as part of the EERE program. Project directions and continuation are based on the scores received in the merit review. The presentations given at the merit review are available to the public.[22]

The research programs include powertrain-system-level and fundamental investigations. Projects range from evaluating the effects of differences in fuel composition on achieving advanced combustion strategies and their impact on proposed aftertreatment systems, to developing comprehensive chemical kinetic mechanisms for non-petroleum-based fuels, primarily bio-derived fuels. In addition, fundamental experiments are being done using advanced optical diagnostics to evaluate the impact of nonpetroleum fuels on the combustion subprocesses known to be important with petroleum fuels, such as fuel spray development, vaporization, and in-cylinder mixing. Differences have been observed and correlated with characteristics of the fuels (Genzale et al., 2010; Kook and Pickett, 2009; Mueller et al., 2009). Also, some of the activities within the 21CTP seem to focus on light-duty gasoline-

[21] The cetane number is a measure of the ignition quality of diesel fuel.

[22] See http://www1.eere.energy.gov/vehiclesandfuels/resources/proceedings/2010_merit_review.html. Accessed on May 18, 2011.

like fuels. To logically include these as part of the 21CTP, one would expect to see projects aimed at investigating gasoline-like fuels for heavy-duty applications. This could be very timely and strategic with the expected excess supply of gasoline in the next 30 years (discussed below). A small part of one research project is investigating dual-fuel use, but this project appears to be in a light-duty engine and states as one of its motivations the increased use of ethanol to help meet the requirements of the Energy Independence and Security Act (EISA) of 2007 (Public Law No. 110-140).

DOD Fuels Program

The military fuels program has an objective of minimizing the amount of fuel used through engine and vehicle efficiency improvements and increasing the amount of nonpetroleum fuels used. The military would prefer to use the same fuel in all of its vehicles, and JP-8 is the desired single fuel. The Army, through TARDEC, has the unique role of qualifying alternative jet fuels for use in tactical/combat vehicles having diesel engines. TARDEC is concerned over fuel lubricity and the wide variations that the Army sees in its worldwide fuel surveys of JP-8, its primary fuel. For example, cetane index varies from 33 to 50, density by about 6 percent, and volumetric energy density by about 5 percent. These variations affect engine operation and vehicle range. Fuel-quality sensors are needed to minimize these adverse effects. For ground vehicle use, the Army would like the cetane number of JP-8 to be at least 50.

Although the Army would like to use less petroleum-derived fuels, it realizes that it will be difficult to do so. It is exploring biodiesel fuels; however, the lower energy density of these fuels is a negative. The military has a very rigorous procedure for qualifying alternative fuels, which by itself could prevent any from qualifying. A recent report from the RAND Corporation stated that the U.S. military would derive no meaningful benefit from increased use of alternative fuels, because the technologies needed were unproven, too expensive, and too far from commercial scale to have any impact in the next decade (Bartis and von Bibber, 2011).

The Army keeps many of its engines for 40 to 50 years, and so there is concern about the compatibility of newer fuels and lubricants in these engines. For example, diesel engine fuel pump life can be an issue with low-lubricity fuels. Also, the newer low-sulfur, phosphorus and ash lubricants will not be compatible with these engines.

Advanced Petroleum Fuels

For many years to come, trucks will be powered by diesel engines, and the fuel for these engines will be diesel-like fuel. This fuel will continue to be made mainly from petroleum, in spite of considerable efforts to develop processes for making biodiesel fuel and to use alternative fuels. For the United States in 2008, 95 percent of transportation fuel was from petroleum, 3 percent was biofuels (essentially all ethanol that is not usable in diesel engines), and 2 percent was natural gas.[23] This distribution will change with time, with the biofuels portion increasing slowly, but petroleum will be the dominant source for many years. For example, in *BP Energy Outlook 2030*, BP predicts that biofuels will supply 3 to 4 percent of world transportation energy by 2030, whereas oil will supply almost 90 percent (BP, 2011).

Thus, it is surprising and disappointing that the DOE efforts on petroleum-based fuels have been eliminated from the FY 2011 budget. For FY 2011, the DOE Fuels Technology Budget request was $11 million, down from the $24 million appropriation for FY 2010, with none of it for petroleum fuels.[24] However, the DOE has remained active in programs that address diesel fuel properties, whether they are petroleum-derived or not. Programs such as Fuels for Advanced Combustion Engines (FACE), which provides a fuels matrix considering variations in cetane number, aromatics content, and T90, are worthwhile and should be continued. Also, the involvement on fuels with the U.S. Coordinating Research Council (CRC) and the American Society for Testing and Materials (ASTM) should be continued.

Diesel fuel properties vary considerably, both for petroleum-derived fuels and for those containing biomass-derived-components. As engines continue to be fine-tuned for improved fuel consumption and reduced emissions, variations in fuel properties, such as cetane number, aromatics and sulfur concentration, and T90, become more important. For example, the Alliance of Automobile Manufacturers' annual diesel fuel-quality survey for 2005 reported that cetane quality varied from under 42 to over 50, with 80 percent of the samples under 46 cetane number and 20 percent over 46.

Truck engine manufacturers, through the Worldwide Fuel Charter, have expressed a desire for a minimum diesel fuel cetane number of 55 for markets with the highest degree of emission control, such as the United States. This cetane number is considerably higher than the current United States average in the mid-40s, and it is typical of cetane quality in European diesel fuel. A higher cetane number would improve engine thermal efficiency and cold starting, reduce engine noise, white smoke, and odor emissions (ACEA, 2006). It would improve the ability of diesel-hybrid trucks to restart in stop and go operations, especially at low temperatures. The 2006 version of the Worldwide Fuel Charter (WFC, 2006) provides sufficient evidence to conclude that there is a very large, still unexploited potential for improvements in road fuels, which will provide major reductions in pollutant emissions both in vehicles already on the road as well as in future dedicated vehicles. However, the committee is not aware of conclusive test results showing the effects of cetane number on the engine thermal efficiency and exhaust emis-

[23] Patrick Davis, DOE, "NAS Review of 21CTP—Phase 2," presentation to the committee, September 8, 2010, Washington, D.C.

[24] Ibid.

sions of modern U.S. heavy-duty diesel engines or engines with advanced-concept combustion systems.

A series of trends needs to be explored in order to make proper decisions on future diesel fuels, whether they are petroleum- or biomass-derived:

1. In the United States and worldwide, future demand for diesel fuel is expected to continue to increase in spite of the forthcoming truck-fuel-consumption standards in the United States and elsewhere (EIA, 2011a).
2. In the United States, demand for gasoline is expected to decline as a result of more stringent fuel economy standards, greater use of ethanol, and the increased use of battery-driven light-duty vehicles. Worldwide, even with higher fuel economy standards, the demand for gasoline may increase because of the growth of vehicle sales in Asia (EIA, 2011b).

A presentation to the committee showed that even with improvements in vehicle fuel consumption, total truck petroleum consumption would continue to increase through 2035, dominated by the increase for heavy-duty trucks.[25] This will have major implications for refineries in the United States, which will see the ratio of gasoline to diesel fuel production change from highly gasoline-biased to highly distillate-fuel-biased, and will necessitate changes in refinery configuration and operation.

The United States for many years has been a large importer of petroleum, mainly to meet the needs of transportation. If demand for diesel fuel continues to increase, reducing gasoline demand might not have the desired favorable impact on petroleum imports because diesel fuel production will become the controlling factor. Continuing to add ethanol to gasoline for use in light-duty vehicles, be it from corn or cellulosic materials as prescribed by the Renewable Fuels Standards (RFS) in the Energy Policy Act of 2005 (Public Law 109-58), might not contribute to reducing petroleum imports, if increasing diesel fuel demand must be met. The major efforts in the United States to develop and commercialize cellulosic ethanol, many supported by the DOE, may be counterproductive. It may be wise to redirect the DOE biofuels programs to the development of hydrocarbons for use in distillate fuels (diesel fuel and jet fuel) and away from oxygenated fuels for use in gasoline. Also, the DOE will also need to consider efforts toward combustion system development, which uses gasoline-like fuels in high-efficiency, heavy-duty engines.

In looking ahead, the government, in cooperation with industry, could look at maximizing the miles driven per barrel of petroleum used. Studies of this type were done more than 40 years ago, but there have been none of note recently. Because one of this nation's major goals is to reduce petroleum imports, the DOE could also explore how the nation might use petroleum in the most efficient manner. Such a study would include the following factors:

1. What is the optimum distribution (gasoline/distillate/other) of the barrel to maximize miles or work done per barrel of petroleum processed?
2. What are the optimum ignition characteristics, cetane quality of diesel fuel and octane quality of gasoline?
3. How many grades of each fuel, at what specific octane and cetane qualities, would be optimum?

Such a study would look at the impacts of the continued use of ethanol in gasoline. It could show the potential for reducing petroleum demand and providing fuels tailored for future optimized internal combustion engines. A cost-benefit analysis should be included to compare the costs of the required refinery modifications, with the savings in reduced automotive fuel use. Transition to such an optimal fuel and engine combination would have to be facilitated through the development of engines that are capable of adjusting to the fuel characteristics.

Biofuels

For many years, biofuels have been held out as the "holy grail" that could reduce petroleum imports and greenhouse gas emissions and provide domestic jobs. The DOE has been heavily involved with developing technology and processes for cellulosic ethanol and biodiesel fuels, with the ultimate goal of commercialization. Biodiesel fuels were envisioned as ideal supplements or replacements for petroleum-derived diesel fuel. Congress established the Renewable Fuels Standards in 2005, which set a goal of 36 billion gallons of biofuels by 2022. Congress has provided tax credits and incentives for biofuels production, including that of ester-based and renewable diesel fuels, and they were recently extended. A recent study conducted by Sandia National Laboratories and General Motors concluded that there are no theoretical barriers to achieving the stated goal of producing 90 billion gallons of ethanol per year by 2030.[26] A number of practical obstacles were identified however. In particular, investment in cellulosic ethanol production needs long-term protection against oil and feedstock price volatility. Capital costs are significant, and investment risk needs to be managed. Technology improvements, particularly in cellulosic conversion yields, are critical and must be sustained over a number of years. Finally, large-scale development of energy crops is necessary. An NAS-NAE-NRC (2009) report concluded that the resources for biomass were available in the United States and that significant biofuels could be produced by 2020. However, the applicabil-

[25] Kevin Stork, DOE, "DOE Fuel and Lubricant Technologies R&D," presentation to the committee, November 15, 2010, Washington, D.C.

[26] Available at http://www.sandia.gov/news/publications/white-papers/90-Billion-Gallon-BiofuelSAND2009-3076J.pdf. Accessed May 18, 2011.

ity of the forecasts in the NAS-NAE-NRC (2009) report for fuels used in heavy-duty diesel engines remains an open issue.

With all of the above, what success has occurred, and what is the outlook for biofuels, especially for biodiesel fuels that could be used in truck diesel engines? To date, the only success has been with corn-based ethanol, which is added to gasoline across the United States for use in gasoline engines. This policy has been clouded with controversy over "food for fuel," tax credits, and the reduction of greenhouse gas emissions. To date, the commercial production of cellulosic-derived ethanol, envisioned in the RFS, has not occurred, and the production of biodiesel (essentially fatty acid methyl ester [FAME] and other esters) and renewable diesel (a biomass-derived feedstock used in refineries for diesel fuel production) has been minimal. For 2011, the RFS targets for all biofuels are these: biomass-derived diesel, 0.8 billion gallons; advanced biofuels, 1.35 billion gallons; cellulosic biofuels, 5 to 17.1 million gallons; total renewable fuels (almost all corn-derived ethanol), 13.95 billion gallons. Compared with Energy Information Agency (EIA) values for annual gasoline consumption in the United States in 2009 of 138 billion gallons, and diesel fuel consumption in 2008 of 42 billion gallons, these amounts are very small (EIA, 2011a,b). By 2022, the RFS requirement is 4 billion gallons of advanced biofuels, which can be just about any renewable fuel except corn-based ethanol. Even if all of this was biodiesel fuel, it still would meet only about 10 percent of diesel fuel demand.

To meet future RFS requirements, significant advancements in technology and reductions in cost are needed for biofuels. Many years are expected to be required to explore the different approaches to producing biofuel, which are currently at the fundamental laboratory research stage, and to transition promising technologies from the laboratory to large-scale production. The DOE has been actively involved in process development, in monitoring the quality of the biofuels produced, and in understanding their use in engines and vehicles. A significant portion of the DOE effort at the National Renewable Energy Laboratory has been in the generation of cellulose-derived fuels, primarily ethanol. Insufficient work is being done in the generation of hydrocarbon fuels, which are better suited for diesel engines.

In looking toward bio-derived hydrocarbons for diesel fuel, there are two options:

1. Generate a biodiesel fuel, such as FAME or other esters, from a specific feedstock, such as soybeans, and blend that into existing diesel fuel.
2. Generate a bio-crude oil (renewable diesel fuel) that can be used at the refinery in the production of diesel fuel.

Much of the effort to develop diesel biofuels has been directed toward the development of ester-based fuels such as FAME. These fuels are now blended, to a limited extent in the United States, commercially in diesel fuel, and to a greater extent in Europe. But, they are not without problems relating to low-temperature operability and deposits.

Bio-mass derived dimethyl ether (DME) has received attention, especially in Europe, as a sulfur-free diesel fuel substitute because of its high cetane number (55) and very low emissions of PM, NO_x, and CO. It would require minor engine modifications, and would need a special system for distribution and storage, which could be a major stumbling block.[27]

Others are working on renewable diesel fuels, which essentially use biomass feedstocks to generate a hydrocarbon blend that is processed at a refinery during the production of diesel fuel. It results in a finished diesel fuel that is essentially the same as a completely petroleum-derived fuel, with very similar properties.

The California Environmental Protection Agency has recently compared FAME with renewable diesel fuel and concluded that renewable diesel fuel is a better option for the following reasons: it is all hydrocarbon and is chemically more like diesel fuel; it is compatible with current diesel fuel infrastructure and engines; and it avoids unwanted effects associated with ester-based biodiesel fuels (FAME), such as lower volumetric energy content, instability, hygroscopicity, injector fouling, and low-temperature operability, among other factors. The U.S. EPA recently approved a prototype renewable diesel fuel as an "advanced biofuel."[28]

The DOE has encouraged production of renewable diesel fuel. On January 20, 2011, DOE Secretary Steven Chu announced[29] a loan guarantee of $241 million to support the construction of a 137 million gallon per year renewable diesel facility that will use animal fats, used cooking oil, and other waste grease as feedstock. Many more of these plants will be needed to make a significant dent in biodiesel fuel supply.

The use of other sources for biodiesel fuel production has not yet been successful. For example, a recent United Nations Food and Agricultural Organization report stated that biodiesel from the Jatropha plant that grows in arid climates could not make a significant contribution toward reducing oil dependence (UNFAO, 2010).

Algae have been receiving much attention as a potential source of biofuels. In the United States, major energy companies such as ExxonMobil and Chevron have invested hundreds of millions of dollars, the DOE's National Renewable Energy Laboratory has an extensive R&D program, and the DOE has invested $24 million in three programs to tackle key hurdles in the commercialization of algae-based

[27] A study of DME as an alternative fuel for diesel engine applications (TP 13788E) is available at http://www.tc.gc.ca/eng/innovation/tdc-summary-13700-13788e-718.htm (accessed July 5, 2011).

[28] See http://www.epa.gov/otaq/fuels/renewablefuels/compliancehelp/triton-determination.pdf. Accessed May 22, 2011.

[29] See DOE Blog; Available at http://blog.energy.gov. Accessed on May 9, 2011.

biofuels.[30] In spite of all of this effort, algae-based biofuels are not on the horizon. A study presented to the National Petroleum Refiners Association (NPRA) meeting in March 2010 (NPRA, 2010) concluded: "significant production seems a minimum of five years off and likely 10 years." The same study said: "even the current algae-oil technology leaders have big cost hurdles facing them, with algae-derived crude liquids likely to cost between $6 and $20 per gallon" (NPRA, 2010). A study from Wageningen University in the Netherlands concluded that the cost of producing biodiesel from algae is 3.5 times more than the cost of producing biodiesel from crude oil, and twice as much as producing it from rapeseed. The study also stated that it would be 10 to 15 years before commercial production would be feasible, and the cost would have to go down by a factor of 10 (Wijffels and Barbosa, 2010). Even the DOE's National Algae Biofuels Technology Roadmap concluded that the technology is at an early stage and will require years of development to reach commercialization (Wijffels and Barbosa, 2010). As stated above, the development time for new biofuels will be long, measured in decades.

The International Energy Agency's report *Status of 2nd Generation Biofuels Demonstration Facilities* in June 2010 concluded: "A high number of projects are being pursued, but only few facilities in the demonstration phase are actually operating" (Bacorsky et al., 2010). An Australian program has reduced the cost of algae-derived biodiesel fuel by a factor of four to $11 per gallon and admits that further reduction is not a simple process (Wards, 2011). A Japanese report concluded that Japan's biofuel industry would have a hard time surviving without government subsidies (Ethanol and Biofuels News, 2010). The situation is similar throughout the world.

Finding 3-7. In spite of efforts to reduce the fuel consumption of light-duty and heavy-duty vehicles and to develop biomass-derived fuels (an effort which, except for corn-based ethanol, has not progressed as much as had been expected), petroleum will remain the primary source of light-duty and heavy-duty vehicle fuel for many years to come. Whereas future U.S. gasoline demand is expected to be flat for the next 20 years, diesel fuel demand is expected to grow, necessitating changes in refinery operations.

Recommendation 3-4. The DOE should reinstate its program for advanced petroleum-derived fuels (they will be transportation's primary fuels for many years to come) with the objective of maximizing the efficiency of their use.

Alternative Fuels

One of the DOE's original 21CTP goals was to replace 5 percent of petroleum fuels with fuels from nonpetroleum sources by 2010. In the NRC Phase 1 report (NRC, 2008), it was stated that this was unlikely to happen, especially for fuels for heavy-duty diesel engines, and this has been the case. Other than ethanol from corn for use in gasoline-fueled engines and the use of natural gas in fleet vehicles, success with alternative fuels has been limited. About 110,000 natural gas vehicles are on U.S. roads today, with about 66 percent of those being transit buses. In 2009, 26 percent of all new transit bus orders were for natural-gas-fueled buses.[31]

Interest remains in utilizing natural gas, either directly or as a feedstock, for the production of diesel fuel. The natural gas resources of the United States can be greatly expanded through the fracturing of shale deposits, providing a greater opportunity for the use of natural gas resources in producing transportation fuels. However, there is concern about the adverse effects of methane emissions in the fracturing process and about the impacts of those emissions on global warming.

There is also interest in coal, oil shale, and tar sands. However, high costs, environmental issues, and greenhouse gas emissions have stymied most of these efforts. Commercial success for coal and shale oil is unlikely in the next 10 to 20 years, especially if the price of petroleum remains relatively low. The production of transportation fuels from Canadian tar sands has been a commercial success, but not without environmental controversy.

Gasoline fuel for heavy-duty applications is currently receiving attention in several areas. First, evidence is emerging that some gasoline engines are replacing some diesel engines in pickup, delivery, and other smaller trucks as a lower-cost alternative to the 2010 emissions-standards-compliant diesel engines. Second, R&D work is under way on gasoline-fueled heavy-duty HCCI engines. Work on this type of engine at the EPA was previously described in this chapter. Other research work on similar gasoline-fueled HCCI engine concepts is also under way.

There are no simple and inexpensive substitutes for petroleum-based liquid energy carriers. However, the global demand and diminishing growth in petroleum reserves mandate that alternatives be pursued. Because of the tremendous utility, high energy density, and specific energy of liquid hydrocarbon fuels, the DOE should continue to follow progress in utilizing alternative resources to generate transportation fuels, especially hydrocarbon fuels for heavy-duty diesel engines. The DOE should focus on processes likely to be commercially successful in the next 10 to 20 years.

Lubricants

Lubricants for truck engines will have to provide reduced fuel consumption, powertrain durability, and emissions-control-system compatibility. The DOE recognizes this.

[30] See http://www.energy.gov/9167.htm. Accessed May 9, 2011.

[31] See http://www.ngvamerica.org/about_ngv/. Accessed April 25, 2011.

Regarding emissions, the DOE has been active in the Collaborative Lubricating Oil Study on Emissions (CLOSE) project. Of particular concern is the effect of sulfur from the fuel and the engine oil and of phosphorus from the engine oil on emission-control-system performance, especially regarding PM and NO_x.

It is well known that reducing powertrain friction, through design changes and the use of more efficient engine oils and transmission fluids, can reduce fuel consumption. Significant reductions in powertrain friction, and improvements in lubricants, especially for light-duty vehicles, have occurred in the past 20 years. The DOE recognizes that not as much progress has been made with heavy-duty vehicles, especially with their lubricants. For instance, the use of 5W engine oils is recommended for most light-duty engines, and 0W engine oils are now recommended for some, whereas 15W engine oils are recommended for most heavy-duty engines, primarily to provide the necessary bearing oil-film thickness, because heavy-duty diesel engines have significantly higher bearing unit loads than do typical light-duty engines.

The DOE has an objective of reducing "parasitic losses in system efficiency by developing improved engine and transmission lubricants."[32] The target benefits are as follows: for 2016, 10 percent engine/15 percent drivetrain friction reduction, leading to approximately 1.5 percent fuel economy benefit; for 2030, 25 percent engine/35 percent drivetrain friction reduction leading to approximately 3 to 4 percent fuel economy benefit. Considering what has already been done to reduce powertrain friction, these are very ambitious targets. To help achieve these goals, the DOE is supporting research to develop and improve the understanding of microfluidic transport, ionic liquids, and lubricant film formation.[33]

The DOE includes the retrofit of improved engine and transmission fluids in its plans and projected fuel economy improvements. This is a problematic strategy for several reasons:

- Before lower-viscosity engine oils are recommended, either for factory fill or in-use lubricants, manufacturers and truck owners will have to be assured that engine durability is not compromised.
- Transmissions are very sensitive to the quality of the transmission fluids. Fluids are not changed frequently, and manufacturers and operators will need great assurance before risking the use of a retrofit transmission fluid.

A critical development goal is for backward-compatible products, which would minimize the complications of supplying multiple products in the industry. A program that relies on retrofit of lubricants absolutely needs to be conducted in conjunction with the engine and transmission suppliers and users, and with the engine oil and transmission fluid developers and suppliers. Without their involvement, it has zero chance of success.

The process for periodic enhancements of heavy-duty diesel truck engine lubricating oil has been initiated, beginning nearly a year ago as the EPA and DOT/NHSTA GHG emissions and fuel efficiency rule was anticipated (EPA/NHTSA, 2010). This primary objective is an improved fuel efficiency contribution through the implementation of a high temperature high shear (HTHS) viscosity property. Early industry testing with HTHS controlled oils has shown promise for fuel consumption reduction while other parameters continue to protect traditional performance. HTHS can be managed relatively independently from traditional viscosity designations (e.g., 10W30 and 15W40 are frequently used diesel oil viscosity specifications). Fuel consumption results are shown to be duty-cycle-specific. The results characterizing over-the-road cycles may experience up to a 1 percent reduction, whereas an urban/suburban pickup and delivery cycle may achieve up to 1.5 percent reduction or more.[34]

Bio-derived lubricants are not receiving much attention. However, the DOT has provided \$370,000 to the National Ag-based Lubricant Center at the University of Northern Iowa to study the feasibility of using readily biodegradable soy-based lubricants in railroad engines (Ethanol and Biofuels News, 2010). Because diesel engines are involved, the DOE should maintain cognizance of this study.

Finding 3-8. The DOE recognizes the importance of reducing truck powertrain friction and the need for improved lubricants that reduce fuel consumption.

Recommendation 3-5. The DOE must work closely with industry in exploring improved lubricants that reduce fuel consumption, especially with regard to using such lubricants in existing truck engines and transmissions.

Three Different Sets of 21CTP Goals for Fuels and Lubricants

To determine the current goals for the 21CTP fuels program, the committee addressed the following comments to the Partnership:

> The 21CTP fuels and lubricants goals stated in the NRC (2008) Phase 1 report were:
>
> - By 2010, identify and validate fuel formulations optimized for use in advanced combustion engines exhibiting high efficiency and very low emissions, and

[32] Kevin Stork, DOE, "DOE Fuel and Lubricant Technologies R&D," presentation to the committee, November 15, 2010, Washington, D.C.

[33] Kevin Stork, DOE, "DOE Fuel and Lubricant Technologies R&D," presentation to the committee, November 15, 2010, Washington, D.C.

[34] Private communication between Gregory Shank, Volvo Powertrain, and the committee, September 8, 2010.

facilitating at least 5 percent replacement of petroleum fuels.

- By 2010, identify and exploit fuel properties that could increase efficiency and reduce overall tailpipe emissions through (1) lower engine-out emissions, including new low-temperature combustion regimes, and (2) enhancement of aftertreatment performance for 2010 emissions regulations.
- By 2013, identify non-petroleum fuel formulations (i.e., renewables, synthetics, hydrogen-carriers) for advanced engines and new combustion regimes for the post-2010 time frame that enable further fuel economy benefits and petroleum displacements while lowering emissions levels to near zero, thus adding incentives for using non-petroleum fuels.

A draft white paper presented to the committee on November 15-16, 2010, provided the following new list of 21CTP fuels and lubricants goals (DOE, 2010a). The new Goal 3 repeats the earlier Goal 1, and the new Goal 4 repeats the earlier Goal 3. The status of each goal is provided in the comments following each goal. Little or no progress has been made on these goals because DOE funding reductions in fuels and lubricants technologies have resulted in no funding for heavy-duty fuels and lubricants R&D.

- *New 21CTP fuels and lubricants Goal 1:* "Establish the influence of fuel and lubricant sulfur on emission-control technologies."
  Status: The 21CTP engine white paper dated August 30, 2010, states: "The sulfur and ash content of lubricants are sufficiently high to be factors in degradation of performance of NO_x adsorber catalysts and to influence the cleaning intervals and regeneration phenomena in DFPs, for example." No progress was reported to the committee for this goal.
- *New 21CTP fuels and lubricants Goal 2:* "Identify and exploit fuel properties that reduce overall tailpipe emissions through lower engine-out emissions, including new low-temperature combustion regimes and enhancement of aftertreatment performance."
  Status: The 21CTP engine white paper dated August 30, 2010, states: "The understanding of fuel property effects on emissions is highly empirical. Similarly, understanding the relation between fuel properties and low-temperature combustion modes is far from well-understood." The DOE reported that high-volatility diesel (HVD) fuel increased efficiency and lowered emissions for PCCI operation according to optical engine studies, which showed that liquid fuel films on the piston are avoided.
- *New 21CTP fuels and lubricants Goal 3:* "By 2010, identify and validate fuel formulations optimized for use in advanced combustion engines exhibiting high efficiency and very low emissions, and facilitating at least 5 percent replacement of petroleum fuels."
  Status: This DOE goal was discussed in detail in the NRC Phase 1 report (NRC, 2008). For this present review, the DOE reported that it contributed to an improved biodiesel ASTM specification, resolving shortcomings of the ASTM D6751 specification issued in 2002. The DOE also reported on studies under way to determine the factors (including radiative heat transfer, reaction rates, and combustion temperatures) that result in increased NO_x emissions when fueling with biodiesel. A set of common research fuels were developed, analyzed, and distributed for understanding low-temperature combustion (the FACE project), although a low-temperature combustion engine for testing these fuels had not been identified. The NRC Phase 1 report did not contain plans for achieving the goal of replacing 5 percent of petroleum fuel with nonpetroleum fuels by 2010. The DOE subsequently reported that zero percent reduction in petroleum consumption for the total heavy-duty truck fleet was achieved in 2010, but forecasted a 5 percent reduction in 2015.[35]
- *New 21CTP fuels and lubricants Goal 4:* "By 2013, identify non-petroleum fuel formulations (e.g., renewables, synthetics, hydrogen-carriers) for advanced engines and new combustion regimes for the post-2010 time frame that enable further fuel economy benefits and petroleum displacements while lowering emissions levels to near-zero, thus adding incentive for using non-petroleum fuels."
  Status: The NRC Phase 1 report (2008) stated that this goal "was intended to emphasize the development of non-petroleum fuel formulations beyond biodiesel," previously addressed by Goal 1. Similar to Finding 3-15 in the NRC Phase 1 report, the DOE provided little insight into the scope and magnitude of the effort to address this goal. The Phase 1 report stated, "It appears unlikely that the fundamental mechanisms that control the formation of HC, NO_x, and particulate emissions in a diesel engine can be dramatically altered with a change in fuel formulation to the extent that the emissions could approach zero."

Another draft 21CTP engine white paper dated February 25, 2011 (DOE, 2011a) states the goal for fuels as follows:

- "Through experiments and models with FACE fuels and other projects, determine the most essential fuel properties, including renewables, needed to achieve 55 percent engine brake efficiency, by 2014."

[35] Kevin Stork, DOE, "Fuel & Lubricant Technologies R&D Overview for NAS Review of 21CTP," presentation to the committee, November 15, 2010, Washington, D.C.

Status: Plans for achieving this goal were not reviewed with the committee. As noted in an earlier section, significant progress in low-temperature combustion has been realized through the use of gasoline or dual gasoline and diesel fuels. The DOE should recognize this progress in defining this goal and developing specific plans for achieving the goal.

Finding 3-9. The DOE established three different sets of goals for the fuels program from 2008 to 2011, which made an assessment of progress against the goals difficult. In total, little progress has been made toward the achievement of these DOE goals, which were not specified goals of the 21CTP.

Recommendation 3-6. The DOE fuel goals should be re-evaluated in line with the FY 2012 budget and the recommendations of this report. Specific plans for achieving these goals should be established.

AFTERTREATMENT SYSTEMS

Introduction

Considerable effort and research funding have been focused on improving aftertreatment systems as part of the 21CTP. This effort is complementary to the development of combustion processes that would minimize or preclude the in-cylinder formation of criteria pollutants. It is most appropriate to think of the combustion processes within the cylinder and the aftertreatment devices within the exhaust as an interconnected system. Minimal emissions leaving the combustion chamber result in reduced demands on the aftertreatment system, allowing for simpler, less expensive aftertreatment systems. Also, highly effective aftertreatment systems can facilitate different engine calibrations for better efficiency, which would otherwise have been precluded because of emission constraints. Aftertreatment research and development must be done with an eye toward the likely progress in combustion system development.

In addition, aftertreatment systems are the subject of extensive research within the technical community. Consequently it is critical for the 21CTP be aware of the broad scope of activities taking place outside of its program and to make sure that the research activities within its purview address fundamental concerns and are not already being done as part of the research and development efforts of the other agencies or industry.

To this end, the research structure of the 21CTP aftertreatment program is well organized. The research programs are built around teams with participation from national laboratories, universities, and relevant stakeholder industries. This is directly stated in the goals of their aftertreatment program. The research focuses on developing new technologies and on gaining an enhanced fundamental understanding of catalysis and governing phenomena limiting the effectiveness of current approaches. Success in these programs could lead to combining multiple aftertreatment approaches into a single unit, developing catalysts with higher resistance to poisoning, and implementation of retrofit systems.

The 21CTP aftertreatment research has already helped manufacturers meet the EPA 2007 and 2010 new-engine emissions standards. The system architecture generally being applied to meet U.S. EPA 2010 emissions is shown in Figure 3-4. The attainment of these emission standards is providing substantial health and environmental benefits by reducing ground-level ozone and fine particulates (mass as well as number), in addition to regional haze. Thus, con-

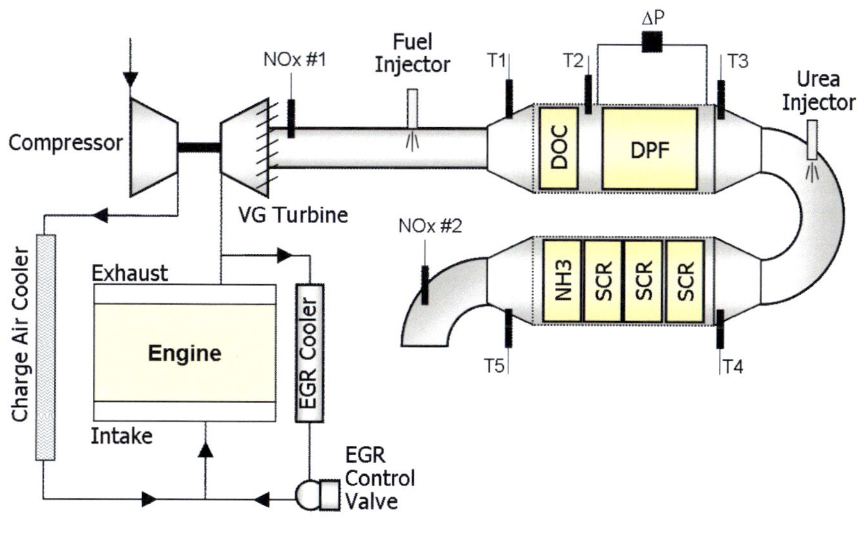

FIGURE 3-4 Emission control system architecture generally being applied to meet 2010 new engine emissions standards of the Environmental Protection Agency. Acronyms are defined in Appendix I. SOURCE: DOE (2011a).

tinued research into improved aftertreatment is extremely important.

Diesel particulate filters have reduced diesel particulate matter by approximately 90 percent. Particulate matter emitted from modern diesel engines is typically in the size range of 10 to 250 nanometers (µm). There are concerns that current regulatory limits for particulate matter from engines, which are in terms of emitted mass, are not a proper measure of the health hazard. One particle of 10 micrometers (µm) diameter has approximately the same mass as 1 million particles of 100 nm diameter. There is evidence that particles smaller than 100 nm can pass through cell membranes and migrate into other organs. Research into the potential health impacts of very small particles is currently being sponsored by the Health Effects Institute (HEI). Proposals for new regulations exist in some countries, with suggestions to limit the particle surface area or the particle number. Although particulate number regulations are not in effect in the United States, they are being considered. Vehicles equipped with particulate filters should have no difficulty meeting newly proposed particle number regulations. However, as new combustion technologies, such as LTC and high injection pressure, lifted flame, mixing controlled combustion, are integrating into the engine map, it may be possible to meet the particulate mass regulation without a filter. Under such situations it will be important to understand the characteristics of the particulate number being emitted from these future combustion systems. Technologies to address this issue will be needed in the future.

Response to Recommendations from NRC Phase 1 Review

Prior to 2007, the aftertreatment goals of the 21CTP were focused on (1) meeting the 2007 heavy-duty engine emission standards and (2) eliminating the need for aftertreatment. In its Phase 1 review of the 21CTP (NRC, 2008), the committee found that no specific goals had been outlined for 21CTP diesel engine aftertreatment systems, but some goals had been set for eliminating aftertreatment. The committee found that the goal of eliminating aftertreatment did not appear to be achievable in the foreseeable future. The committee also found that the Crosscut Lean Exhaust Emissions Reduction Simulations (CLEERS), Diesel Crosscut Team (DCT), and CRADAs had contributed to many successful projects and programs. The NRC Phase 1 review recommended that specific goals should be set for aftertreatment systems (improved efficiency, lower fuel consumption, lower cost of substrates, lower-cost catalyst, etc.) and that the 21CTP should continue with the CLEERS, DCT, and CRADA activities for aftertreatment systems. The 21CTP accepted the recommendations of the NRC (2008) review and has adjusted its aftertreatment program to address them.

Goals

The 21CTP responded to the recommendations of the NRC Phase 1 review with a critical evaluation of the role of aftertreatment in meeting its overall program goals and has written a white paper specifying the role of aftertreatment within its program (DOE, 2011a). Formally the 21CTP has identified six goals specifically related to aftertreatment technologies. These goals are as follows with comments from the committee on the work supporting these goals.

- *21CTP Goal 1 related to aftertreatment technologies:* "Improve performance and durability of NO_x control technology through improved combustion and aftertreatment processes, sulfur management, reductant strategies, and improved materials."[36]
  Comment: This goal would appear to be met by further reduction of sulfur in the diesel fuel, and developments of SCR catalysts, lean NO_x traps, DEF reductants, low temperature combustion, and iron and copper zeolite materials.
- *21CTP Goal 2 related to aftertreatment technologies:* "Develop and apply advanced fuel injection, engine control strategies, new combustion regimes, air-handling, and aftertreatment for emissions reduction, with modeling, simulation and controls integrated in the approach."
  Comment: This goal would appear to be met through the various CRADAs and research projects discussed in this chapter.
- *21CTP Goal 3 related to aftertreatment technologies:* "Develop and implement cost effective retrofit emission control technologies."
  Comment: OEM systems for PM control and NO_x control have been supported, and these have made their way into retrofit (aftermarket) systems, but no direct support of retrofit systems or programs are evident.
- *21CTP Goal 4 related to aftertreatment technologies:* "Determine the best configuration and controls for NO_x and particulate matter (PM) reduction through engine/aftertreatment integration."

[36] Fundamental work is being done to improve the reduction efficiency (performance) of NO_x aftertreament systems, with particular emphasis on high mileage (durability) where compliance with applicable standards is required. The higher the efficiency of the aftertreatment system the larger the cylinder-out NO_x can be. The efficiency of the NO_x aftertreatment system ends up constraining the allowable NO_x leaving the cylinder. Constraining the cylinder-out NO_x constrains what one can do to improve the efficiency of the engine—especially at high load. If the NO_x reduction efficiency were 100 percent, the engine engineers would have great leeway in improving the engine efficiency—which they currently do not have. The closer one can get to this ideal the better. The objective of the research work in the 21CTP is to continually improve the NO_x aftertreatment reduction efficiency at the least possible cost; thus the statement to improve performance and durability of the NO_x control technology.

TABLE 3-11 Department of Energy 21CTP Supported Aftertreatment Research Programs

Project Title	Organization	FY 2009 DOE Funding	FY 2010 DOE Funding	Link to Most Recent EERE Merit Review
CLEERS Coordination and Joint Development of Kinetics for LNT and SCR	ORNL	$200,000 $450,000	$200,000 $500,000	http://www1.eere.energy.gov/vehiclesandfuels/pdfs/merit_review_2010/emissions_control/ace022_daw_2010_o.pdf
Development of Advanced Diesel Particulate Filtration (DPF) Systems	ANL	$500,000	$500,000	http://www1.eere.energy.gov/vehiclesandfuels/pdfs/merit_review_2010/emissions_control/ace024_lee_2010_o.pdf
CLEERS: Aftertreatment Modeling and Analysis	PNNL	$750,000	$750,000	http://www1.eere.energy.gov/vehiclesandfuels/pdfs/merit_review_2010/emissions_control/ace023_lee_2010_o.pdf
Experimental Studies for DPF and SCR Model, Control System, and OBD Development for Engines Using Diesel and Biodiesel Fuels	Michigan Technological University	NA	$583,000	http://www1.eere.energy.gov/vehiclesandfuels/pdfs/merit_review_2010/emissions_control/ace028_johnson_2010_o.pdf
Combination and Integration of DPF-SCR Aftertreatment Technologies	PNNL	$200,000	$400,000	http://www1.eere.energy.gov/vehiclesandfuels/pdfs/merit_review_2010/emissions_control/ace025_rappe_2010_o.pdf
Development of Optimal Catalyst Designs and Operating Strategies for Lean NO_x Reduction in Coupled LNT-SCR Systems	University of Houston	NA	$637,000	http://www1.eere.energy.gov/vehiclesandfuels/pdfs/merit_review_2010/emissions_control/ace029_harold_2010_o.pdf
TOTAL		NA	$3,570,000	

NOTE: Acronyms are defined in Appendix I. NA, not available.

Comment: This goal is being met through the support of many programs discussed in this Chapter. The "best" configurations are the systems developed by the OEMs.

- *21CTP Goal 5 related to aftertreatment technologies:* "Achieve production-feasible, life-cycle effective, emission control system(s) that will meet NO_x and PM regulations phasing in starting 2007, also with reductions of unregulated "toxic" emissions."
  Comment: This goal was achieved, as discussed in the Health Effects section of this chapter.
- *21CTP Goal 6 related to aftertreatment technologies:* "Research pathways to post 2010 regulations for emissions, such as toxics and carbon dioxide."
  Comment: The regulation for carbon dioxide emissions from heavy-duty trucks will entail research into pathways for meeting this regulation. For a potential future regulation for particulate number, research will need to be carried out to develop an understanding of the characteristics of the particulate number being emitted from future diesel combustion systems.

Aftertreatment Projects

Aftertreatment research in the 21CTP is overseen by the DOE, EPA, and 21CTP industry partners. Approximately $37 million has been spent for heavy-duty-truck aftertreatment research for the past 7 years and $3.6 million was the funding for FY 2010.

The development of effective exhaust gas aftertreatment systems and the optimal integration of these technologies as part of the engine-powertrain system are a critical aspect of meeting the goals of the 21CTP. Of specific interest to the 21CTP is the reduction of NO_x within a lean exhaust environment and effective capture and regeneration of particulate matter from the exhaust stream. This happens to be the focus of the combustion research and development community at large, so the 21CTP not only contributes to, but also benefits from, the research activities of a much larger technical community.

As part of the Phase 2 review, the DOE supplied to the committee a list of research programs that should be considered as part of the 21CTP. The names of these projects, the lead agency performing the work, their DOE funding level, and a link to the most recent EERE merit review presentation are presented in Table 3-11.

The aftertreatment research projects currently supported by the 21CTP, listed in Table 3-11, cover a wide range of technologies. Included in the funding is support for the organizational structure called Crosscut Lean Exhaust Emission Reduction Simulations. CLEERS, which is managed through the Oak Ridge National Laboratory, acts as a coordinating body and a disseminator of the newest kinetic schemes and models for aftertreatment system simulation. This is done through its website.[37]

[37] Available at http://www.cleers.org.

More specifically, CLEERS is a collaborative effort among the national laboratories ORNL, ANL, and the Pacific Northwest National Laboratory (PNNL), and industry and academia; CLEERS works to identify, prioritize, and coordinate R&D needs within industry to expedite the development of detailed technical data necessary to simulate lean NO_x trap (LNT) and SCR. Through their efforts, LNT and SCR kinetics have been published, the toluene poisoning mechanism for SCR has been identified using surface spectroscopy, and new catalyst formulations for LNTs that are more sulfur-resistant have been developed. Most of the major catalyst suppliers and engine and vehicle manufacturers participate with national laboratories and academia in these activities.

Within the aftertreatment program, efforts are under way at the ANL to develop advanced diesel particulate filters. This effort involves collaboration with Corning, Caterpillar, the University of Illinois–Chicago, and the University of Wisconsin–Madison; it is focused on studying the oxidation processes to enable better control systems and optimization of regeneration strategies.

The PNNL is leading an effort on aftertreatment modeling and analysis. This work is being done through a new CRADA involving DOW, PACCAR, Ford, GM, and Cummins. The participants are working together to enhance the scientific understanding of DPF, SCR, and LNT technologies. Recent accomplishments include the identification of which hydrocarbons have detrimental effects to NO_x reduction, why water can be an inhibitor to HC storage, and the role of the BaO/alumina interface in LNT performance with ceria as a supplement. These results feed into models that are important in aftertreatment design and control strategy optimization.

Under 21CTP funding, Michigan Technological University is leading an activity with Cummins, Navistar, John Deere, Johnson Matthey, Watlow, ORNL, and PNNL to perform an experimental assessment of DPF and SCR models. This effort is being done with diesel and biodiesel fuels. The importance of particulate maldistribution within the DPF, NH_3 loading in the SCR system, and the ability of different sensors in conjunction with state estimation models to monitor the state of the aftertreatment system effectively are being evaluated.

Funding for a research program being led by the University of Houston is exploring optimal catalyst design and operating strategies for lean NO_x reduction in a coupled LNT/SCR system. Participants in this program include the University of Kentucky's Center for Applied Energy, Ford, BASF Catalysts, and ORNL. The mechanisms of NO_x reduction in LNT/SCR coupled systems are not understood. The program participants are also exploring catalyst synthesis with better desulfation and durability. Recently it has been determined the non-NH_3 mechanisms may be important in SCR kinetics for Fe-zeolite systems.

Finally, a research program is being led by the PNNL with PACCAR and DAF Trucks looking at an advanced aftertreatment system consisting of an integrated DPF and SCR for simultaneous particulate and NO_x control from the same device. The current focus of this work is to determine the optimal SCR catalyst loading that will maximize NO_x reduction while minimizing the pressure drop across the DPF.

These projects have several global aims: to evaluate and address emission control technology barriers; address deficiencies in modeling capabilities and basic understanding; develop the understanding of degradation from sulfur in fuels; address the high platinum metal content and high cost; work to expand the effectiveness of catalytic systems to a broader temperature range; eliminate the inefficient use of fuel for regeneration of diesel particulate filters and desulfurization of NO_x reduction systems, as well as poor reductant utilization (lean NO_x catalyst, LNC); improve inadequate sensors for processing control diagnostics; and address cost and packaging constraints.

Finding 3-10. The research agenda of the 21CTP is focused on improving the NO_x reduction performance of SCR and LNT systems, improving the efficiency of and reducing the fuel consumption associated with PM filter regeneration, and improving the ability to model aftertreatment systems. The DOE CLEERS program does a good job of coordinating the aftertreatment research programs within the 21CTP and disseminating the results to the technical community at large.

Finding 3-11. The demands on the aftertreatment system and its performance are intimately linked to the combustion process taking place within the cylinder. Consequently, the aftertreatment system must be developed and its performance evaluated in conjunction with the combustion system. The 21CTP realizes this, and its new goals for the aftertreatment program specifically state this.

Recommendation 3-7. The aftertreatment program within the 21CTP should be continued, and the DOE should continue to support the activities of CLEERS that interface with the activities of the aftertreatment technical community at large.

Finding 3-12. Particulate size distribution is not a problem with current diesel-type combustion using DPFs. However, as new combustion processes, possibly using different fuels ranging from petroleum-derived fuels to biofuels and synthetics, are integrated into future engine operating maps, it is important to assess particulate size distribution characteristics if particulate filter designs are changed or if DPFs are not used.

Recommendation 3-8. In light of the progress being made with new combustion technologies, which show potential for very low cylinder-out NO_x and particulate emissions, the 21CTP should incorporate studies of particulate number emissions into their research portfolio.

TABLE 3-12 Projects Attributed to Health Effects Studies Receiving Funding from the Department of Energy as Part of 21CTP

Title	Organization	FY 2009 DOE Funding	FY 2010 DOE Funding	Link to Reference
Advanced Collaborative Emission Study (ACES)	HEI, Lovelace Respiratory Research. Institute, CRC	$600,000	$600,000	http://www1.eere.energy.gov/vehiclesandfuels/pdfs/merit_review_2010/health_impacts/ace044_greenbaum_2010_o.pdf
Measurement and Characterization of Unregulated Emissions from Advanced Technologies	ORNL	$475,000	$450,000	http://www1.eere.energy.gov/vehiclesandfuels/pdfs/merit_review_2010/health_impacts/ace045_storey_2010_o.pdf
Collaborative Lubricating Oil Study on Emissions (CLOSE Project)	NREL	FY 2006-FY 2010 (DOE) $892,000		http://www1.eere.energy.gov/vehiclesandfuels/pdfs/merit_review_2010/health_impacts/ace046_lawson_2010_o.pdf

NOTE: Funding levels represent DOE funds only.
SOURCE: Information based on DOE Vehicle Technologies Program Merit Review (DOE, 2010b).

EMISSIONS AND RELATED HEALTH EFFECTS

Introduction

The health effects research within the 21CTP is coordinated through the Health Effects Institute. The HEI is a nonprofit corporation chartered in 1980 as an independent research organization to provide high-quality, impartial, and relevant science on the health effects of air pollution. Typically the HEI receives half of its core funds from the U.S. Environmental Protection Agency and half from the worldwide motor vehicle industry. The 21CTP's support to the HEI represents a strong collaboration between the HEI, DOE, EPA, California Air Resources Board, Coordinating Research Council, Engine Manufacturers Association (EMA), American Petroleum Institute (API), engine and aftertreatment manufacturers, and lubricant suppliers. The objective of the research is to provide a sound scientific basis underlying any potential health hazards associated with the use of new powertrain technologies, fuels, and lubricants in transportation vehicles. Furthermore, the program endeavors to ensure that vehicle technologies being developed by the DOE Vehicle Technologies Program for commercialization by industry will not have adverse impacts on human health through exposure to toxic particles, gases, emanation of electromagnetic fields,[38] and other effects generated by these new technologies. This is being done by characterizing the emissions from vehicles using advanced technologies and also screening these emissions for toxicity. In selected cases if necessary and where possible, the team will work to identify the components responsible for toxicity and engineer solutions to reduce the toxic components.

Three main efforts are under way as part of 21CTP and the Vehicle Technologies Program to evaluate potential health concerns related to heavy-duty diesel vehicles. These efforts aim to undertake high-risk mid- to long-term research; utilize unique national laboratory expertise and facilities; help create a national consensus; and work cooperatively with industry and other agencies. The three efforts are presented in Table 3-12.

Project Overviews

Advanced Collaborative Emissions Study

The combination of advanced-technology, compression-ignition engines, aftertreatment systems, reformulated fuels, and oils developed to meet the 2007 and 2010 emissions standards are demonstrating reduced emissions. Substantial public health benefits are expected from these reductions. However, with any new technology it is prudent to conduct research to confirm benefits and to ensure that there are no adverse impacts on public health and welfare. This is the overall objective of the Advanced Collaborative Emissions Study (ACES) (DOE, 2010b; Greenbaum et al., 2010).

There are three phases to ACES:

1. In phase 1 (completed), the Southwest Research Institute (SwRI) characterized emissions from four 2007 engines. A final report was issued in June 2009.[39]
2. In phase 2, emissions from 2010 model year engines are being characterized.
3. In phase 3, health effects testing is being conducted by the Lovelace Respiratory Research Institute. In this phase, short-term biological screening and long-term

[38] The World Health Organization website includes the statement: "To date no adverse health effects from low level, long-term exposure to radio frequency or power frequency fields have been confirmed but scientists are actively continuing to research this area." Available at http://sho.int/peh-emf/about/what is EMF/en.index1.htm.

[39] Available at http://www.crcao.org/reports/recentstudies2009/ACES%20Phase%201/ACES%20Phase1%20Final%20Report%2015JUN2009.pdf.

health effects tests are being conducted on the emissions from 2007 model year engines.

The ACES project funding totals $15.5 million, with the DOE contribution in FY 2009 and FY 2010 being $600,000 each year. The project will continue until mid-2013.

To date, the reported results from the 2007 engines include the demonstration of emission reductions below the 2007 standards as follows: PM, 89 percent below the 0.01 g/hp-h limit; CO, 98 percent below the 15.5 g/hp-h limit; NMHC, 95 percent below the 0.14 g/hp-h limit; and NO_x, 10 percent below the 1.2 g/hp-h average limit. Unregulated emissions are 90 percent below those of a 2004 technology engine. NO_2 emissions increased over those from the 2004 engines but are expected to go down in 2010; no results have been published yet. Short-term animal studies are complete, and a report is being prepared for review in 2011. The long-term exposure study is under way, with results expected in 2013.

Measurement and Characterization of Unregulated Emissions from Advanced Technologies

The project on the Measurement and Characterization of Unregulated Emissions from Advanced Technologies is largely focused on light-duty vehicles. However, some analysis of SCR systems is being done for the ACES project. Results reported to date have included detailed particulate characterization from direct-injection spark-ignition engines with gasoline and ethanol blends. In addition, the HEI is undertaking a review of the literature on ultra-fine particulates (UFPs). The review will encompass information on the contribution of mobile sources to atmospheric UFPs, health effects of UFPs, and the potential for environmental exposure leading to potential health effects in humans. Animal studies are being considered.

The presentation by the DOE's James Eberhardt to the committee titled "Overview of DOE Health Impacts Research" provided the following observations that may indicate the need for further study:[40]

- "NO_2 emissions have increased significantly at two sampling locations in the South Coast Air Basin since the introduction of 2007 MY trucks. (Starting in 2007, CARB introduced regulations to limit the increase in NO_2 emissions from DPF equipped diesel engines.)
- Certain liquefied natural gas (LNG)-powered trucks are emitting large amounts of ammonia."

[40] James Eberhardt, DOE, "Overview of DOE Health Impacts Research," presentation to the committee, November 15, 2010, Washington, D.C. See http://www1.eere.energy.gov/vehiclesandfuels/pdfs/merit_review_2010/health_impacts/ace00d_eberhardt_2010_o.pdf.

Collaborative Lubricating Oil Study on Emissions (CLOSE) Project

The objective of the CLOSE project is to quantify the relative contributions of fuel and engine lubricating oil to motor vehicle particulate matter and semi-volatile organic compound emissions through the extensive chemical and physical characterization of emissions under a variety of engine operating conditions (Eberhardt and Louison, 2010). At the time of this review, testing from the HD compressed natural gas and diesel buses has been completed, and detailed analyses and source apportionment of exhaust from all vehicle samples are now under way.

Response to Recommendations from NRC Phase 1 Review

Recommendation 3-19 in the NRC (2008) Phase 1 report stated that the committee endorses the DOE multiparty effort to characterize the emissions and assess the safety and potential health effects of new, advanced engine systems, aftertreatment, fuels and lubricants (ACES), and recommends that this continue for the remainder of the study until results become available in the 2012-2013 time period. The DOE response to this recommendation was that it appreciates the endorsement of the ACES program and agrees with the NRC committee on its value.

Finding and Recommendation

Finding 3-13. The Advanced Collaborative Emissions Study, the Collaborative Lubricating and Oil Study on Emissions, and the project on Measurement and Characterization of Emissions from Advanced Technologies are comprehensive and cooperative projects that are investigating important issues related to heavy-duty diesel engine health effects. Based on the activities reported, the committee finds a high degree of collaboration among government agencies, national laboratories, and industry stakeholders.

Recommendation 3-9. The DOE should continue funding the Advanced Collaborative Emissions Study, the Collaborative Lubricating Oil Study on Emissions, and the project on Measurement and Characterization of Unregulated Emissions from Advanced Technologies until results are finalized and reported for all three studies.

HIGH-TEMPERATURE MATERIALS

Introduction

Current heavy-duty diesel engines are extremely durable, in most cases performing reliably for more than 1 million miles in Class 8 truck applications. However, modern diesel engines have pushed the performance of materials to the limit. As the 21CTP develops the next generation of clean and efficient engines, new, higher-performance, lightweight, and cost-effective materials will be needed, as well as manufacturing and inspection methods and appropriate standards. An example of this need for materials is that the thermal efficiency of the diesel engine is enhanced with the ability to run the engine at higher peak cylinder pressures. Higher cylinder pressures and temperatures will challenge the current mechanical property limitations of many engine components, so new materials will be needed to achieve the engines' efficiency potential. A second example is in the potential to reduce fuel consumption through the use of fuel injection systems with higher injection pressure, finer spray control, and multiple injection events. To realize these new fuel injection systems, new materials with higher strength, dimensional stability, and erosion resistance are needed. Lowering the rotation mass in valve trains and air-handling systems has the potential to improve engine response and thermal efficiency and to lower emissions. To capitalize on these potential performance improvements, cost-effective, lightweight material with superior mechanical properties is needed for valve train and air-handling components. Aftertreatment limitations include an incomplete understanding and optimization of catalysts, inadequate test methods for rapid age testing of catalysts, and inadequate sensors for process control or diagnostics. Other barriers to improved fuel efficiency are tribological limits of current lubricants.

The DOE's Heavy Vehicle Propulsion Materials Team has been helpful in identifying commercial materials solutions introduced in 2007 engines. A number of these materials have been identified as enablers of higher-efficiency engines that are being developed for future engine technology. The Heavy Vehicle Propulsion Materials program has been instrumental in developing materials technologies for future engine technologies.

In general, advanced material needs include the following:

- *Major engine components:* cost-effective materials with higher strength and fatigue resistance (cylinder blocks, cylinder heads, pistons, cylinder liners, camshafts, crankshafts, bearings);
- *Fuel injection systems:* materials with higher strength, better dimensional stability, and erosion resistance;
- *Valve train:* cost-effective materials with lower reciprocating mass and greater wear resistance;
- *Air-handling:* corrosion-resistant materials for EGR systems; higher-strength and creep-resistant materials for turbocharger components;
- *Exhaust systems:* materials with higher strength and better creep resistance (DOE, 2011a).

As an example of advanced materials needs, cast components such as cylinder heads and engine blocks are limited at peak temperature and pressure by material tensile strength. Current engines with a peak cylinder pressure of 190 bar are approaching the design limit for traditional cast iron, with a tensile strength of 44,000 pounds per square inch. As cylinder pressures increase to 260 bar for engines with higher efficiency, midterm goals for tensile strength will be approaching 75,000 pounds per square inch, which exceeds the 65,000 pounds per square inch tensile strength for compacted graphite cast iron. Longer-term goals for tensile strength may approach 90,000 pounds per square inch as cylinder pressures approach 300 bar.[41]

Propulsion Materials Programs

There are a number of individual research programs in the area of propulsion materials for both light- and heavy-duty engines.[42] This work is being conducted at the ORNL, ANL, PNNL, and LLNL. The majority of these programs are at the ORNL. Most are in cooperation with industrial partners of which many of the programs are CRADAs. The overall 21CTP budget for heavy vehicle propulsion materials had averaged about $5 million for the past decade. It was $4.86 million in FY 2009 and is $5.66 million in FY 2010. For FY 2011, the total budget for materials for internal combustion engines is expected to be $8.71 million, with an industry cost share of $5.15 million. Research areas include the following:[43]

- Sensors,
- Internal engine components,
- Friction reduction,
- Turbomachinery,
- Fuel injection,
- Alternative fuel compatibility,
- Exhaust aftertreatment,
- Modeling.

It is beyond the scope of this report to review each program individually, but a few are highlighted and listed in Appendix D. A complete list of 21CTP projects in the area of high temperature materials, conducted prior to 2007, can be found in project quad sheets for February 2007 (ORNL, 2007).

The DOE's Heavy Vehicle Propulsion Materials Team also has been helpful in identifying commercial materials solutions that were introduced in 2007 engines. CF8C-Plus stainless steel was developed in an ORNL/Caterpillar CRADA. CF8C-Plus steel received a U.S. patent, and Caterpillar has filed foreign patent applications to facilitate commercial licensing. CF8C-Plus steel also received an

[41] Response by 21CTP to committee question 52, Advanced Materials, received March 1, 2011.

[42] Answers submitted by the 21CTP to committee questions.

[43] Jerry Gibbs, DOE, "Materials Support for the 21st Century Truck: Lightweighting and Propulsion Materials," presentation to the committee, November 15, 2010, Washington, D.C.

ASTM new-alloy designation. Caterpillar commercialized CF8C-Plus as the burner housing for the Caterpillar Regeneration System (CRS) on all CAT on-highway truck engines in January 2007. Deployed in the fall of 2006, more than 500 tons of CF8C-Plus steel have been cast to produce more than 35,000 CRS units.

The project continues in 2011 to explore thin-sections for turbocharger and manifold applications (Maziasz and Pollard, 2007).

Additional examples of successful projects that led to commercialization were presented on February 17, 2011, to committee members John Johnson and David Merrion.[44] Examples include the following:

- Development of high-performance inconel, titanium, and silicon nitride engine valves;
- Development of a piezoelectric control valve actuator for fuel injectors; and
- Exhaust aftertreatment research leading to improved catalyst durability, improved diesel particulate filter durability, and improved EGR cooler performance.

Finding 3-14. The propulsion materials program is addressing a broad range of materials issues associated with heavy-truck propulsion systems. Many of the initiatives are funded as CRADAs with significant industry cost sharing, showing strong support by industry for this area of work.

Recommendation 3-10. The DOE should fund programs in the areas outlined in its "21CTP White Paper on Engines and Fuels" (February 25, 2011) in the section "Approach to Reaching Goals" covering materials R&D for valve trains, major engine components, air-handling systems (turbochargers and EGR systems), and exhaust manifold sealing materials.

High Temperature Materials Laboratory

The High Temperature Materials Laboratory was established more than 20 years ago as a National User Facility to provide specialized, and, in some cases, one-of-a-kind instruments for materials research and characterization. The laboratory is located at the Oak Ridge National Laboratory housing six centers:

- Materials Analysis;
- Mechanical Characterization and Analysis;
- Residual Stress;
- Thermography and Thermophysical Properties;
- Friction, Wear, and Tribology;
- Diffraction.

Additional information on the HTML can be found in the NRC (2008) Phase 1 report, Chapter 3.

Capabilities of HTML that are of current value to the 21CTP through the Vehicle Technologies Program's HTML User Program include the following (DOE, 2010d):[45]

- *The Spallation Neutron Source (SNS)*—a particle accelerator that provides neutron streams in short bursts, enabling the study of materials that are not in equilibrium and are changing dynamically;
- *VULCAN*—a diffractometer, which enables the investigation of stress formation in a material as it solidifies from the molten state as an alloy component (e.g., brake disk rotor, engine block);
- *VENUS*—which makes measurements of the absorption of neutrons by the various nuclei in the material. By doping some ^{6}Li in with the predominant ^{7}Li isotope, VENUS can create an image illustrating the flow of Li ions in a working Li-ion battery. Other applications of VENUS include viewing the lubricant flow in a working engine.

During FY 2009, HTML had 11 industry participants, 14 university participants, and 3 national laboratory participants. Projects focused on such topics as the following:

- Determine the effect of machining parameters on residual stress in ceramic diesel engine exhaust valves,
- Study the residual stresses resulting from piercing truck frame rails,
- Characterize the atomic structure of thermoelectric materials,
- Examine the plastic behavior of wrought magnesium alloys, and
- Determine the mechanical and thermal properties of fibrous diesel particulate filter materials.

These and other projects were described in detail in a presentation to committee members John Johnson and David Merrion on February 17, 2011.[46]

Perhaps just as important as the direct support of the 21CTP is the extensive benefit to the broader research and development community that comes from the research conducted at HTML. This research covers a wide range of challenging problems, for which solutions require the unique instrumentation at HTML as well as the expertise of the knowledgeable DOE researchers who oversee and operate the facility. The fact that many academic researchers, as well as industry research specialists, seek collaboration with

[44] Ron Graves, ORNL, "Material Technology for the 21st CTP Program," presentation to a subgroup of the committee, February 17, 2011, Oak Ridge, Tenn.

[45] See 21CTP response to committee questions, submitted to the committee on November 12, 2010.

[46] Edgar Lara-Curzio, ORNL, "Materials Characterization Capabilities at the High Temperature Materials Laboratory and HTML User Program Success Stories," presentation to John Johnson and David Merrion, February 17, 2011, Oak Ridge, Tenn.

HTML speaks to the value of the facility to the advancement of knowledge on many fronts.

Finding 3-15. The High Temperature Materials Laboratory continues to be a valuable resource for materials research for the 21CTP, providing specialized and in many cases unique instrumentation and professional expertise. The expertise of those who oversee the laboratory, and therefore the value of HTML to all users, is enhanced by the participation of the HTML staff in the research.

Recommendation 3-11. The DOE should continue to provide 21CTP researchers and other potential users access to HTML, and it should make every effort to maintain support for HTML and to maintain the cutting-edge capability of the facility. Moreover, the DOE should provide sufficient funding for HTML, and for the research specialists who oversee and operate the facility, to enable continued research collaboration with the academic community, other government laboratories, and industry. In particular, HTML support should not be reduced to a level that allows only maintenance of the equipment for paying users.

REFERENCES AND BIBLIOGRAPHY

ACEA (European Automobile Manufacturers Association). 2006. Worldwide Fuel Charter, Fourth Edition, Brussels, Belgium, September. Available at http://www.acea.be/index.php/files/worldwide_fuel_charter_september_2006/.

ANL (Argonne National Laboratory). 2009. Nanofluid Development for Engine Cooling Systems. Available at http://www1.eere.energy.gov/vehiclesandfuels/pdfs/merit_review_2009/vehicles_and_systems_simulation/vssp_21_timofeeva.pdf. Accessed on May 5, 2011.

ANL. 2010. CRADAs: Cooling Boiling in Head Region-PACCAR, Integrated Underhood Thermal and External Aerodynamics-Cummins, Routbrot. VSS004. June 8.

ANL. 2010a. Erosion of Radiator Fluids by Nanofluids. Available at http://www1.eere.energy.gov/vehiclesandfuels/pdfs/merit_review_2010/propulsion_materials/pm008_routbort_2010_o.pdf. Accessed on May 5, 2011.

ANL. 2010b. Friction Modeling for Lubricated Engine and Drivetrain Components. June. Available at http://www1.eere.energy.gov/vehiclesandfuels/pdfs/merit_review_2010/propulsion_materials/pm026_ajayi_2010_p.pdf. Accessed on June 1, 2010.

ARB (Air Resources Board). Final Regulation Order to Reduce Greenhouse Gas Emissions from Heavy-Duty Vehicles. Available at http://www.arb.ca.gov/regact/2008/ghghdv08/ghgfro.pdf. Accessed on May 5, 2011.

ATDynamics. 2010. Four East Coast Trucking Fleets Put TrailerTails® on Tractor-Trailers to Reduce Fuel Consumption in Advance of the Holiday Season. Available at http://www.atdynamics.com/trailertail_faq.htm. Accessed January 1, 2011.

Bacorsky, D., M. Dallos, and M. Worgetter. 2010. Status of 2nd Generation Biofuels Demonstration Facilities in June 2010. A report to the International Energy Agency Bioenergy Tack 39. Available at http://www.iea.org/papers/2010/second_generation_biofuels.pdf.

Bartis, J.T., and L. von Bibber. 2011. Alternative Fuels for Military Applications. Santa Monica, Calif.: RAND Corporation.

BP (British Petroleum). 2011. BP Energy Outlook 2030. London, England. Available at http://www.bp.com/liveassets/bp_internet/globalbp/globalbp_uk_english/reports_and_publications/statistical_energy_review_2008/STAGING/local_assets/2010_downloads/2030_energy_outlook_booklet.pdf. Accessed May 19, 2011.

Brusstar, M.J., and C.L. Gray, Jr. 2007. High Efficiency with Alcohol Fuels in a Stoichiometric Medium Duty Spark Ignition Engine. SAE Paper No. 2007-01-3993.

Cooper, K.R. 2004. Commercial Vehicle Aerodynamic Drag Reduction: Historical Perspective as a Guide. In: The Aerodynamics of Heavy Vehicles: Trucks, Buses, and Trains. Springer.

DOE (U.S. Department of Energy). 2006. 21st Century Truck Partnership Roadmap and Technical White Papers. Doc. No. 21CTP-003. December. Washington, D.C.: Office of FreedomCAR and Vehicle Technologies.

DOE. 2009. Annual Merit Review. June. Washington, D.C: Office of Vehicle Technologies.

DOE. 2010a. Updated 21st Century Truck Partnership Roadmap and Technical White Papers. Working draft, September 1, 2010. Washington, D.C.: Office of Vehicle Technologies.

DOE. 2010b. Annual Merit Review. June. Washington, D.C: Office of Vehicle Technologies.

DOE. 2010c. Vehicle System Simulation and Testing. Multi-Year Program Plan, 2011-2015. December. Washington, D.C.: Vehicle Technologies Program.

DOE. 2010d. Multi-Year Program Plan, 2011-2015. Section 2.5-3. Vehicle Technologies Program. December.

DOE. 2011a. 21st Century Truck Partnership White Paper on Engines and Fuels. February 25. Washington, D.C.: Office of Vehicle Technologies.

DOE. 2011b. National Energy Technologies Laboratory, FY2011 Vehicle Technologies Program Wide Funding Opportunity Announcement. Funding Opportunity Number DE-FOA-0000239. Announcement Type: Amendment No. 000002. CFDA Number: 81086 Conservation Research and Development. Issue date: December 16, 2010. Amendment issue date: February 2, 2011.

Eberhardt, J., and D.R. Lawson. 2010. Overview of the DOE Health Impacts Research. Available at http://www1.eere.energy.gov/vehiclesandfuels/pdfs/merit_review_2010/health_impacts/ace00d_eberhardt_2010_o.pdf.

EIA (Energy Information Administration). 2011a. Liquid Fuels Supply and Disposition, Reference Case. Available at http://www.eia.gov/oiaf/aeo/tablebrowser/#release=AEO2011&subject=0-AEO2011&table=11-AEO2011&region=0-0&cases=ref2011-d020911a.

EIA. 2011b. Light-Duty Vehicle Energy Consumption by Technology Type and Fuel Type, Reference Case. Available at http://205.254.135.24/oiaf/aeo/tablebrowser/#release=AEO2011&subject=0-AEO2011&table=47-AEO2011&region=0-0&cases=ref2011-d020911a.

EPA (Environmental Protection Agency). 2010. Greenhouse Gas Emissions Standards and Fuel Efficiency Standards for Medium- and Heavy-Duty Engines and Vehicles. EPA-HQ-OAR-2010-0162. NHTSA-2010-0079. FRL-9219-4. November 30.

EPA. 2011. SmartWay Transport Verified Technologies. Available at http://www.epa.gov/smartwaylogistics/transport/what-smartway/verified-technologies.htm. Accessed April 20, 2011.

EPA/NHTSA (EPA/National Highway Traffic Safety Administration). 2010. Greenhouse Gas Emissions Standards and Fuel Efficiency Standards for Medium- and Heavy-Duty Engines and Vehicles. October 25. Available in Dockets No. EPA-HQ-OAR-2010-0162 and No. NHTSA-2010-0079 at http://www.regulations.gov.

EPA/NHTSA. 2011. Greenhouse Gas Emissions Standards and Fuel Efficiency Standards for Medium- and Heavy-Duty Engines and Vehicles. Federal Register, Vol. 76, No. 179, September 15. Available at http://www.nhtsa.gov/staticfiles/rulemaking/pdf/cafe/2011-20740.pdf.

Ethanol and Biofuels News. 2010. Japan Biofuels Exist Only Because of Subsidies: Study. Vol. XXII, Issue 30. August 4. McLean,Va.: Hart Energy Publishing.

Genzale, C., L.M. Pickett, and S. Kook. 2010. Liquid Penetration of Diesel and Biodiesel Sprays at Late-Cycle Post-Injection Conditions. SAE Paper 2010-04012.

Greenbaum, D., J. Mauderly, et al. 2010. Advanced Collaborative Emissions Study (ACES). Available at http://www1.eere.energy.gov/vehiclesandfuels/pdfs/merit_review_2010/health_impacts/ace044_greenbaum_2010_o.pdf.

Kokjohn, S., R.M. Hanson, D. Splitter, and R.D. Reitz. 2009. Experiments and Modeling of Dual-Fuel HCCI and PCCI Combustion Using In-Cylinder Fuel Blending. SAE Paper 2009-01-2647.

Kokjohn, S. R. Hanson, D. Splitter, J. Kaddatz, and R. Reitz. 2011. Fuel Reactivity Controlled Compression Ignition (RCCI) Combustion in Light-and Heat-duty Engines. SAE Paper 2011-01-0357.

Kook, S., and Pickett, L.M. 2009. Effect of Fuel Volatility and Ignition Quality on Combustion and Soot Formation at Fixed Premixing Conditions. SAE Paper 2009-01-2643.

Lawson, D.R., and J. Eberhardt. 2010. Collaborative Lubricating Oil Study on Emissions (CLOSE Project). June 2010. Available at http://www1.eere.energy.gov/vehiclesandfuels/pdfs/merit_review_2010/health_impacts/ace046_lawson_2010_o.pdf.

Maziasz, P.J., and M.J. Pollard. 2007. CF8C-Plus: A New Cast Stainless Steel for High Temperature Diesel Exhaust Components. Available at http://www1.eere.energy.gov/vehiclesand fuels/pdfs/deer_2007/session9/deer07_maziasz.pdf. Accessed April 5, 2011.

Mueller, C.J., A.L. Boehman, and G.C. Martin. 2009. An Experimental Investigation of the Origin of Increased NO_x Emissions When Fueling a Heavy-Duty Compression-Ignition Engine with Soy Biodiesel. SAE Paper 2009-01-1792.

NAS-NAE-NRC (National Academy of Sciences-National Academy of Engineering-National Research Council). 2009. Liquid Transportation Fuels from Coal and Biomass: Technological Status, Costs, and Environmental Impacts. Washington, D.C.: The National Academies Press.

Navistar. 2010. Navistar Pursues Natural Gas for MaxxForce 13 Engines. Warrenton, Ill.: Navistar Corporation.

Navistar. 2008. Navistar Awarded Department of Energy Funding to Develop Aerodynamic Trailers. Available at http://ir.navistar.com/releasedetail.cfm?ReleaseID=354505.

Nelson, C. 2010. Exhaust Energy Recovery—Program Summary Presentation. Prepared as Final Report for Contract to Cummins. DOE DE-FC 26-05NT42419. May.

NESCCAF/ICCT (Northeastern States Center for a Clean Air Future/International Council on Clean Transportation). 2008. Reducing Heavy-Duty Long Haul Combination Truck Fuel Consumption and CO_2 Emissions. P. Miller, ed. October 28.

New York Times. 2010. DOE's National Algae Biofuels Technology Roadmap, June 29, 2010.

NPRA (National Petroleum Refiners Association). 2010. Algae's Potential as a Major Fuel Producer. Paper AM-10-156. Washington, D.C.

NRC (National Research Council). 2008. Review of the 21st Century Truck Partnership. Washington, D.C.: The National Academies Press.

NRC. 2010. Technologies and Approaches to Reducing the Fuel Consumption of Medium-and Heavy-Duty Vehicles. Washington, D.C.: The National Academies Press.

ORNL (Oak Ridge National Laboratory). 2007. Propulsion Materials Quad Sheets. Available at www.ornl.gov/sci/propulsionmaterials/pdfs/21CTP_QuadSheets_2007.pdf. Accessed April 5, 2011.

ORNL. 2009. Effect of Wide-Based Single Tire on Class-8 Combination Fuel Efficiency. Oscar Franzese, Helmut Knee, and Lee Slezak. ORNL. October 28.

Ra, Youngchul, R. Reitz, M. Andrie, R. Krieger, D. Foster, R. Durrett, V. Gopalakrishnan, A. Plazas, R. Peterson, and P. Szymkowicz. 2010. Study of High Speed Gasoline Direct Injection Compression Ignition (GDICI) Engine Operation in the LTC Regime. SAE Paper 2011-01-1182.

Splitter, D., R. Hanson, S. Kokjohn, and R. Reitz. 2010. Reactivity Controlled Compression Ignition (RCCI) Heavy Duty Engine Operation at Mid- and High-Loads with Conventional and Alternative Fuels. SAE Paper 2011-01-0363.

Stanton, D. 2010. High Efficiency Clean Combustion for Super Truck. 2010 DEER Conference, September 29, Detroit, Mich.

Sun, R., R. Thomas, and C.L. Gray, Jr. 2004. An HCCI Engine: Powerplant for a Hybrid Vehicle. SAE Paper 2004-01-0933.

Sun, H., D. Hanna, L. Hu, J. Zhang, E. Krivitzky, and C. Osborne. 2010. Analyses Guided Optimization of Wide Range and High Efficiency Turbocharger Compressor. DEER Conference, September 29, 2010, Detroit, Mich.

TIAX. 2009. Assessment of Fuel Economy Technologies for Medium- and Heavy-Duty Vehicles. Report prepared for the National Academy of Sciences by TIAX LLC, Cupertino, Calif., July 31, 2009.

UNFAO (United Nations Food and Agricultural Organization). 2010. Jatropha: A Smallholder Bioenergy Crop. Integrated Crop Management, Vol. 8. Rome, Italy. Available at http.www.fao.org/docrop/012/i1219e/i1219e.pdf.

Wards. 2011. Algae-derived biodiesel fuel, Wards Auto.com, January 19, 2011.

WFC (Worldwide Fuel Charter). 2006. European Automobile Manufacturers Association, Alliance of Automobile Manufacturers Association. Available at http://www.jama.or.jp/eco/wwfc/pdf/WWFC_Sep2006_Brochure.pdf.

Wijffels, H., and M.J. Barbosa. 2010. An Outlook on Microalgal Biofuels. Science, Vol. 329, No. 5993, Pages 796-799.

4

Medium- and Heavy-Duty Hybrid Vehicles

INTRODUCTION

The 21st Century Truck Partnership (21CTP) focuses on "research and development of advanced heavy-duty hybrid propulsion systems that will reduce energy consumption and pollutant emissions" (DOE, 2011). A heavy-duty hybrid vehicle has an internal combustion engine, an energy storage system, and a means for absorbing or delivering torque from the drivetrain. Hybrid vehicles provide improvements in fuel consumption by several means, including the conversion of kinetic energy into a storable form of energy for later use, the downsizing of the internal combustion engine, engine shutoff, and accessory electrification. A mild hybrid provides idle-stop functionality and regenerative braking. Idle stop-start functionality is made possible by having the electric motor quickly restart the engine. A full hybrid design provides traction capability and higher rates of brake energy regeneration as well as idle-stop functionality.

The two major types of heavy-duty hybrid vehicles discussed in this chapter are hybrid electric vehicles (HEVs), which are the primary focus of the Department of Energy (DOE) in the 21CTP, and hydraulic hybrid vehicles, which are the focus of the Environmental Protection Agency (EPA). In addition to HEVs, similar technologies are being applied to plug-in hybrid electric vehicles (PHEVs) and battery electric vehicles (EVs). Hybrid electric vehicles are considered "a key technology that will help the 21CTP achieve its goals" by enabling manufacturers to achieve reduced fuel consumption and emissions (DOE, 2011). The various electric hybrid architectures have been well described elsewhere (NRC, 2010, 2011). Heavy-duty hybrid electric vehicles provide additional benefits from their unique capability to "creep" in queues and idle without operating the main engine. For Class 8 long-haul trucks, heavy-duty hybrid systems can recharge large-capacity batteries during daytime operation to provide for hotel loads (powering support services for the driver in the truck tractor), eliminating overnight idling of the main engine.

The 21CTP partners work together on technical and commercial programs related to the reduction of fuel consumption by heavy-duty hybrid trucks and the commercialization of these more-fuel-efficient vehicles. In addition to the DOE, the U.S. Department of Defense (DOD), the U.S. Department of Transportation (DOT), EPA, and several original equipment manufacturers (OEMs) and key suppliers, the 21CTP has strategic alliances with the Engine Manufacturers Association (EMA), the Truck Manufacturers Association (TMA), and the Hybrid Truck Users Forum (HTUF). The national laboratories assist the 21CTP in conducting research and development (R&D) involving hybrid technologies. Missing from these partnership alliances are companies in the electric machines, energy storage, and semiconductor-device industries. Such relationships should be fostered through 21CTP R&D programs with industry.

During the 2007-2010 period, the 21CTP project priorities in the area of medium- and heavy-duty hybrid vehicles have been the following:

- Simulation and modeling,
- Subsystem R&D, and
- Vehicle demonstrations.

Progress in addressing each of these priorities is discussed later in this chapter. Programs in these priority areas have been pursued by the DOE through the Argonne National Laboratory (ANL), the National Renewable Energy Laboratory (NREL), and DOE's Office of Vehicle Technologies; by the DOD through the U.S. Army's Tank-Automotive Research, Development and Engineering Center (TARDEC); by the DOT through its Federal Transit Agency's National Fuel Cell Bus Program; and by the EPA through its National Vehicle and Fuel Emissions Laboratory.

Differences Between Heavy-Duty and Light-Duty Hybrids

Requirements for heavy-duty hybrid vehicles are significantly different from those of light-duty (LD) hybrid vehicles, and they necessitate unique solutions. In this chapter, the term *heavy-duty vehicles* refers to both medium heavy-duty trucks and heavy-duty trucks, as defined in Figure 1-3 in Chapter 1 of this report. Many technologies that apply to light-duty vehicles do not apply to heavy-duty vehicles. The heavy-duty truck and light-duty vehicle hybrid technologies leverage each other only at the most basic level. Consequently, the 21CTP hybrid program is needed to address the unique technology needs of heavy-duty vehicles.

Unlike LD hybrid vehicles, the broad class of vehicles comprising the heavy-duty fleet is very diverse and includes tractor-trailer, refuse, dump and utility trucks, package delivery vehicles, buses, and large pickups. These vehicles have highly differentiated mission profiles, which make it difficult to establish architectures or performance metrics that are commonly applicable to this broad range of heavy-duty vehicles. Key differences between heavy-duty trucks and light-duty vehicles (LDVs) include the following:

- *Volume:* Annual sales volume for heavy-duty trucks is about 5 percent of that for LDVs, and the former can be bought in a thousand times more configurations than the latter.
- *Buying criteria:* Heavy-duty truck buyers prioritize reliability and cost of ownership, whereas LDV buyers prioritize a variety of attributes including cost, functionality, reliability, performance, and styling.
- *Weight:* A heavy-duty truck can weigh up to a hundred times more than an LDV and has peak horsepower up to twice that of LDVs.
- *Life expectancy and driving cycles:* Heavy-duty vehicles have longer life expectancy and more demanding duty cycles than those for LDVs. Heavy-duty vehicles have expected lifetime mileages nearly 10 times greater than those of LDVs.

Driven by these vehicle differences, factors differentiating hybrid systems for heavy-duty vehicles and those for LDVs are power rating, energy storage capacity, number of relevant driving cycles, and economics. Unlike cars and light trucks, which are available in only a relatively restricted range of sizes and weights and whose missions have been characterized by a few standardized driving cycles, heavy-duty trucks span a size range from 8,500 lb (Class 2b) to greater than 33,000 lb (Class 8), with gross vehicle weights (GVWs) of up to 200,000 lb (DOE, 2011). As a result, heavy-duty hybrid vehicles require high energy storage density, much like an EV, as well as high power density for acceleration and deceleration, like light-duty hybrids.

Functional differences between hybrid systems for heavy-duty trucks and those for LDVs result in substantial economic differences. Light-duty vehicles are fundamentally similar. Their weight range is relatively limited (up to Class 2a, under 8,500 lbs GVW), and their driving schedules are characterized by a small number of driving cycles. They are manufactured in high volumes, and their expected lifetime mileage is up to 150,000 miles. Hybrid systems for these vehicles can be manufactured in volumes large enough to benefit from the economies of scale.

Heavy-duty trucks by contrast vary widely in both tare (empty) and gross weights. Their missions vary from daily runs with frequent stops (e.g., the work of a delivery van) to 24-hour–a-day, multiple-day long hauls of tractor-trailers up to 200,000 lb GVW. The total fleet of heavy-duty vehicles is 10.99 million, and the average life of a vehicle is up to 1 million miles, resulting in a small market with low turnover and a challenge to making an economic argument for hybrid systems generally applicable to the heavy-duty fleet. A conventional hybrid design applied to vehicles whose missions incorporate a lot of stop-and-go driving, such as delivery vans, urban transit buses, or refuse trucks, has the potential to be economically sound (i.e., to result in a favorable payback period) by providing substantial fuel economy benefits of 20 to 40 percent (17 to 29 percent reduction in fuel consumption) (Greszler, 2009).[1] A further benefit of hybridization in these vehicles is the reduced brake wear and maintenance resulting from regenerative braking.

In contrast to medium-duty delivery vans, urban transit buses, and refuse trucks, a heavy-duty, Class 8 long-haul truck makes few stops, maintains a relatively constant speed, and requires high power for long periods of grade climbing. According to the 21CTP, there are three primary reasons to consider hybridizing a Class 8 long-haul powertrain (Greszler, 2009):

1. *Reduced engine idle time,* through the hybrid energy storage and use of electric auxiliaries;
2. *Reduced fuel use,* through the electrification of components, thereby improving efficiency; and
3. *Reduced fuel usage during cruise,* through energy management with traffic-induced speed variation and in rolling terrain.

An additional reason to consider hybridizing a long-haul truck is that, as mentioned above, the large-capacity batteries can be recharged during daytime operation to provide for hotel loads so that overnight idling of the main engine can be eliminated.

Hybrid Technology for the SuperTruck Program

In FY 2010, the DOE announced the establishment of the SuperTruck program, with an overall goal to develop and demonstrate a 50 percent improvement in freight efficiency

[1] Answers provided by Ken Howden, DOE Office of Vehicle Technologies, to committee questions 9(a) and 42.

expressed in ton-miles per gallon (33 percent reduction in the proposed fuel consumption standard expressed in gallons per 1,000 ton-miles) for Class 8 long-haul trucks (see Chapter 8). The three project teams selected for the SuperTruck program are Cummins-Peterbilt, Daimler Trucks North America, and Navistar, Inc. Two of the teams, Daimler and Navistar, are using hybrid technology, with Navistar specifying a dual-mode (series/parallel) electric hybrid. The Cummins-Peterbilt team is using waste heat recovery (WHR) and a solid oxide fuel cell auxiliary power unit (APU) for idle reduction. All three teams are planning to use some electrically driven accessories as part of their hybridization and idle-reduction systems.

Impact of Duty Cycle

Fuel consumption improvements in a heavy-duty hybrid vehicle are highly dependent on the duty cycle, because the duty cycle determines the amount of kinetic energy available to be recovered, the time available for engine shutoff at idle, and the benefit of the electrification of accessories operating on demand and at constant speed versus operating full time at speeds proportional to engine speeds. There are no industry standards yet for heavy-duty hybrid vehicle testing. The duty cycle used to measure fuel consumption is typically determined by the way that a certain type of vehicle is used in the targeted application and the creation or selection of an appropriate cycle based on measured data taken to characterize the vehicle usage.

The Argonne National Laboratory has conducted modeling studies of the effect of duty cycle on a Class 8 long-haul truck with a mild hybrid (50 kW motor and 5 kWh battery) and a full hybrid (200 kW motor, a 50 kW starter/generator, and a 25 kWh battery) (NRC, 2010). Both were pre-transmission hybrids, that is, hybrids in which the electric drive motor is located between the clutch and the transmission, allowing torque multiplication through the transmission and electric-mode-only operation during low power demand. Five drive cycles were evaluated; they included three highway cycles: HHDDT 65 (Heavy Heavy-Duty Diesel Truck), HHDDT Cruise, and HHDDT High Speed; and two transient/urban cycles: HHDDT Transient and UDDS (Urban Dynamic Driving Schedule) Truck (NRC, 2010). Key observations from this ANL analysis as described in the National Research Council (NRC) report were as follows:

- On the highway cycles, fuel savings were less than 10 percent for the full hybrid and less than 5 percent for the mild hybrid.
- Neither hybrid system had enough electrical storage to contribute to cruise power demand for any significant length of time.
- Fuel savings on the transient/urban cycles were greater, although the mild hybrid showed significantly lower savings than the full hybrid, peaking at 14 percent, while the full hybrid showed savings over 40 percent.

Other simulation studies of the fuel consumption benefit from a variety of hybrid truck configurations have been conducted and reported (NRC, 2010). The qualification for these studies is relevant and is repeated here. "While indicative of the range of potential benefits, it should be noted that simulations are carried out under ideal conditions—hence results typically represent best-case scenarios. Real-world savings in fuel consumption are likely to be lower, because of off-design duty cycles and practical production vehicle constraints" (NRC, 2010, p. 83).

Limited test results for different duty cycles are available for medium- and heavy-duty hybrid vehicles. Regarding one set of results, the EPA conducted tests on a series hydraulic hybrid delivery truck over a number of different test duty cycles. The results indicated that the fuel economy improvements ranged from negligible for the Highway Fuel Economy Test (HWFET) to more than 100 percent improvement (50 percent reduction in fuel consumption) on the Manhattan Bus Cycle.[2]

Commercialization

Hybrid truck technology is currently available in demonstration vehicles as well as commercial vehicles. Heavy-duty hybrid electric trucks are commercially available from several manufacturers, including Freightliner/Daimler Trucks, International/Navistar, Kenworth, Peterbilt, Ford, and GMC. Details of the specific models of commercially available heavy-duty hybrid trucks are listed in Appendix E of this report. Commercially available battery electric trucks are also listed in Appendix E. Heavy-duty hydraulic hybrid trucks are becoming commercially available, with the hydraulic hybrid systems primarily supplied by Eaton, Parker Hannifin, and Bosch-Rexroth. A summary of the status as of 2009 of the wide variety of hybrid system architectures currently available in the market is provided in Table 4-1.

Hybrid Truck Users Forum (HTUF)

The HTUF, which was briefly described in the NRC Phase 1 report (NRC, 2008), is a North American, user-driven program to speed the commercialization of heavy-duty hybrid and high-efficiency technologies. It is operated by CALSTART, a member-supported organization with headquarters in California and dedicated to supporting a growing clean transportation industry, in partnership with the U.S. Army's National Automotive Center (NAC), with project support from the Hewlett Foundation, which makes grants to solve social and environmental problems, and the DOE. The HTUF focuses on developing the commercial hybrid industry

[2] John Kargul, EPA, "Clean Automotive Technology, Cost Effective Solutions for a Petroleum and Carbon Constrained World," presentation to a subgroup of the committee, October 26, 2010, Ann Arbor, Michigan.

TABLE 4-1 Hybrid Vehicle Architectures, Their Status as of 2009, and Primary Applications

Architecture	Technology Status	Primary Applications
Medium-duty/heavy-duty parallel HEV	Available now: Eaton, Azure, Volvo	Refuse, urban pickup, and delivery (P&D)
Medium-duty/heavy-duty parallel HEV w/e PTO	Available now: Eaton, Volvo	Bucket truck
Parallel gasoline or diesel HEV bus	Available now: ISE, Enova, BAE	Transit bus
Two-mode diesel HEV bus	Allison	Transit bus
Series gasoline or diesel HEV bus	Available now: ISE	Transit bus
Parallel hydraulic hybrid	Introduced in 2009: Eaton, Parker Hannifin, Crane Carrier	Refuse, urban P&D
Series hydraulic hybrid	Demo vehicles	Refuse, urban P&D
Parallel Class 2b	Demo vehicles	Class 2b pickups and vans
Two-mode Class 2b	Demo vehicles	Class 2b pickups and vans
Line-haul dual-mode HEV	Demo vehicles	Line-haul tractor trailer.
Line-haul parallel HEV	Demo vehicles	Line-haul tractor trailer, motor coach

NOTE: Acronyms are provied in Appendix I. See Appendix E, Table E-1, for a list of medium- and heavy-duty vehicle models.
SOURCE: TIAX (2009).

through increasing user-driven volumes in key platforms to provide the benefits of reduced fuel use and lowered emissions. HTUF efforts have accelerated the market "pull" that assisted in helping to launch the first production of hybrid trucks.

Ten years ago, the U.S. Army and collaborative partner CALSTART launched an initiative to promote hybrid and high-efficiency dual-use technologies for the commercial trucking industry and military platforms. The HTUF held its first annual conference at that time. In September 2010, the 10th annual HTUF conference was held in Dearborn, Michigan, with 700 attendees. It was the largest national conference in the program's history, signaling the expanding interest in hybrid trucks and their growing market. During a panel discussion at this conference, efforts to secure purchase incentives (through an expansion and extension of the hybrid truck tax credit) and R&D funding (through H.R. 3246, the Advanced Vehicle Technology Act) were discussed, but both of these efforts have stalled.

The prime role of the HTUF with its military partners was to help create the commercial industrial capability to support military hybrid advanced technology and vehicle needs. To date, commercial hybrid use has surpassed that in pure military applications. When HTUF user specifications were developed, they included functionality (stealth mode/silent watch, power generation) of importance to the military. Therefore, most commercial systems can be easily adapted to military needs. Military hybrid truck adoption has been slowed owing to the prolonged and deployed status of the military since September 11, 2001. The military has been in a replace-and-repair mode with its vehicles since that time. The Army has had to delay or halt new vehicle development efforts that included hybrid vehicles. The HTUF is focusing its near-term efforts on the following areas for the military: (1) increasing deployments of commercial hybrids on military bases, (2) identifying and demonstrating hybrid and advanced systems in off-road commercial construction equipment of use to the military, and (3) assisting with identifying and testing commercial-based systems in tactical vehicle applications.

FUNDING

The only heavy-duty hybrid truck funding available since 2007 has been used to conduct in-use and laboratory evaluations on hybrid vehicles, mostly delivery trucks and transit bus applications. None of this funding was for hybrid R&D. Since 2007, the $1 million to $1.5 million per year for heavy-duty R&D has been focused on aerodynamics, thermal management, and friction and wear reduction. Although there were no congressional requests for FY 2007 through FY 2010 that specifically targeted 21CTP heavy-duty hybrids, the work cited above was completed under the DOE's Office of Vehicle Technologies, which has included both light-duty and heavy-duty advanced vehicle testing and R&D activities, with some of these being hybrid-specific. Project funding since 2007 is shown in Table 4-2.

HYBRID ELECTRIC VEHICLES

The various electric hybrid architectures have been well described elsewhere (NRC, 2010, 2011). The three major components of a hybrid system are the electric machines (motor/generator), the electronic controller, and the electric energy storage system (typically a battery or battery/ultracapacitor combination). In addition, the hybrid powertain system includes an internal combustion engine, a transmission system linking the motor and engine, and electromechanical controls that determine the electrical and mechanical power directions and their paths. The electrical machine(s),

TABLE 4-2 Heavy-Duty Hybrid Funding for FY2007-FY2010

	FY 2007	FY 2008	FY 2009	FY 2010
HD Simulation and Technology Focused HEV R&D	$200,000	$250,000	$200,000	$350,000
HD HEV Fleet Demonstrations	$400,000	$450,000	$300,000	$1,500,000
Subsystem R&D	$0	$0	$0	$0

NOTE: Acronyms are defined in Appendix I.
SOURCE: Lee Slezak and Kevin Walkowicz, NREL, "Hybrid Team Progress on Past Goals," presentation to the committee, November 15, 2010, Washington, D.C.

its coupling to the driveline, and the necessary power and control electronics are generally referred to as the drive unit. Of all the components, the energy storage system presents the greatest economic challenge for the truck application. Trucks in stop-and-go driving and with long lives will require long-life energy storage systems, as identified in the goal for energy storage systems discussed in the next section.

Goals and Status

The overall goal of the 21CTP for the heavy-duty hybrid system is to develop and demonstrate system technologies that deliver substantial reductions in fuel consumption for both urban start-stop use and for higher mileage applications. The top-priority hybrid electric vehicle R&D areas requiring support are the following:

1. Drive-unit optimization,
2. Drive-unit cost,
3. Energy storage system reliability,
4. Energy storage system cost, and
5. The ability to attain fuel consumption improvements.

Three hybrid technology goals for 2012 were established in 2006 by the 21CTP.[3] The progress toward achieving each of these goals is discussed below. The revised, stretch goals for heavy-duty hybrids issued by the 21CTP in 2011 are summarized in the section below titled "Revised Goals Issued in 2011."

21CTP Goal 1: Develop a new generation of drive unit systems that have higher specific power, lower cost, and durability matching the service life of the vehicle. Develop a drive unit that has 15 years of design life and costs no more than $50/kW by 2012.

The drive unit consists of the motor, controller, and packaging. No R&D work for drive-unit systems technology for heavy-duty hybrid vehicles was funded by the DOE for FY 2007 through FY 2010. The NRC Phase 1 report (NRC, 2008) stated that "very little evidence was presented to the committee to substantiate any significant progress made by 21CTP-funded researchers toward achieving the desired reliability target of 15 years design life for the hybrid propulsion equipment. In fairness, the number of prototype heavy-duty hybrid trucks currently in the field is very low, making it particularly difficult to gather any meaningful reliability data."

The DOE did not discuss the status of costs of the drive-unit system with the present committee. The NRC Phase 1 Report (NRC, 2008) stated that "meeting the aggressive cost target of $50/kW is proving to be one of the most difficult challenges for the developers of heavy hybrid propulsion systems."

The Advanced Power Electronics and Electric Motors (APEEM) program is one of the DOE's Office of Vehicle Technologies' two basic building blocks for electric propulsion drive vehicles; it includes power electronics, electric motors, thermal management, and packaging. The Office of Vehicle Technologies addresses the needs of both light- and heavy-duty vehicles. A principal objective of the APEEM R&D program is to reduce component and subsystem costs so that a customer can recover the additional cost for advanced electric vehicles in 3 years through fuel savings (DOE, 2010). In addition to cost, APEEM research focuses on components that are significantly smaller, lighter, and more reliable than those of existing systems. The American Recovery and Reinvestment Act of 2009 (ARRA) awarded $500 million in grants to U.S.-based manufacturers to accelerate the development of the U.S. manufacturing capacity for electric drive components. These R&D programs are described in the DOE's "Multi-Year Program Plan 2011-2015" (DOE, 2010, pp. 2.1-20 to 2.1-25). The specific applicability of these projects to heavy-duty vehicles was not provided to the committee.

21 CTP Goal 2: Develop an energy storage system with 15 years of design life that prioritizes high power rather than high energy, and costs no more than $25/kW peak electric power rating by 2012.

The energy storage system consists of the battery, its thermal control system, and packaging. No R&D work for energy storage technology for heavy-duty hybrid vehicles was funded by the DOE for FY 2007 through FY 2010. The NRC Phase 1 report (NRC, 2008) provided the status of energy storage systems for the light-duty FreedomCAR program. As noted earlier, heavy-duty hybrids are significantly different from light-duty hybrids. The DOE indicated to the committee that the state-of-the-art energy storage systems are typically rated for 8 years of useful life.[4] The DOE did

[3] Lee Slezak and Kevin Walkowicz, NREL, "Hybrid Team Progress on Past Goals," presentation to the committee, November 15, 2010, Washington, D.C.

[4] Answers provided by Ken Howden, DOE Office of Vehicle Technolo-

not discuss the status of costs of heavy-duty hybrid energy storage systems with the committee.

The DOE's 2010 cost goal for light-duty hybrid energy storage systems is $750 to $900/kWh (nameplate capacity).[5] However, the committee was informed by one manufacturer that the current cost for heavy-duty hybrid energy storage systems was approximately $2,000/kWh (nameplate capacity). Drivers of the increased cost relative to light-duty energy storage systems are the operating environment (e.g., vibration) in the truck and lower production volumes over which tooling, design, and validation costs can be amortized.

Electrical energy storage has improved in the past decade, but further gains are required for truck applications. The more promising technologies are the nickel metal-hydride battery, lithium technology batteries, and ultracapacitors. The EV requires a high-energy electrical storage capacity for a long driving range. Heavy-duty hybrid vehicles require high energy storage density, much like an EV, as well as high power density for acceleration and deceleration, like light-duty hybrids. In recent years batteries have been the major focus for electrical energy storage for LDVs, and results provide an opportunity for cross-application to medium-duty and heavy-duty hybrid development. Several companies (A123 Systems, Hitachi, and Valence Technologies, Inc.) are producing batteries for transit operations. Since 2007 the DOE has funded seven heavy-duty hybrid projects. These have all been vehicles with a high level of stop-and-go behavior in their mission profiles—for example, delivery vehicles and school and transit buses. Observed fuel economy improvements for hybrid electric vehicles ranged from 0 to 30 percent (0 to 23 percent reduction in fuel consumption), whereas a plug-in hybrid electric school bus exhibited an improvement up to 57 percent (36 percent reduction in fuel consumption), but results were inconsistent because of incomplete battery-charging data.[6]

Opportunity exists for further fuel consumption reductions and cost reductions with the development and application of components designed specifically for the heavy-duty hybrid vehicles. For example, a greater fraction of braking energy could be regenerated if the power electronic and energy storage systems were more efficient and able to transfer energy at higher powers. Other areas for improvement are implementing the system-wide electrification of accessories and increasing the temperature range of operation for lithium batteries. The key challenges identified by the DOE for the heavy-duty hybrid energy storage system are as follows:

- Both procurement costs and life–cycle costs,
- Weight and space claim,
- Life expectancy (in a heavy-duty drive cycle),
- Energy and power capacity for a heavy-duty hybrid application,
- Suitability for the heavy-duty vehicle environment and cooling techniques,
- Architecture and modularity,
- Safety and failure modes,
- Maintainability,
- Management and equalization electronics and algorithms, and
- Supplier base for the storage elements.

On March 19, 2009, the DOE launched a $2 billion competitive grant program under the ARRA to promote the manufacturing of the batteries and parts for electric vehicles. The objective of these grants is to establish a complete "value chain" for lithium battery manufacturing, from material supply, to cell production, to pack assembly. Twelve domestic manufacturers are developing battery materials, production and recycling capabilities, as shown in Table 4-3. DOE funding for each of the manufacturers is listed in the table. Nine domestic U.S. battery manufacturing facilities for cell production and for pack assembly, together with total investment in these facilities, are shown in Table 4-4. The committee was not provided with information on the applicability of the manufacturing facilities shown in Tables 4-3 and 4-4 to light-duty versus heavy-duty vehicles.[7] However, the committee believes that these facilities would be capable of supplying batteries for medium- and heavy-duty trucks, although their shock and vibration requirements are more severe than for LDVs.

As with the APEEM program mentioned above, battery energy storage is one of the DOE's Office of Vehicle Technologies' two basic building blocks for electric propulsion drive vehicles; it includes a wide range of battery technology R&D under the U.S. Advanced Battery Consortium (USABC) that addresses the needs of both light- and heavy-duty vehicles. The battery R&D is engaged in topics ranging from fundamental materials research through battery development and testing and includes the following:[8]

- Seventy laboratory and university projects to address the cost, life, and safety of lithium-ion batteries and to develop next-generation materials; and

gies, to committee question 42.

[5] Patrick B. Davis, DOE, "U.S. Department of Energy Vehicle Technologies Program Overview," presentation to the committee, September 8, 2010, Washington, D.C. Note that these cost goals for electric energy storage are different from the peak power cost goal in Goal 2 of $25/kW.

[6] Answers provided by Ken Howden, DOE Office of Vehicle Technologies, to the committee's written questions to 21CTP representatives, Question 9a.

[7] James Miller, DOE, "Status/Prospects for Energy Storage Technology for Medium-Duty and Heavy-Duty Hybrids," presentation to the committee, January 31, 2011, Washington, D.C.

[8] James Miller, DOE, "Status/Prospects for Energy Storage Technology for Medium-Duty and Heavy-Duty Hybrids," presentation to the committee, January 31, 2011, Washington, D.C.

TABLE 4-3 Electric-Vehicle Battery Materials, Production, and Recycling Capabilities Being Developed by 12 Domestic Manufacturers, with Amount of Department of Energy Funding

Company	Location	Funding	Material	Description
BASF The Chemical Company	Elyria, OH	$50 M	Cathode	Production of nickel-cobalt-metal cathode material for Li-ion batteries
TODA AMERICA	Midland, MI	$70 M	Cathode	Production of nickel-cobalt-metal cathode material for Li-ion batteries
Pyrotek	Sanborn, NY	$23 M	Anode	Production of carbon powder anode material for Li-ion batteries
FutureFuel Chemical Company	Batesville, AR	$25 M	Anode	Production of high-temp anode material for Li-ion batteries
Novolyte technologies	Zachary, LA	$41 M	Electrolyte	Production of electrolytes for Li-ion batteries
Honeywell	Buffalo, NY & Metropolis, IL	$55 M	Electrolyte	Production of electrolyte salt for Li-ion batteries
Celgard	Charlotte, NC	$101 M	Separator	Production of polymer separator material for lithium-ion batteries
Chemetall	Silverpeak, NV & Kings Mtn., NC	$60 M	Lithium	Production of battery-grade lithium carbonate and lithium hydroxide
enerG2	Albany, OR	$28 M	Carbon	Production of high-energy density nano-carbon for ultracapacitors
H&T Group	Holland, MI	$10 M	Cell Casing	Manufacturing of precision aluminum casings for cylindrical cells
TOXCO	Lancaster, OH	$19 M	Recycling	Hydrothermal recycling of Li-ion batteries
ENTEK	Lebanon, OR	JCI Partner	Separator	Production of battery separators for HEVs and EVs

SOURCE: James Miller, DOE, "Status/Prospects for Energy Storage Technology for Medium-Duty and Heavy-Duty Hybrids," presentation to the committee, January 31, 2011, Washington, D.C.

- Thirty-five industry contracts to design, build, and test battery prototype hardware, to optimize materials and processing specifications, and to reduce cost.

Notable accomplishments made by the USABC battery development partners for light-duty vehicles include the following:

- Johnson Controls-SAFT (JCS) is supplying lithium-ion batteries to BMW and Mercedes for hybrids;
- A123 Systems is supplying batteries to Ford EVs and hybrid buses; and
- Compact Power/LG Chem is supplying lithium-ion cells to GM for the Chevrolet Volt.

Progress in heavy-duty batteries is expected to benefit from the development of light-duty vehicle batteries.

Finding 4-1. Although 2012 has been established as the deadline for 21CTP Goals 1 and 2 for hybrid vehicle technology, it is unlikely that these will be met, because there has been no funding for either goal. However, with regard to Goal 2, the DOE's Office of Vehicle Technologies' battery R&D program is supporting a large number of programs addressing issues ranging from fundamental materials research through battery development and testing. Significant progress has been made in developing domestic manufacturing facilities for battery materials production and recycling, cell production, and pack assembly. Although the applicability of these programs to heavy-duty applications was not provided to the committee, the committee believes that these developments are supportive of the needs of medium- and heavy-duty hybrid applications.

21CTP Goal 3: Develop and demonstrate a heavy hybrid propulsion technology that achieves a 60% improvement in fuel economy (38% reduction in fuel consumption), on a representative urban driving cycle, while meeting regulated emissions levels for 2007 and thereafter.

No R&D work to reduce the fuel consumption of heavy-duty hybrid vehicles was funded by the DOE for FY 2007 through FY 2010. However, fleet testing sponsored by the DOE has shown that, for "urban-based, start and stop driving cycles" (e.g., FedEx route, transit bus, etc.), demonstrations in the 20 to 40 percent range (17 to 29 percent reduction in fuel consumption) have been achieved, based on measured

TABLE 4-4 Nine Domestic Manufacturing Facilities for Battery Cell Production and Pack Assembly and Amount of Total Investment

Company	Location	Total Investment	Cell Manu.	Pack Assembly	Description
Johnson Controls	Holland, MI Lebanon, OR	$600 M	✓	✓	Li-Ion: Nickel Metal Cobalt
A123 Systems	Romulus & Brownstown, MI	$500 M	✓	✓	Li-Ion: Iron Phosphate
cpi compact power inc (LG Chem)	St. Clair & Holland, MI	$302 M	✓		Li-Ion: Mixed Manganese
GM	Brownstown, MI	$236 M		✓	Battery Pack Assembly
Saft	Jacksonville, FL	$191 M	✓	✓	Li-Ion: Nickel Metal Cobalt
Dow Kokam	Midland, MI	$320 M	✓	✓	Li-Ion: Manganese Spinel
EnerDel	Indianapolis, IN	$236 M	✓	✓	Li-Ion: Nickel Metal Cobalt
East Penn manufacturing co., inc.	Lyon Station, PA	$98 M	✓	✓	Advanced VRLA and the Ultra Batteries
Exide Batteries	Bristol, TN & Columbus, GA	$70 M	✓	✓	Spiral Wound AGM and Flat Plate Batteries

SOURCE: James Miller, DOE, "Status/Prospects for Energy Storage Technology for Medium-Duty and Heavy-Duty Hybrids," presentation to the committee, January 31, 2011, Washington, D.C.

fuel usage in the field and testing in the laboratory.[9] These data are typically determined by measuring how the particular type of vehicle is used in the specific application, and creating or selecting an existing heavy-duty cycle based on the measured data characterizing the vehicle usage. There are no industry standards yet for heavy-duty hybrid vehicle testing. The demonstrated fuel economy improvement of 20 to 40 percent is somewhat reduced from the finding of 35 to 47 percent improvement (26 to 32 percent reduction in fuel consumption) from the NRC Phase 1 report (NRC, 2008), perhaps because of the additional data that were obtained in the intervening time.

DOE vehicle simulation tools have been used to evaluate the fuel consumption benefits of advanced technologies to meet the goal of 60 percent improvement in fuel economy (38 percent reduction in fuel consumption). Specific advanced technologies and their capabilities of reducing fuel consumption were not discussed by the DOE, although the DOE's modeling and simulation of heavy-duty hybrid vehicles, discussed in the next section, is focused on identifying technologies for improving the fuel consumption of heavy-duty hybrid vehicles.

[9] Answers provided by Ken Howden, DOE Office of Vehicle Technologies, to committee's written questions to 21CTP representatives, Question 42.

Finding 4-2. The DOE did not receive any funding for heavy-duty hybrid R&D in FY 2007 through FY 2010. Consequently, no progress was reported toward the 21CTP's three heavy-duty hybrid goals, primarily focused on R&D, for achieving 15 years of design life, achieving cost goals for drive-unit systems and energy storage systems, and achieving a 60 percent improvement in fuel economy (38 percent reduction in fuel consumption). During this period, the DOE made progress in developing heavy-duty hybrid simulations and models and conducting fleet testing and evaluations of heavy-duty hybrid vehicles.

Recommendation 4-1. The DOE should provide an up-to-date status with respect to the heavy-duty hybrid goals. The DOE should partition the available hybrid funds between heavy-duty and light-duty hybrid R&D technology to promote the R&D required for the development of heavy-duty hybrid technologies, because heavy-duty hybrid requirements are significantly different from light-duty requirements.

Simulation and Modeling

Modeling and simulation work was begun in 2007 at the Argonne National Laboratory and is currently ongoing using the ANL's simulation tool, Autonomie, to support original equipment manufacturer (OEM) hybrid activities as well as those of government agencies (EPA). State-of-the-art component data characterizing heavy-duty vehicle components (engine, transmission, and vehicle aerodynamics) and specific control strategies have been implemented in Autonomie. Several applications were validated using test data provided by the EPA. The close collaborations with OEMs allow the national laboratories and industry users to accelerate the development of heavy-duty hybrid vehicles.

Modeling and simulation work began in 2008 at the National Renewable Energy Laboratory and continues today with simulation and modeling techniques being used to analyze how medium-duty hybrid vehicles are used in a broad array of fleet applications. Measured fuel consumption results were used to validate fuel consumption values derived from dynamic models of the vehicles. A matrix of 120 component sizes, usages, and cost combinations were analyzed to minimize fuel consumption and vehicle cost. The results illustrated the dependency of component sizing on the drive cycle and daily distance driven.

Vehicle Development and Demonstrations

Hybrid electric drive systems are being tested in heavy-duty vehicles through the NREL's fleet testing and evaluation team. The results provide unbiased, third-party assessment of real-world operation and are used to focus future efforts to improve performance and cost-competitiveness. Heavy-duty hybrid testing and evaluation project results since 2007 include the following:

- *GM Allison's HEV transit bus in Seattle, Washington (2007)*—Improved on-road fuel economy by 27 percent (21 percent reduction in fuel consumption) during a 12-month period.
- *ISE's series gasoline HEV transit bus in Long Beach, California (2008)*—Improved on-road fuel economy (based on diesel-equivalent energy content per gallon) by 8 percent (7 percent reduction in fuel consumption) over a 24-month period.
- *BAE's HEV system in New York City, New York (2009)*—Improved fuel economy by 28 percent (22 percent reduction in fuel consumption) over a 12-month period.
- *Eaton Corporation's HEV system in the UPS van fleet in Phoenix, Arizona (2009)*—Improved fuel economy by 29 percent (22 percent reduction in fuel consumption) over a 12-month period.
- *Enova's Plug-In HEV system in IC Corporation's school bus (2009)*—Improved fuel economy up to 57 percent (36 percent reduction in fuel consumption) were observed, but results were inconsistent because of incomplete battery-charging data.
- *Azure's hybrid system in the FedEx Los Angeles delivery fleet (2010)*—No statistically significant improvement in fuel economy was observed during a 12-month evaluation, although 21 percent improvement (17 percent reduction in fuel consumption) was measured on a dynamometer test cycle.
- *Eaton Corporation's HEV system in a Class 8 tractor in the Coca Cola Enterprise beverage delivery fleet in Miami, Florida (2010)*—Chassis dynamometer testing showed 0 to 30 percent improvement in fuel economy (0 percent on mostly highway cycle, 30 percent on urban cycle; 0 to 23 percent reduction in fuel consumption).

The DOT has been involved in the funding of heavy-duty vehicle hybrids through support of hybrid buses for public transport. These programs include the following:

- *National Fuel Cell Bus Program*—$49 million over 4 years, starting in FY 2006;
- *Transit Investments for Greenhouse Gas and Energy Reduction (TIGGER)*—$100 million in ARRA funding in round 1 and $75 million of discretionary funds in round 2; and
- *Emission Certification Support for Hybrid Buses*—carried out at West Virginia University.

These DOT-supported programs have distinct goals and benefits that complement the work of other agencies.

From 2006 to 2010, the DOD has supported heavy-duty hybrid vehicle programs and has taken leadership in the areas of heavy-duty hybrids for combat vehicles and nontactical trucks. The U.S. Army, acting through TARDEC and its subordinate organization, the National Automotive Center, has sponsored heavy-duty hybrid programs. A current program is the Oshkosh ProPulse diesel electric drive system, which provides electric propulsion power as well as 100 kW of clean exportable alternating current (ac) power. This technology became part of the Heavy Expanded Mobility Tactical Truck (HEMTT) A3 platform from Oshkosh, which is a series-electric hybrid design. Recently, ADA Technologies announced a $70,000 contract from the U.S. Army for Phase I research into the development of advanced electrochemical ultracapacitor systems for use in hybrid electric vehicles for high-power military applications (Global and Refining Fuels Today, 2011).

Progress in Commercializing Hybrid ElectricVehicles

The committee obtained insight into industry's progress in commercializing heavy-duty hybrid electric vehicles during a visit to the Eaton Corporation on January 14, 2011. Commercialization at Eaton is focused on developing flexible architectures to accommodate many diverse applications for hybrid systems and developing designs that can provide significant cost reductions through modularization and reusability.

Eaton has been developing hybrid power systems for more than a decade, having developed its first prototype in

2000. In 2002, Eaton began working with FedEx to develop a medium-duty hybrid system. Since that time, Eaton has sold more than 3,500 hybrid systems globally. Although Eaton's largest application for hybrid systems has been for city buses in China, its largest application in the United States has been Class 4 and 5 FedEx delivery vans. Approximately 200 to 300 units have been sold for each class. Eaton's second largest application for hybrid systems in the United States has been for Class 6 and 7 utility trucks. In these applications, Eaton has acted as the system integrator. One of Eaton's challenges has been adapting its hybrid systems to many applications with diverse needs and many niche applications. In addition, there is increased emphasis on all-electric driving, especially in European cities and for port applications in the United States (see the following section, "Plug-in Hybrid and Battery Electric Vehicles").

Eaton is developing two different parallel hybrid systems: (1) an in-line hybrid system with the engine, motor/generator, and transmission arranged in-line, and (2) a power take-off (PTO) input hybrid system for a utility truck, which has the hybrid drive system connected to the PTO of the transmission and located parallel to the driveshaft. To accommodate the proliferation of applications for hybrid systems, Eaton is developing a flexible hybrid system architecture. A base hybrid drive system common to all applications would be interfaced with a flexible architecture extension to provide an energy management system that could be adapted to individual, specialized needs. To facilitate this architecture, Eaton is developing scalable hybrid drive systems that would include scalable motors and batteries for specific applications. Power electronics are being integrated into a module to eliminate cables and simplify packaging in the vehicle (Eaton Corporation, 2009).

Eaton's pursuit of hybrid vehicle cost reductions is focused on (1) smaller motors operating at higher speeds, (2) integration of the hybrid drive system with the transmission at the PTO, and (3) flexible architectures with standard interfaces. To realize standard interfaces, Eaton indicated that a committee of the Society of Automotive Engineers is needed to help drive a standard for this flexible architecture to accelerate the availability of compatible components.

PLUG-IN HYBRID AND BATTERY ELECTRIC VEHICLES

The 21CTP did not have any goals for plug-in hybrid electric vehicles or battery electric vehicles. However, much of the technology for HEVs, including electric machines and energy storage systems, is expected to be directly applicable, or nearly so, to the PHEVs and EVs.

Plug-in hybrids recharge their energy storage system through the electric grid, typically at night, when electricity prices and demand are low. The NRC reported in 2010 that commercial medium- and heavy-duty plug-in hybrid vehicles are being delivered in the United States to numerous companies (NRC, 2010, p. 74). The DOE, under ARRA-Transportation Electrification, is funding the development of four additional PHEV and EV platforms. More than 1,800 commercial vehicles are being funded by the DOE to aid in the development and demonstration of PHEVs and full EVs in fleets around the United States. The objectives of these programs are (1) to accelerate the development of U.S. manufacturing capacity for batteries and electric drive components, (2) to accelerate the deployment of electric drive vehicles, and (3) to help reduce costs and meet the 21CTP goals set forth in 2006. Selected heavy-duty vehicle projects with energy storage being funded by the ARRA include the following:[10]

- *Electric Vehicles:*
  —*Smith Electric Vehicles*—An all-electric Class 3 box truck is being developed with a 120 kW liquid-cooled motor and the option of either a 40 kWh, 80 kWh, or 120 kWh lithium battery pack. The battery packs are assembled in Kansas by Smith Electric Vehicles using modules assembled from Valence Technologies, Inc. cells manufactured in China. Smith Electric Vehicles received a $32 million award from the DOE to develop and deploy 100 of these vehicles.
  —*Navistar*—A battery electric Class 2b/Class3 delivery truck is being developed. Navistar received a $39.2 million award to develop and deploy 400 of these vehicles. Battery packs will be supplied by A123 Systems.
- *Plug-in Hybrid Electric Vehicles:*
  —*Eaton Corporation and Azure Dynamics*—The DOE awarded the South Coast Air Quality Management District $45.4 million, with an additional $45 million that will come from several sources, including $5.5 million from Eaton, to develop and deploy PHEV demonstration trucks and shuttle buses over a 3-year period, starting in August 2009. The 245 utility aerial trucks are based on the Ford F550 chassis equipped with an Eaton PHEV system. The 35 shuttle buses are based on the Ford E450 chassis equipped with an Azure Balance PHEV system. Battery packs will be supplied by Compact Power, Inc. Following the delivery of these vehicles in the summer of 2011, the demonstration period will run for at least 2 years. Additional details of the Eaton program are provided below.
  —*Ford Motor Company*—The DOE awarded Ford $30 million to deploy 150 PHEVs, including 130 Ford Escape PHEVs and 20 E450 Van PHEVs in partnership with 15 utilities. The battery packs will be supplied by Johnson Controls.

Eaton provided the following overview of two heavy-duty PHEV programs that were under way during the committee's January 2011 visit to Eaton:

[10] James Miller, DOE, "Status/Prospects for Energy Storage Technology for Medium-Duty and Heavy-Duty Hybrids," presentation to the committee, January 31, 2011, Washington, D.C.

1. *Ford F550 Diesel PHEV Utility Aerial Truck*—Eaton has begun its program to develop a Ford F550 diesel PHEV utility aerial truck with a 15-mile EV range. Eaton is the system integrator, and Altec manufactures the utility aerial addition for the trucks and will be the manager of the vehicles. The drive unit has a specification of 65 kW and 620 Newton-meters (Nm). An electric air conditioning (A/C) compressor will be used. The control strategy will provide the following modes: (1) maximization of engine-off for job site operations, (2) range-extended EV mode, and (3) blended EV mode.
2. *Class 8 PHEV Day Cab Tractor (for port operations)*—This program is being conducted for the Texas Environmental Research Consortium, whose main interest is in reducing oxides of nitrogen (NO_x) at shipping ports. The Class 8 tractor has a hybrid powertrain, electrified accessories, and engine stop-start management. The control strategy provides the following modes: (1) EV mode, (2) HEV with engine stop/start management, and (3) conventional HEV/engine-on operation. Preliminary results show 44 to 73 percent reduction in NO_x emissions and 13 to 160 percent increase in fuel economy (12 to 62 percent reduction in fuel consumption) over two test routes—the Eaton Port Route with a maximum speed of 35 mph and the City Suburban Heavy Vehicle Route. An electric range of 10.6 miles was achieved at 25 mph and 9.7 miles at 33 mph for the 74,000 lb vehicle test weight. This program was completed in November 2010.

In FY 2009, the Technology Acceleration and Deployment Activity in the DOE Office of Vehicle Technologies provided a $10 million grant to Navistar to fund the development and deployment of the next-generation PHEV school bus. Improvements from Navistar's first-generation design include electric accessories to enable engine-off operation and improved strategies as well as improved energy storage approaches. This vehicle is expected to have a 30-mile range at 45 mph and will use an 80 kWh battery pack. Navistar is currently finalizing the design and build of two PHEV school bus designs. These vehicles will be evaluated, and one of the designs will be selected for the final build of the vehicles.

Finding 4-3. More than 1,800 commercial vehicles are being funded through the ARRA by the DOE to aid in the development and demonstration of plug-in hybrid electric vehicles and battery electric vehicles in fleets. One of the objectives of this program is to develop U.S. manufacturing capacity for all-electric drive components (energy storage, drive motors, power electronics, etc.). However, in at least one of these projects, the battery cells are being manufactured outside the United States.

Recommendation 4-2. The DOE should determine what is needed for the battery cells and other electric drive components in the ARRA-Transportation Electrification programs aimed at development and manufacturing in the United States, as specified in the objectives of these programs.

HYDRAULIC HYBRID VEHICLES

Although the 21CTP did not have any goals for heavy-duty hydraulic hybrid vehicles, the EPA has been conducting R&D of heavy-duty hydraulic hybrid vehicles since early in the 21CTP program. The EPA has supported Cooperative Research and Development Agreements (CRADAs) to help speed the commercialization of federally developed technology. Heavy-duty hydraulic hybrid vehicles were mentioned briefly in the NRC Phase 1 report (NRC, 2008). The NRC (2010) report reviewed the EPA's hydraulic hybrid vehicle program in greater detail, indicating that the EPA had shifted its focus from parallel hydraulic hybrid technology to series hydraulic hybrid technology to realize further improvements in fuel efficiency.

In contrast to an electric hybrid vehicle utilizing batteries for energy storage, a hydraulic hybrid vehicle provides high specific power, but low specific energy, as shown in Figure 4-1. As a result, hydraulic hybrids are effective in providing short-duration launch assist rather than extended-duration power assist for cruise conditions. Delivery trucks and refuse trucks with frequent starts and stops are potential applications for hydraulic hybrids.

Series Versus Parallel Hydraulic Hybrid

The NRC report on medium- and heavy-duty vehicles illustrates the differences between parallel and series hydraulic hybrid systems (NRC, 2010, Figures 4-13 and 4-8, respectively). In a parallel hydraulic hybrid system, the engine and transmission are connected to the differential and drive wheels as in a conventional vehicle. A hydraulic pump-motor unit is geared to the output shaft of the engine and transmission to assist in driving the wheels, or capturing braking energy to recharge the hydraulic accumulator tank. In contrast, in a series hydraulic hybrid system, there is no mechanical connection between the engine/transmission and the differential/drive wheels. A hydraulic pump-motor unit is mounted to the output of the engine, and another pump-motor unit is mounted on the input to the differential for the drive wheels. The respective pump-motor units acting as a pump can charge the hydraulic storage tanks using either engine power or deceleration energy from the wheels. The respective pump-motor units acting as a motor can be used to drive the differential for the drive wheels or to restart the engine. During cruise conditions, the engine-mounted pump-motor unit acting as a pump will provide hydraulic fluid with some buffering by the hydraulic accumulator system to the pump-motor unit acting as a motor mounted

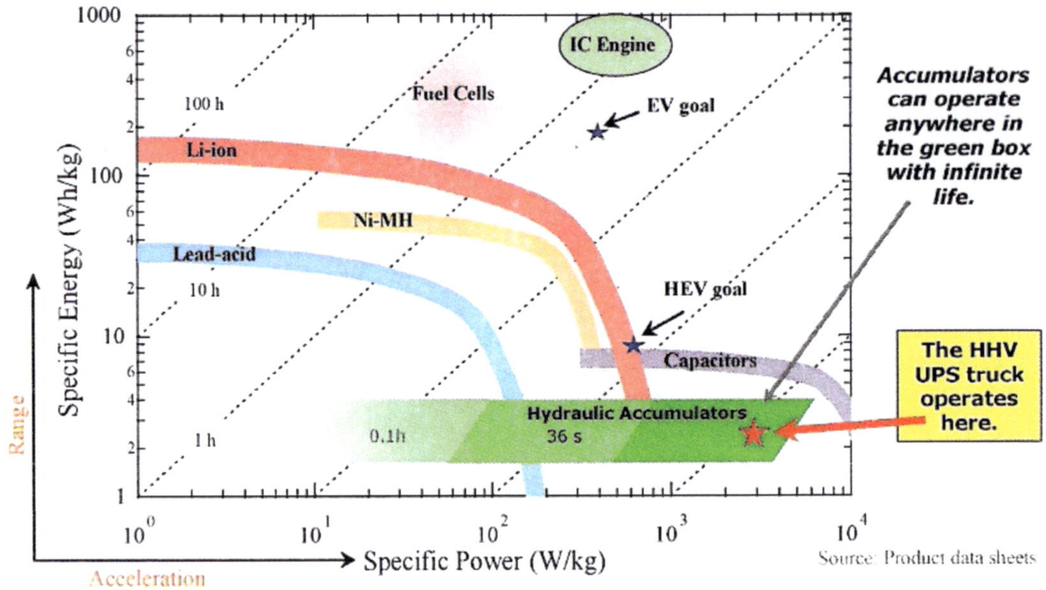

FIGURE 4-1 Relative performance of energy-storage systems. Acronyms are defined in Appendix I. SOURCE: John Kargul, EPA, "Clean Automotive technology, Cost Effective Solutions for a Petroleum and Carbon Constrained World," presentation to a subgroup of the committee, October 26, 2010, Ann Arbor, Michigan.

on the differential for the drive wheels. In a series hydraulic hybrid, the engine is decoupled from the wheels, enabling it to be operated at best-efficiency conditions. Buffering by the hydraulic hybrid accumulator system eliminates the need for quick engine transients. The EPA's program began by investigating parallel hydraulic hybrid systems but switched to focus on series hydraulic hybrid systems.

System Components

One of the key components in the series hydraulic hybrid vehicle is the pump-motor unit. In the EPA's program, a bent-axis pump-motor unit was developed. The bent-axis pump-motor unit is a barrel arrangement pump in which the head of the pump can be pivoted off the centerline of the driveshaft. As the yoke angle increases, the pumping or motoring power increases. A 110 cubic centimeter (cc) pump-motor assembly can provide 300 hp at a pressure of 5,000 pounds per square inch (psi). This pump-motor is small enough to be held in one's hand.

Two hydraulic storage tanks are required. The high-pressure accumulator tank operates between 2,000 and 7,000 psi and the low-pressure accumulator tank operates between 50 and 200 psi. A Class 6 truck would use at least two 22-gallon carbon-fiber accumulator tanks, which are approximately 5 feet long and 12.5 inches in diameter. Hydraulic accumulator tanks can be either bladder or piston type. In the bladder type, adding hydraulic fluid to the accumulator compresses the nitrogen-filled bladder. In the piston type, adding hydraulic fluid to the accumulator compresses the piston into the nitrogen-filled shell.

Fuel Economy Improvements

The potential fuel economy improvements projected for parallel and series hydraulic hybrid trucks are shown in Table 4-5 (TIAX, 2009). Transient start-stop driving cycles, such as those of refuse trucks and delivery vans, are required to realize these potential improvements.

Eaton Corporation is working to improve the performance of parallel hydraulic hybrids in special applications such as refuse trucks. Eaton's first-generation hydraulic hybrid, also referred to as a hydraulic launch assist (HLA) system, provided 10 to 30 percent improvement (9 to 23 percent reduction in fuel consumption) in fuel economy. The EPA showed similar results for its early experimental parallel

TABLE 4-5 Fuel Economy Improvements for Parallel and Series Hydraulic Hybrid Trucks

Type of Truck	Fuel Economy Improvement % (mpg) Fuel Consumption (FC) Reduction (%)
Parallel hydraulic hybrid	25 to 33% (mpg) 20 to 25% reduction in FC
Series hydraulic hybrid	67 to 100% (mpg) 40 to 50% reduction in FC

SOURCE: TIAX (2009).

hydraulic hybrid trucks. Eaton's R&D shows that a parallel hydraulic hybrid with an efficient transmission can improve fuel economy by 35 to 45 percent (26 to 31 percent reduction in fuel consumption) in refuse trucks over the baseline. Eaton's next-generation hydraulic hybrid, or HLA, is currently undergoing fleet testing in the refuse industry with about 50 systems.

EPA experiments with series hydraulic hybrid trucks have shown fuel economy improvements of 60 to 70 percent (38 to 41 percent reduction in fuel consumption) with proper matching of hydraulic components and driving the power steering and brake assist off the accumulator pressure. Although these improvements in fuel economy can be obtained with a system tuned for a particular drive cycle, these conditions generally do not exist in the real world. Such gains are feasible only when there are almost no electrical loads on the system at idle, very few engine starts, and no high-speed travel.[11]

Progress in Commercializing Hydraulic Hybrid Vehicles

The EPA initially partnered with Eaton Corporation to develop series hydraulic hybrid technology in a UPS delivery vehicle. Eaton, which developed a first-generation, parallel hydraulic hybrid refuse truck, or HLA, has more than 100 trucks in the field today. These first-generation vehicles provided a 10 percent improvement in fuel economy (9 percent reduction in fuel consumption) and four times longer brake life. Eaton's next-generation HLAs will replace the Allison transmission with Eaton's Ultra Shift Plus transmission with the hydraulic unit attached to the transmission's PTO and will use hydraulic power to fill in torque during shifts. Eaton expects that this next-generation HLA will provide five times longer brake life and 30 percent better fuel economy (23 percent reduction in fuel consumption) than the baseline Peterbilt refuse truck. Eaton believes that a refuse truck provides an ideal duty cycle for the hydraulic hybrid vehicle and estimates that the North American refuse truck market is up to 10,000 vehicles per year.

Delivery trucks, refuse trucks, yard hustlers, and shuttle buses have been the focus of the EPA's series hydraulic hybrid program over the past 5+ years. EPA technology transfer partners are moving to preproduction and/or full production of hydraulic hybrid vehicles in the 2010-2012 time frame. The following time lines for these programs was provided by the EPA:

- *Refuse Vehicles:*
  —*2008-2011:* Fleet testing of 20 pilot vehicles.
  —*2010-2011:* Purchases of 89 early/preproduction vehicles announced.
- *Delivery Vehicles:*
  —*2008-2011:* Fleet testing of 6 pilot vehicles.
  —*2010-2012:* Purchases of 46 early/preproduction vehicles announced.

The EPA has worked with Parker Hannifin Corporation, Freightliner, and FedEx in developing a hydraulic hybrid delivery vehicle. Currently, Parker Hannifin (in an Autocar refuse truck), Eaton (in a Peterbilt refuse truck), and Bosch-Rexroth (in an American LaFrance refuse truck) are progressing toward full production status with their hydraulic drive systems. Refuse trucks were a logical application, because they already utilize extensive hydraulic systems.

On November 23, 2010, Parker Hannifin Corporation announced that it had developed several hydraulic hybrid technologies that improve the performance of refuse and delivery vehicles to levels that exceed proposed new federal fuel-efficiency and criteria emission standards (Parker Hannifin, 2010). Parker Hannifin stated that its advanced hydraulic hybrid system is unique because it disconnects the engine from the rear wheels of the vehicles, indicating a series hybrid configuration. The hydraulic hybrid truck has the potential to reduce brake maintenance significantly owing to regenerative braking. Brake life can be extended two to eight times relative to the baseline vehicle, which requires three to four brake replacements per year, at approximately $2,000 each. This reduction in brake maintenance can assist in reducing the payback period. Parker Hannifin stated that the technology is already in use on 11 refuse vehicles in three South Florida communities. UPS and FedEx have become the first companies to order a variation of the technology for use in delivery vehicles, scheduled to be on the road in 2011. Parker Hannifin is currently developing further advancements in the technology for use on an advanced bus platform that is targeting a 45 percent reduction in fuel usage over average diesel powertrains.

REGULATORY CONSIDERATIONS

Regulatory considerations for heavy-duty hybrid vehicles include hybrid truck tax credits, credits for fuel-efficiency standards, and emission certification. Each is discussed in the sections that follow.

Hybrid Truck Tax Credits

Tax credits, which are an important incentive for hybrid vehicles, have expired, but they are described below as background information for future consideration of credits (ACEEE, 2009). The Energy Policy Act of 2005 provided tax credits for heavy-duty hybrid vehicles in the period January 1, 2006, through December 31, 2009. The credit amount depended on the weight class of the vehicle, its fuel economy relative to a comparable conventional vehicle, and the incremental cost. By December 31, 2009, the following 10 manufacturers had certi-

[11] Information provided by Mihai Dorobantu, Eaton Corporation, to NRC staff, James Zucchetto, by e-mail, January 31, 2011.

fied tax credits through the Internal Revenue Service (IRS) for at least one heavy-duty hybrid truck or bus model:

Azure Dynamics
Freightliner (Daimler Trucks)
Freightliner/Eaton
International Truck/Eaton
Kenworth/Eaton
Navistar/Eaton
Navistar/IC Bus
Peterbilt/Eaton
Workhorse/Eaton
Odyne/Navistar

A hybrid vehicle's incremental cost was defined in the Energy Policy Act of 2005 as the amount by which the manufacturer's suggested retail price of the hybrid vehicle exceeded that of a "comparable" vehicle, but the incremental cost was capped by GVW class, as shown in Table 4-6.

The percentage of the incremental cost (up to the maximum allowed) that the credit covered was determined by the improvement in city fuel economy, as shown in Table 4-7. For example, the most efficient 20,000 lb hybrid truck could receive a credit of $6,000 ($15,000 maximum incremental cost × 40 percent [hybrid credit for improvement in city fuel economy of at least 50 percent]).

Also, under the Energy Policy Act of 2005, the vehicle had to meet a threshold value of "maximum available power," a measure of the percentage of total vehicle power available from the rechargeable energy storage system of the vehicle. For a vehicle of 8,500 to 14,000 lb gross vehicle weight, the requirement was 10 percent of the maximum available power; over 14,000 lb, the requirement was 15 percent. This requirement was to ensure that a qualifying vehicle incorporated substantial hybrid technology. The language of the act did not specify the type of hybrid; both electric and hydraulic hybrids, for example, could qualify.

TABLE 4-6 Maximum Incremental Cost of a Hybrid Truck Qualifying for a Tax Credit

Gross Vehicle Weight (GVW) Rating	Maximum Qualified Incremental Cost
8,501 to 14,000 lb	$7,500
14,001 to 26,000 lb	$15,000
> 26,000 lb	$30,000

TABLE 4-7 Hybrid Truck Tax Credit as a Function of Fuel Economy

Improvement in City Fuel Economy	Hybrid Credit as Percentage of Qualified Incremental Cost
At least 30% and under 40% (23 to 29% reduction in fuel consumption)	20%
At least 40% and under 50% (29 to 33% reduction in fuel consumption)	30%
At least 50% (At least 33% reduction in fuel consumption)	40%

In 2007, the IRS issued guidance setting out procedures for manufacturers to use in certifying a vehicle as an eligible heavy-duty hybrid, and the amount of the credit for that vehicle.[12] The credit depended on the hybrid's fuel economy relative to that of a comparable conventional vehicle, but there was no standard fuel economy test procedure in place for heavy-duty vehicles. Therefore, the IRS guidance directed manufacturers to assign fuel economies to their vehicles using an approach "substantially similar" to one used by the federal government or the state of California to measure fuel economy or emissions. Once the IRS acknowledged a vehicle credit, the manufacturer could certify it to purchasers. By the end of this program, 10 manufacturers were eligible for credits. Specific models for each manufacturer can be found on the IRS website.[13]

In February 2011, Senator Debbie Stabenow of Michigan introduced the Charging America Forward (S. 298) bill, which calls for an extension of the credits through December 31, 2014, and sets new qualifying incremental cost levels for which the purchaser can receive a portion as a tax credit. As of March 2011, this bill had been referred to the Senate Finance Committee.

Fuel Efficiency Standards

The EPA and the DOT/National Highway Traffic Safety Administration (NHTSA) introduced proposed standards for greenhouse gas (GHG) emissions standards and fuel efficiency standards for medium- and heavy-duty engines and vehicles on October 25, 2010 (EPA/NHTSA, 2010) and issued final standards on September 15, 2011 (EPA/NHTSA, 2011). The agencies (a term used throughout this section to indicate the EPA and DOT/NHTSA) noted that, although the standards are not premised on the use of hybrid powertrains, certain vocational vehicle applications may be suitable candidates for use of hybrids owing to the greater frequency of stop-and-go urban operation and their use of power take-off (PTO) systems. As an incentive, the agencies are providing credits for the use of hybrid powertrain technology. There is an important distinction between the two types of credits for hybrid vehicles discussed above. The credits provided by the fuel-efficiency standards could be used to meet any of the heavy-duty fuel-efficiency standards and are not restricted to the averaging set generating the credit. In contrast, the credits for hybrid vehicles, provided by the U.S. Energy Policy Act of 2005, were direct tax credits for the purchasers of hybrid

[12] See Internal Revenue Service Notice 2007-46. Available at http://www.irs.gov/irb/2007-23_IRB/ar08.html. Accessed December 3, 2010.

[13] See http://www.irs.gov/businesses/article/0,,id=175456,00.html.

trucks that deliver improved fuel economy. Because the fuel-efficiency standards provide procedures for defining the fuel economy improvements of hybrid vehicles, these procedures could provide a means, in lieu of the previously generated IRS guidance for truck purchasers, to obtain tax credits, if such credits are reinstated.

Two different classes of emissions, GHG emissions and "criteria pollutants," are referred to in this section. The first class of emissions, GHG emissions (referring primarily to CO_2), will be controlled by the new GHG emissions standards. The new GHG emissions standards also address hydrofluorocarbon emissions from air-conditioning systems and includes nitrous oxide (N_2O) and methane (CH_4) emissions standards. The second class of emissions, criteria pollutants, refer to hydrocarbons (HC), carbon monoxide (CO), NO_x, and particulate matter (PM) exhaust emissions that have been regulated by the EPA for many years and require a certificate of conformity for engines based on tests conducted on heavy-duty transient engine dynamometers.

The agencies' new fuel consumption and CO_2 emissions standards are to be tailored to each of three regulatory categories of heavy-duty vehicles:

1. Heavy-duty pickup trucks and vans,
2. Vocational vehicles, and
3. Combination tractors.

The new rules for the fuel consumption and CO_2 standards contain provisions for obtaining credits as an incentive for applying hybrid powetrain technology. The approach to account for the use of a hybrid powertrain when evaluating compliance with the standards is described in Appendix F.

The procedure requires testing heavy-duty pickup trucks and van hybrids on the light-duty federal test procedure (FTP) and highway fuel economy test (HWFET) with suitable adjustments to the test procedures. For vocational vehicles and combination tractors incorporating hybrid powertrains, two methods are specified: chassis dynamometer or engine dynamometer. Each method requires a comparison test (hybrid versus baseline) of the actual physical product (engine, powertrain, or vehicle), because the agencies are not aware of analytical models that can assess the technology. The measured fuel consumption advantage of a hybrid vehicle versus the baseline is multiplied by the production volume and the useful life to provide the fuel consumption credits for the specific vehicle line that is applied to the manufacturer's production volume-weighted annual fleet average standard.

Emission Certification

The NRC Phase 1 report noted the lack of an emissions test and certification procedure for heavy-duty hybrid trucks (NRC, 2008). The Phase 1 report noted that the EPA was developing a procedure to measure directly the emissions of complete heavy-duty vehicles, including hybrids. Unfortunately, no progress has been reported on this effort by the EPA. Even though the EPA and the NHTSA introduced their NPRM on October 25, 2010, providing test procedures for CO_2 emissions and fuel consumption for hybrid heavy-duty trucks, a manufacturer still needs a certificate of conformity that the vehicle's internal combustion engine meets the EPA criteria emissions standards for heavy-duty engines, a procedure that does not recognize hybrid heavy-duty trucks. Therefore, the potential reduction in emissions with a hybrid heavy-duty truck cannot be recognized, and so reductions in the size or simplification of the emission control system cannot be considered.

The California Air Resources Board (CARB) is currently drafting vehicle-level test procedures for heavy-duty hybrid vehicles. California is coordinating this effort with the EPA's efforts to establish GHG standards for heavy-duty vehicles. The CARB's work is aimed at quantifying the improvement in fuel consumption and emissions due to hybridization. However, at this time, procedures are in draft form, not finalized (DOE, 2011).

Finding 4-4. The EPA and the DOT's National Highway Traffic Safety Administration (NHTSA) issued their final rules on September 25, 2011, for "Greenhouse Gas Emissions Standards and Fuel Efficiency Standards for Medium- and Heavy-Duty Engines and Vehicles." Although these standards contain test procedures for determining fuel consumption for heavy-duty hybrid trucks, a manufacturer still needs a certificate of conformity showing that a vehicle's internal combustion engine meets the EPA criteria emission standards for heavy-duty engines (a procedure that does not recognize hybrid heavy-duty trucks). The CARB is currently drafting vehicle-level test procedures for heavy-duty hybrid vehicles.

Recommendation 4-3. As partners of the 21CTP, the EPA and DOT/NHTSA should work with the CARB to develop test procedures for the certification process for criteria emissions so that the emissions benefits of hybridization will be recognized, allowing the reduction in size or simplification of the emission control system of heavy-duty hybrid vehicles to be realized.

HYBRID COSTS AND PAYBACK PERIODS

The NRC report on medium- and heavy-duty vehicles (NRC, 2010) contains extensive information and projections on benefits and costs of hybrid truck systems. Because this information is contained in NRC (2010) in many tables with a variety of other information, a consolidated table presenting the information exclusively for hybrid truck systems is provided here. The fuel consumption benefit, capital cost, annual mileage, and typical fuel economy in miles per gallon for technologies projected to be available in the 2015-2020 time period are shown in Table 4-8. The capital costs are

TABLE 4-8 Hybrid Trucks—Break-even Cost Analysis (Future 2015-2020 Technology)

Category	Description	Fuel Consumption Benefit (%)	Forecasted Capital Cost ($)[a]	Annual Mileage[b]	Typical MPG[b]	Payback Period to Breakeven Years[c]	References[d]
Class 2b Pickups and Vans	Parallel electric hybrid	18	$9,000	27,500	12.5	7.6	Table 6-9
Class 3 to 6 Straight Box Truck	Parallel electric hybrid	30	$20,000	41,250	9.4	5.1	Table 6-7
Class 3 to 6 Bucket Truck	Parallel electric hybrid with electric power takeoff	40	$30,000	13,300[e]	9.4	17.7	Table 6-8
Class 8 Tractor Trailer Truck	Mild parallel electric hybrid with idle reduction	10	$25,000	137,500	5.75	3.5 (not directly comparable because of inclusion of idle reduction feature)	Table 6-5
Urban Transit Bus With federal subsidy of incremental cost	Series electric hybrid	35	$220,000 ($22,000)	137,500	6.0	9.1 0.9	Table 6-16
Class 8 Refuse Hauler	Parallel hybrid electric	30	$39,000[f]	50,000	4.25	3.7	Table 4-9
Class 8 Refuse Hauler	Parallel hydraulic hybrid	25	$30,000	50,000	4.25	3.4	Table 6-10
Class 8 Refuse Hauler	Series hydraulic hybrid	50	N/A	50,000	4.25	N/A	Table 4-9

NOTE: The break-even period is calculated on the basis of a fuel cost of $3.00/gal. For a fuel cost of $4.00/gal, the break-even period would be reduced by 25 percent.
[a]Costs for 2015-2020 are forecast to be reduced up to 47% from 2013 to 2015 costs.
[b]Average values of ranges from NRC (2010), Table 2-1.
[c]Breakeven Time = Capital Cost/(Annual Mileage/MPG × % Improvement × $/gal).
[d]Tables 6-5 through 6-16 provide future costs for 2015-2020. All tables are in NRC (2010).
[e]NRC (2010), page 138.
[f]NRC (2010), page 141.

projections for 2015-2020 contained in NRC (2010). Using this consolidated information, a simple calculation of the payback period to break even, defined as the time to achieve fuel cost savings equal to the initial capital cost of the hybrid system, was made assuming a $3.00/gal fuel cost.[14] Maintenance costs and discount rates were not included. The results are also shown in Table 4-8. It is important to note that the capital costs listed in Table 4-8 are forecasted to be reduced by up to 47 percent from the 2013-2015 costs, which, in turn, are expected to be lower than current costs for hybrid systems. Thus, the capital costs shown are forecasts containing reductions of well over 47 percent from current costs.

The 21CTP acknowledges that current break-even times without subsidies, based on current costs and fuel consumption improvements, are typically twice as long as the 5 years that fleets normally require for a return on investment in new hardware for cost savings. The 21CTP stated that "heavy-hybrid technology for commercial trucks and buses needs significant research and development (R&D) before it will be ready for widespread commercialization at prices that can be borne by the vehicles' operators" (DOE, 2011). The 21CTP further stated, "HD [heavy duty] hybrid technology is far from mature.... Many of today's HD hybrid vehicles have used components that are commercially available but were not designed or optimized for on-road HD hybrid vehicles. Some HD hybrid components cannot be found elsewhere and must be custom designed for the application. These will be costly due to low production volumes that have not justified the development of high volume manufacturing tools and processes to produce them economically." A very large investment is needed by the developers of the heavy-duty hybrids, the engine manufacturers, and the truck manufacturers to fully develop and implement the technologies in commercial vehicles. The capital costs shown in Table 4-8 reflect the significant reductions that will be required for HEV systems to become commercially viable.

Fuel economy improvements of 20 to 40 percent (17 to 29 percent reduction in fuel consumption) have been achieved for hybrid trucks over urban-based, start-and-stop driving cycles. DOE vehicle simulation tools are currently being used to evaluate fuel consumption benefits of advanced technologies to meet the goal of 60 percent improvement in fuel economy (38 percent reduction in fuel consumption).

[14] For a fuel cost of $4.00/gal, the breakeven period would be reduced in length by 25 percent.

By using these forecasted capital costs, most of the technologies listed in Table 4-8 achieve payback, or break even, within approximately 5 years, except for the Class 2b pickups and vans, Class 3 to 6 bucket trucks, and transit buses without the 90 percent federal subsidy for the hybrid system incremental cost. The Class 2b pickups and vans need to achieve greater fuel consumption improvements and/or capital cost reductions to achieve payback in 5 years. The Class 3 to 6 bucket trucks have exceptionally low mileage that extends the payback period well beyond 5 years. No remedy for this shortfall appears to be available at this time. The continued application of hybrid technology to transit buses will require the continuation of federal subsidies.

Finding 4-5. The 21CTP acknowledges that current heavy-duty hybrid vehicle break-even times, without subsidies, based on current costs and fuel consumption improvements, are typically twice as long as the 5 years that fleets normally require for a return on investment on new hardware for cost savings. Heavy-duty hybrid components tend to be costly because they are not designed or optimized for the application and are produced in low volumes. Fuel-economy improvements of heavy-duty hybrid vehicles have not achieved the 60 percent improvement goal (38 percent reduction in fuel consumption).

Recommendation 4-4. Dual paths should be pursued to achieve a break-even time of 5 years for heavy-duty hybrid vehicles. First, the DOE should use its vehicle simulation tools to determine the advanced technologies needed to meet the goal of 60 percent improvement in fuel economy (38 percent reduction in fuel consumption), from the current status of 20 to 40 percent improvement (17 to 29 percent reduction in fuel consumption) and initiate R&D programs to develop these technologies. Second, manufacturers should be encouraged to explore modular, flexible designs, which could yield higher production volumes and thus achieve significant reductions in capital costs of hybrid systems.

REVISED GOALS ISSUED IN 2011

In February 2011, the 21CTP issued revised stretch goals for heavy-duty hybrids; they are similar to the goals provided to the committee in November 2010, but the revised goals provide more background and/or greater specificity. These revised stretch goals are summarized below; the complete version is available in the February 28, 2011, Hybrid Propulsion White Paper (DOE, 2011).

Summary of Revised Heavy-Duty Hybrids Goal 1—Electric Machines: Develop advanced motor technology that will deliver electric machines with improved durability, lower cost, better power density, and alternatives to rare-earth permanent magnets.

—Greater than 1 million miles (Class 8 line-haul application) or 15 years of life (vocational applications).

—Power density for some motor designs today is at approximately 0.5 kW/kg. The objective is to nearly double the power density to approximately 1 kW/kg. A cost target of $50/kW by 2016 has been established.

—Motors and generators have efficiencies typically at approximately 94 percent today. The objective is 96 to 97 percent by 2016.

—Demonstrate a nonpermanent magnet motor technology in a commercial vehicle application that would equal or meet current hybrid system requirements by 2013.

- Permanent magnet motors are used in several hybrid systems today. Concern is increasing that higher hybrid volumes will create significantly higher demands for rare-earth magnet material that is predominantly supplied by China. The development of alternate motor designs that do not depend on a commodity supplied primarily by a single foreign power is deemed to be in the best interests of the United States.

Summary of Revised Heavy-Duty Hybrids Goal 2—Inverter Design/Power Electronics: Develop technologies that will improve the cycle life of critical components within the inverter and other power electronics within the hybrid system.

—Develop an improved switching device (insulated gate bipolar transistors [IGBTs] or other) that has a broader operating temperature range higher than today's 50°C limit and offers improved system life and durability. Develop this improved switching system and demonstrate benefits by 2016.

- IGBTs are today one of the main components of the inverters used in heavy-duty hybrid systems. Their cycle life is limited owing to thermal expansion and contraction that occurs with varying power transmission levels, especially at the IBGT-to-device package interface. In addition, the allowable temperature range of these devices is limited, with recommendations to maintain operating temperatures below 50°C. Technology developments focused on these areas will improve system life, simplify cooling requirements, and drive down total life–cycle cost.

—By 2016, reduce the overall weight of inverter designs by 20 percent through higher efficiency of switching devices with higher operating temperatures and potential integration with engine cooling systems.

Summary of Revised Heavy-Duty Hybrids Goal 3—Energy Storage Systems: Develop an energy storage system with 15 years of design life, a broader allowable temperature operating range, improved power density and energy density, and significantly lower cost.

—Develop a system that can provide a cycle life of 5,000 full cycles, which should achieve the target of 1 million miles (on highway) or 15 years (vocational). Current state-of-the art energy storage systems are typically rated for 8 years of life.

—Expand the acceptable operating temperature range for lithium-ion batteries, currently at 0 to 55°C, by 2017.

—Develop battery technologies that will significantly increase power and energy densities.

—Proposed cost targets:

- $45/kW and/or $500/kWh for an energy battery by 2017;
- $40/kW and/or $300/kWh for a power battery by 2020; and
- By 2016, the cost of the overall battery pack should not exceed the cost of the cells themselves by more than 20 percent.

—Establish an "end-of-life" strategy for advanced batteries and provide the necessary funding related to either the remanufacturing or recycling of batteries by 2017.

Summary of Revised Heavy-Duty Hybrids Goal 4—Hybrid System Optimization, Medium Duty: To develop and demonstrate medium-duty hybrid system technology that can deliver substantial increases in fuel economy, beyond what is available with today's systems.

—Potential applications for demonstration include medium-duty shuttle buses, vocational trucks, or on/off highway medium-duty work trucks.

—A vehicle demonstration program that provides a platform for developing these medium-duty technologies (similar to the SuperTruck program for heavy-duty technologies) is one potential approach, with development and demonstrations to be completed by 2017.

Summary of Revised Heavy-Duty Hybrids Goal 5—Hybrid System Optimization, Heavy Duty: An overarching goal is to develop and demonstrate heavy-duty hybrid system technology that can deliver substantial increases in fuel economy.

—For urban, heavy start-and-stop driving cycles, a stretch goal of 60 percent (38 percent reduction in fuel consumption) has been identified.

—For regional haul and line-haul applications, the percentage improvements would be more modest, with a stretch goal of 25 percent (20 percent reduction in fuel consumption).

—Additional review and development need to be considered for those vehicles that would possess alternative anti-idling devices that could be provided without additional infrastructure changes.

Summary of Revised Heavy-Duty Hybrids Goal 6—Electrified Power Accessories: Develop robust, durable, efficient electric power accessories for use with medium- and heavy-duty hybrid systems.

—Electrifying accessories such as power steering, air compressors, and air-conditioning compressors can achieve significant reductions in parasitic losses by powering them "on demand."

—Targeted availability of such improved accessories: 2016.

Comments on New Hybrid Goals

Some of the metrics from the three 2007-2010 goals discussed earlier in this chapter have been carried over to these revised new goals. Specifically,

- The new revised Goal 1 for electric machines carries over the 15-year design life and cost of no more than $50/kW for the drive-unit system from the 2006 Goal 1.
- The new revised Goal 3 for energy storage systems carries over the 15-year design life for the energy storage systems. However, the revised Goal 3 specifies a higher cost of $40/kW peak electric power, whereas the 2006 Goal 2 specified a lower cost of no more than $25/kW peak electric power.
- The new revised Goal 5 for Hybrid System Optimization, Heavy Duty, specifies a stretch goal of 60 percent improvement in fuel economy (38 percent reduction in fuel consumption), which is the same as the 2006 Goal 3.

The committee agrees with the white paper (DOE, 2011) that these goals are indeed "stretch goals." They appear to be reasonable technical goals, although the 21CTP did not provide the rationale for developing them. The cost and design-life objectives in the 2006 goals had been identified earlier by the 21CTP as necessary for achieving commercially viable heavy-duty hybrid vehicles. Although the 21CTP did not provide specific plans for achieving the six new revised goals, a significant budget is expected to be required through the target dates specified in the goals. Such a significant budget would require a significant increase from the zero budget for heavy-duty hybrid R&D over the past 3 years (FY 2007 to FY 2010).

Finding 4-6. Six new stretch technical goals have been established by the 21CTP for heavy-duty hybrid vehicles. The committee agrees with the 21CTP that these are indeed stretch goals. Specific plans for achieving these new goals, some of which were carried over from the previous three goals that had been set for hybrids, were not provided to the committee. Nor was the rationale provided for these new goals, although they are appropriately focused on fuel consumption reductions, cost reduction, and a 15-year design life for the technologies. They appear to be reasonable technical goals. The cost and design life objectives in the previous goals had been identified earlier by the 21CTP as being necessary for achieving commercially viable heavy-duty hybrid vehicles. It is expected that a significant budget would be required through the target dates specified in the new goals, and a significant increase from the zero budget for heavy-duty hybrid R&D over the past 3 years would be required.

Recommendation 4-5. The 21CTP should establish plans and develop realistic budgets for accomplishing the six new stretch goals for heavy-duty hybrid vehicles in accordance with the committee's findings, explain the rationale behind the new goals, and provide the current status of the applicable technology for each of the goals so that the magnitude of the tasks for each can be assessed.

RESPONSE TO RECOMMENDATIONS IN NRC PHASE 1 REPORT

The DOE concurs with the recommendations regarding hybrid powertrains made in the NRC Phase 1 report (see Recommendations 4-1 through 4-9 in NRC, 2008). In the energy storage area, the DOE recognizes that current light-duty battery work may not necessarily apply to heavy-duty applications. A DOE-sponsored meeting on energy storage, held in January 2011, provided a crosscut of light-duty/heavy-duty work (see Appendix C in this volume, presenting the 21CTP Response to Finding 4-2 and Recommendation 4-2 on hybrids). The DOE plans activities to increase codes and standards work for hybrid components. It has plans to explore additional power density needs for heavy-duty applications, and two of the SuperTruck projects include hybridization of Class 8 trucks. One recommendation (Recommendation 4-7 in NRC, 2008) addressed the structure of the partnership between the DOD, the DOT, and the DOE. Given the separate funding mechanisms of these agencies, the DOE acknowledges their central role in fostering cross-agency communications. The DOE plans to explore strategic alliances with stakeholder organizations.

REFERENCES

ACEEE (American Council for Energy-Efficient Economy). 2009. Tax Credits for Heavy-Duty Hybrids. Available at http://www.aceee.org/blog/2009/01/tax-credits-heavy-duty-hybrids. Accessed December 1, 2010.

DOE (U.S. Department of Energy). 2010. Multi-Year Program Plan 2011-2015. Washington, D.C.: Office of Vehicle Technologies, December.

DOE. 2011. Hybrid Propulsion White Paper, February 28, 2011. Submitted by the DOE Office of Vehicle Technologies to the committee on behalf of the 21st Century Truck Partnership.

Eaton Corporation. 2009. Hybrid Solutions for MD Commercial Vehicles. ERC Symposium, University of Wisconsin, Madison, June 10, 2009.

EPA/NHTSA (U.S. Environmental Protection Agency/National Highway Traffic Safety Administration). 2010. Greenhouse Gas Emissions Standards and Fuel Efficiency Standards for Medium- and Heavy-Duty Engines and Vehicles. Dockets No. EPA-HQ-OAR-2010-0162 and No. NHTSA-2010-0079. October 25. Available at http://www.regulations.gov.

EPA/NHTSA. 2011. Greenhouse Gas Emissions Standards and Fuel Efficiency Standards for Medium- and Heavy-Duty Engines and Vehicles. Federal Register, Vol. 76, No. 179, September 15. Available at http://www.nhtsa.gov/staticfiles/rulemaking/pdf/cafe/2011-20740.pdf.

Global Refining & Fuels Today. 2011. ADA Technologies Receives US$70K for R&D from U.S. Army. February 8, 2011, Vol. 3, Issue 25. Available at http://www.worldfuels.com/newsletters/Global-Refining-Fuels-Today/3/25/.

Greszler, A. 2009. Presentation by Anthony Greszler to the House Committee on Science and Technology Subcommittee on Energy and Environment. March 24.

NRC (National Research Council). 2008. Review of the 21st Century Truck Partnership. Washington, D.C.: The National Academies Press.

NRC. 2010. Technologies and Approaches to Reducing the Fuel Consumption on Medium- and Heavy-Duty Vehicles. Washington, D.C.: The National Academies Press.

NRC. 2011. Assessment of Technologies for Light-Duty Vehicles. Washington, D.C.: The National Academies Press.

Parker Hannifin. 2010. Parker's Hydraulic Hybrid Technology Helps Truck Owners Exceed Proposed Federal Fuel Efficiency and Emissions Standards. Press Release. November 23, 2010. Available at http://www.parker.com. Accessed November 30, 2010.

TIAX. 2009. Assessment of Fuel Economy Technologies for Medium- and Heavy-Duty Vehicles. Report prepared for the National Academy of Sciences by TIAX, LLC. for the Committee to Assess Fuel Economy Technologies for Medium- and Heavy-Duty Vehicles. July 31. Cupertino, Calif.

5

Vehicle Power Demands

INTRODUCTION

The net power output of a truck engine is consumed by numerous forces opposing the truck's movement on the road. These forces in the case of a heavy-duty truck derive from a number of subsystems or components that consume energy by various forms of friction, by unfavorable pressure differentials, or by energy conversion to power a device. The 21st Century Truck Partnership (21CTP) has identified the areas of energy consumption for a typical Class 8 vehicle operating on a level road at a constant speed of 65 mph with a gross vehicle weight (GVW) of 80,000 lb. In this case, the engine losses are about 322 horsepower (hp), 60 percent of the fuel energy. The truck power demands consume the remaining 40 percent, as summarized in Table 5-1.

Under the described operating conditions, aerodynamic drag and tire rolling resistance are major contributors to energy loss. Although all trucks would benefit from aerodynamic drag reduction, those with high average speed and miles traveled would benefit most. Class 8, over-the-highway, tractor-trailer combination trucks best fit that operational description.

TABLE 5-1 Heavy-Duty Truck Power Consumption (each hour)

Operating Load	Power Consumed (hp)	Power Consumed (%)
Aerodynamic	114	53
Rolling resistance	68	32
Auxiliaries	20	9
Drivetrain	12	6
Total	214	100

NOTE: Horsepower consumed each hour is an energy term (hp-h).
SOURCE: NRC, 2010, Table 5-4.

The resisting aerodynamic horsepower is proportional to

$C_d \times A \times V^3$, where C_d = coefficient of drag, A = frontal area, and V = forward velocity (NRC, 2010, p. 28).[1] This illustrates the important role of vehicle speed on aerodynamic horsepower consumption. The relationship is shown in Figure 5-1.

An analysis of the consumptions shown in Table 5-1 has illustrated that a 20 percent reduction in C_d results in about a 10.5 percent reduction in fuel consumption. This supports the industry's rule of thumb that for high–road-speed tractor-trailers, the fuel consumption change is one-half of the drag change.

Although it is typical in truck aerodynamic improvement practice to report fuel consumption at 65 mph, rarely is an average speed of 65 mph achieved without frequently exceeding the common Interstate highway speed limit. In the case of many real-world trucks' duty cycles with perhaps an average 55 mph, the aerodynamic fuel-consumption reduction is about 9 percent rather than the 10.5 percent associated with a 20 percent C_d reduction.

Also illustrated in Figure 5-1 is the tire-rolling-resistance power consumption.[2] Figure 5-1 indicates that these two sources of power consumption are equal at about 50 mph, while rolling resistance power consumption is about twice that because of aerodynamic drag at 37 mph, when the tractor-trailer is equipped as a 2008 model pre-SmartWay[3] combination vehicle.

[1] C_d = coefficient of drag = Force/(1/2 ρ A V^2), which is dimensionless; F = force the vehicle must overcome due to air, ρ = density of the air, A = projected area perpendicular to the direction of travel, and V = velocity of the vehicle relative to the fluid (air).

[2] "Rolling resistance" is the force at the axle in the direction of travel required to make a loaded tire roll. The coefficient of rolling resistance (C_{rr}) is the value of the rolling resistance force divided by the wheel load.

[3] SmartWay is an Environmental Protection Agency (EPA) voluntary certification program for tractor-trailer combinations. The EPA notes that the

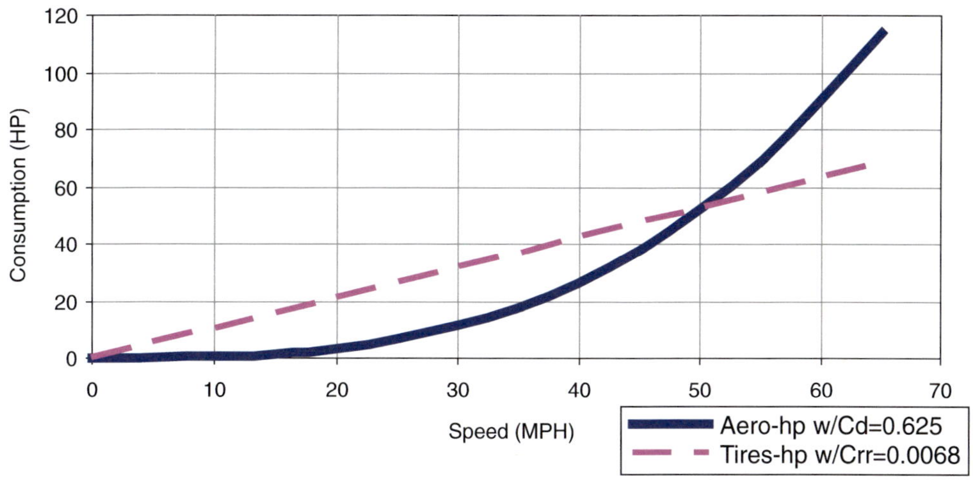

FIGURE 5-1 Aerodynamic and tire power consumption for tractor-trailer combination. SOURCE: NESCCAF/ICCT (2009).

REDUCTION GOALS FROM THE PARTNERSHIP WHITE PAPERS

The goal of the 21CTP is to conduct research, provide hardware demonstrations, and validate and deploy cost-effective, reliable, and durable technologies that reduce vehicle power demands. The Partnership will continue to utilize a vehicle system approach to continually track the benefits of individual technologies on overall vehicle efficiency and performance. Five primary technology goals applicable to the target tractor-trailer truck are to be achieved over the next 10-year period (DOE, 2011).

1. *Goal 1 (reference level of 53 percent aerodynamic power consumption):* Reduce the aerodynamic drag coefficient by 20 percent (from a C_d of 0.69). Evaluate a stretch goal of 30 percent reduction in aerodynamic drag. (The baseline is from the Environmental Protection Agency/National Highway Traffic Safety Administration [EPA/NHTSA] final rule "Greenhouse Gas Emissions Standards and Fuel Efficiency Standards for Medium- and Heavy-Duty Engines and Vehicles" for a conventional Class 8 tractor with high-roof sleeper. [EPA/NHTSA, 2011; NRC, 2010].) (Also, see footnote 3 in this chapter.)
2. *Goal 2 (reference level of 32 percent rolling resistance power consumption):* Reduce tire rolling resistance by 35 percent from tractors equipped with dual-tire drive wheels. (The baseline C_{rr} is 0.0082 from the EPA/NHTSA rule for tires of the dual tire drive wheels, only, of a Class 8 tractor [EPA/NHTSA, 2011; NRC, 2010].) (Also, see footnote 3 in this chapter.)
3. *Goal 3 (reference level of 9 percent auxiliaries' power consumption):* Reduce essential auxiliary loads by 50 percent. (The baseline for this goal is a Class 8 highway tractor-trailer with sleeper, operating 5 days in over-the-highway operations at 80,000 lb gross combined vehicle weight [GCVW].)
4. *Goal 4:* Reduce tare weight by 10 percent, and for the long-term stretch goal by 20 percent, from a 34,000 lb base tractor-trailer capable of an 80,000 lb GCVW operation, and comprised of a tractor (19,500 lb), trailer (13,500 lb), and fuel (1,500 lb).
5. *Goal 5: Thermal Management, and Friction and Wear Reduction*
   a. *Thermal Management Systems:* Increase heat-load rejected by 20 percent without increasing radiator size.
   b. *Friction and Wear (reference level of 6 percent drivetrain power consumption):* Reduce powertrain and drivetrain consumptions by 50 percent (NRC, 2010).[4]

The committee appreciates these carefully formulated goals statements. The committee believes that these goals are achievable within the 10-year period specified, but only if adequate research and development efforts are expended. Typically, achieving some goals will be found more problematic than achieving others. In the more problematic

SmartWay brand identifies products and services that reduce emissions, such as greenhouse gas emissions. Certification using EPA test methods allows carriers, manufacturers, and shippers to apply the SmartWay logo. See the EPA website at http://www.epa.gov/smartwaylogistics/basic-information/index.htm and NRC (2010) for more detailed information.

[4] Notes to Goals 1 and 2 were prepared to compare these current numerical goals to those stated in the DOE (2006) White Paper Roadmap, as well as to compare certain C_ds and C_{rr}s corresponding to other frequent descriptions (current available, SmartWay, Advanced SmartWay). See Appendix G in this report. The baseline for these goals is a Class 8 highway tractor-trailer with sleeper operating at a steady 65 mph at 80,000 lb GCVW.

areas, typically of higher technology risk and cost, 21CTP processes are very important. The ability of the Department of Energy (DOE) to assist in understanding how to achieve the goals through its Vehicle Technologies Program, including specifically the Vehicle Systems Simulation and Testing (VSST) protocols (DOE, 2010b), should be very beneficial and is particularly encouraged.

AERODYNAMICS

Aerodynamic drag arises principally from the pressure differentials on fore and aft body surfaces. Surface friction is a much less significant issue if surfaces are substantially smooth, which is a common but not yet a universal design feature; a notable example is container boxes (DOE, 2010a). Full-truck on-road testing following Society of Automotive Engineers (SAE) protocols (e.g., SAE J1263 coast-down testing; SAE J1321, Type II over-the-road testing with control truck) is relatively imprecise for evaluating the aerodynamic effect of design changes. Wind tunnel tests, also following SAE protocols (e.g., SAE J1252), are an improvement in precision and, importantly, allow an evaluation of the effects of off-axis forces (yaw) due to prevailing wind. Computational fluid dynamics (CFD) is often useful for fine-tuning component design to take account of the aerodynamic contribution, but it is in limited use for full-truck behavior.

Historically, the truck manufacturers have not reported C_d values for tractors. This situation is likely due to the competitive nature of those values, especially in light of the imprecision of prevailing standard test procedures.

Four regions of the tractor-trailer combination truck are amenable to aerodynamic design improvements. These regions include (1) the various tractor-related "aero" details, (2) the tractor-trailer gap, (3) the trailer skirt (or underbody), and (4) the trailer "base" fairing, which are all illustrated in Figure 5-2, along with the approximate fuel-consumption reductions related to each region that are estimated to be achievable in the near term.

The DOE funded a $4.2 million study, initiated in 2007, on trailer aerodynamic improvements to reduce fuel consumption and presented a project status report to the committee (DOE, 2010a). The authors provided the committee a pre-publication draft of their results and analysis. The study combined CFD studies with a large battery of full-size tractor-trailer wind tunnel tests at 65 mph in the huge National Full-Scale Aerodynamics Complex (NFAC) at the National Aeronautics and Space Administration (NASA) Ames Research Center. This research was supported by numerous industrial partners. These data are reported as yaw-wind averaged drag, which is acquired only in a wind tunnel, and which most aerodynamicists agree represents the best on-road performance figure. The study clearly is an excellent one, with very thorough evaluation of numerous candidate design improvements for all three of the trailer regions in Figure 5-2. The analysis is quite insightful, providing very helpful commentary to clarify why certain design details yield the observed results. The best performance observed for the case of a 2008 long sleeper tractor and straight frame (LS/SF) tractor-trailer was a C_d reduction of 23 percent. The draft did not proffer a base case C_d for this LS/SF combination. However, if a C_d of 0.63—used in this chapter for a 2006-2008 pre-SmartWay tractor—is assumed, a fuel-consumption reduction of about 12 percent would be expected. This can be compared to the average fuel consumption of 9.3 percent for a full package described in the section below titled "TIAX Summary" and Figure 5-3.

EPA SmartWay Transport Program

In 2004 the EPA developed and implemented SmartWay, an organized effort to specify a collection of current and emerging technologies for creating fuel-efficient tractor-

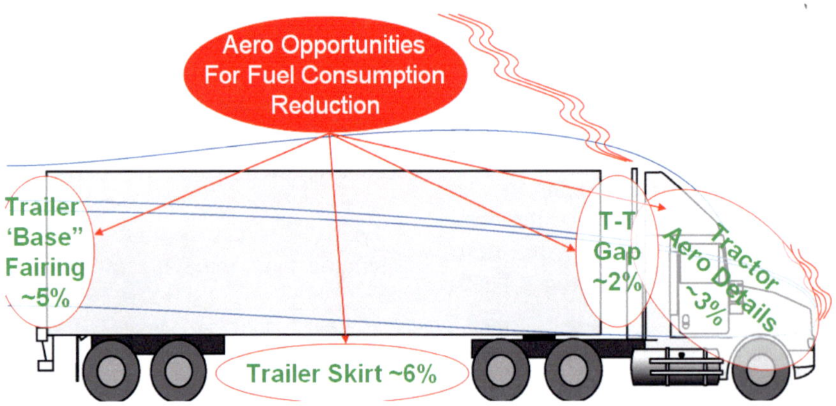

FIGURE 5-2 Tractor-trailer (T-T) combination truck showing areas of energy-saving opportunities. NOTE: Percentage changes refer to fuel consumption with base C_d = 0.625. SOURCE: Personal communication, Richard M. Wood, Solus, LLC.

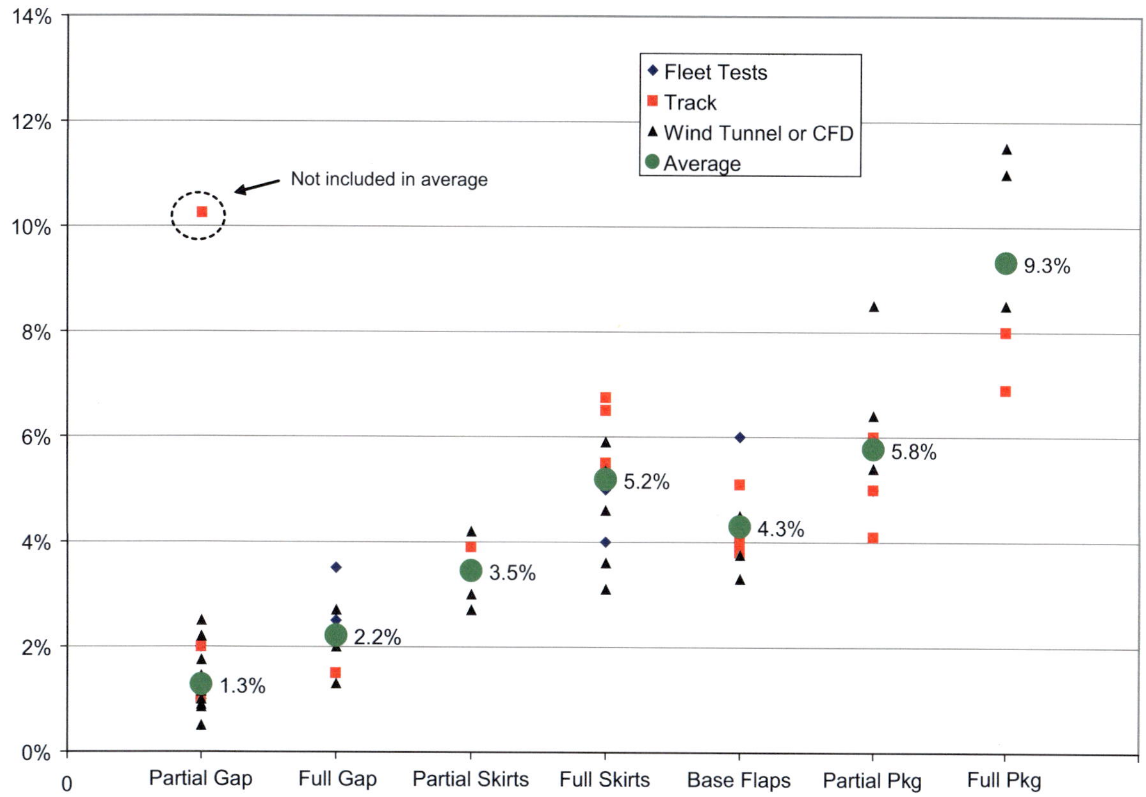

FIGURE 5-3 Summary of trailer aerodynamic device fuel consumption reduction (baseline C_d of 0.625). CFD, computational fluid dynamics. SOURCE: TIAX (2009).

trailer combinations with the best environmental performance in terms of both air pollution and emissions of greenhouse gases (GHGs) (EPA, 2011; NRC, 2010). SmartWay tractors must have an aerodynamic profile that includes a high-roof sleeper, integrated roof fairings, cab side extenders, fuel tank side fairings, and aerodynamic bumpers and mirrors. They must be powered by a 2007 or newer engine, with a SmartWay-approved option for idle reduction. The tires must be SmartWay-approved, low-rolling-resistance tires. All six major U.S. truck manufacturers have one or more complying tractors. Thousands of carriers, shippers, and manufacturers are SmartWay members or affiliates, attributing to the excellent success of this voluntary efficiency program.

SmartWay dry van trailers are 53 ft long and must have side skirt fairings, a front gap or rear fairing, and SmartWay-approved low-rolling-resistance tires. Trailer fairings and tires can be either provided by the original equipment manufacturer (OEM) on new trailers or retrofitted to older trailers by more than a dozen manufacturers that have developed these aerodynamic components.

The SmartWay program has provided an important visibility in the heavy-duty truck industry to the benefits of adapting manufacturer-developed aerodynamic trailer features and low-rolling-resistance tires. SmartWay combinations consume a minimum of 8 percent less fuel than pre-SmartWay specifications by virtue of their aerodynamic and tire-rolling-resistance changes alone.

California Regulation Based on SmartWay Program

One of the most significant consequences of the SmartWay certification program is action taken by the California Air Resources Board (CARB) to reduce the state's GHG emissions to 80 percent of 1990 levels by 2020 (ARB, 2008). In response to the California legislature's Global Warming Solutions Act of 2006, the CARB required all tractor-trailers to implement SmartWay technologies.

Beginning in January 2010, with the 2011 model year, all sleeper cab tractors on California highways that pull 53-ft or longer box van trailers must be SmartWay-certified. Day cab tractors must have SmartWay-approved low-rolling-resistance tires. At the same time, in model year 2011 and beyond, all 53-ft van trailers must also be SmartWay-certified.

TIAX Summary

TIAX provided data on the performance of a collection of aerodynamic modifications to trailers, including packages of various combinations, using various testing methods.

These are reported in Figure 5-3. The full package includes the partial gap filler, full or partial trailer skirt, and base flaps (base fairings and boat tails). TIAX also reported that next-generation aerodynamic tractors would improve fuel consumption by 3 to 4 percent in the 2013-2015 time period (TIAX, 2009).

It might be helpful to note here that a 0.63 C_d base was used in several earlier documents, including the previous Partnership White Papers (DOE, 2006), the performance-improvement estimates in Figure 5-2, and the TIAX (2009) study. Further, the current Partnership white paper (DOE, 2011) adopted the proposed EPA/NHTSA GHG rule C_d base of 0.69 for application to the new goal. This difference in goal between the current and previous white papers represents a fuel consumption penalty increment of about 4 percent for the tractor-trailer operating condition described for Table 5-1.

Adjusting these TIAX data to the 0.69 new C_d base, the committee believes that in the 2013-2015 time frame, a 13 percent reduction in fuel consumption is achievable at 65 mph when utilizing the trailer "full package" average of 9.3 percent reduction. Note that this performance represents a C_d reduction of about 25 percent, compared to the expressed goal of 20 percent.

In the 2015 to 2020 time period, if it is assumed that the performance of the trailer individual devices at the 75th percentile would eventually be achieved, then the combined full package (of partial gap, full skirt, and base flaps) would be 16 percent (C_d base adjusted to 0.69). Combining that trailer performance with the forecast next-generation tractor performance would yield a 19 percent reduction in fuel consumption (a C_d reduction of about 38 percent) achievable at 65 mph, in the 2015-2020 time period. This result is derived through a method of multiplication of fuel-consumption reductions (DOE, 2010a). Take particular note that these packaged solutions include a boat tail or similar device that might interfere with trailer docks, and must be appropriately accommodated. At least one manufacturer of such devices reported that its boat tail folds flat onto the doors in 6 seconds (ATDynamics, 2010).

Navistar, with partners Kentucky Trailer and Freight Wing, initiated a Cooperative Research and Development Agreement (CRADA) with the DOE beginning in December 2008 to bring to market a tractor-trailer combination and tire package that can reduce the fuel consumption of a heavy-duty vehicle by at least 15 percent. After completion of the project, the team members will make this fuel-efficient technology package available for sale (Navistar, 2008).

All three of the SuperTruck projects will investigate tractor and trailer aerodynamic features (see Chapter 8 in this report). These projects will provide appropriate next steps under the 21CTP for many of the vehicle power demand topics.

Findings and Recommendations

Finding 5-1. Aerodynamic improvement studies need to become increasingly integrated, because individual component improvements are typically not additive. Appropriately, the perspective of the 21CTP for the SuperTruck projects is to utilize a vehicle systems approach for the validation of research and development results.

Finding 5-2. The aerodynamic test procedures may not be sufficiently precise, and only wind tunnel testing accounts for important yaw effects, so that competitive pressures discourage truck-tractor manufacturers from publishing C_d figures. Recommendation 5-1[5] from the National Research Council's 2010 report titled *Technologies and Approaches to Reducing the Fuel Consumption of Medium- and Heavy-Duty Vehicles* provided good suggestions for standardizing C_d reporting.

Finding 5-3. The proposed EPA/NHTSA greenhouse gas emissions standards rule chose not to regulate trailer operational efficiency. Regardless of the reasons, this seems a significant omission, because both trailer aerodynamic devices and low-rolling-resistance tires that are currently production-available can provide an immediate, combined fuel consumption reduction of about 13 percent (compared to the rule's baselines).

Finding 5-4. Aerodynamic design packages are expected to improve tractor-trailer fuel consumption by 19 percent at 65 mph when fully developed in the 2015-2020 time period. This reduction corresponds to a C_d reduction of nearly 40 percent (from the newly adopted 0.69 C_d baseline).

Recommendation 5-1. The Partnership should consider setting an aerodynamic drag stretch goal of 40 percent instead of 30 percent.

TIRE ROLLING RESISTANCE

When tires roll under a truck's weight, their shape changes cyclically, deforming and recovering with inherent hysteresis. That deformation energy is converted into heat and the tire's environment dissipates the heat accumulation, and the heat dissipated is due to the tire's rolling resistance. The power consumed by the tires is proportional to truck weight, speed, and the coefficient of rolling resistance (C_{rr}) (NRC, 2010, p. 28). Rolling resistance is influenced by many factors, including design and construction features, materials, internal pressure, and tread design, which may be different according to intended tire use at steer, drive, or trailer locations. The change from bias to radial ply in the 1970s and 1980s reduced C_{rr} by 25 percent.

A common 18-wheeler highway tire specification for today (pre-low rolling resistance) will have an average C_{rr}

[5] "Regulators should require that aerodynamic features be evaluated on a wind-averaged basis that takes into account the effects of yaw. Tractor and trailer manufacturers should be required to certify their drag coefficient results using a common industry standard" (NRC, 2010).

of about 0.0068. This results from the tire combination of 2 steers, 8 drives, and 8 trailers from a range represented by data such as in Figure 5-4. (See Appendix G of this report for the tire weighting values used in this averaging.) Similarly, the EPA SmartWay 18-wheeler average threshold C_{rr} is 0.0063 based on the values from Figure 5-4, and might be achieved with low-rolling-resistance dual tires.

The tire industry has taken additional initiative, spurred by competitive market forces, to develop next-generation wide-base single tires (NGWBSTs), which have substantially reduced C_{rr} and are becoming the current state of the art; some fleets are replacing dual tires with the NGWBST systems when new trucks are ordered. Figures for C_{rr} in the low 0.004s are available for some tire positions (trailers). A currently available set of tractor-trailer tires, including wide-base low-rolling-resistance designs, provide a C_{rr} of 0.0055.

As noted above in Goal 2, the 21CTP has recently adopted the C_{rr} baseline used by the EPA for tractor tires on dual drive wheels, from the proposed truck GHG/fuel efficiency rule promulgated by the EPA and NHTSA (EPA/NHTSA, 2010). To put the Goal 2 drive tires' C_{rr} baseline of 0.0082 in perspective for an 18-wheeler, one must create a combination that includes the EPA/NHTSA rule's selection for steer tires, 0.0078, along with a committee-assumed trailer tire value of 0.0072. That tractor-trailer combined weighted average C_{rr} then is about 0.0077. Now, by comparison, the current NGWBST combined weighted average C_{rr} is a reduction of nearly 30 percent, to 0.0055, and yields a fuel-consumption reduction of about 10.5 percent. (Note that 18-wheelers become 10-wheelers when their current 8 pairs of duals are replaced by single wide-base tires.)

This level of reduced fuel consumption has been validated in an extensive on-highway test series conducted by the Oak Ridge National Laboratory (ORNL) in cooperation with a carrier, Schrader Trucking, and using Michelin-supplied tires (ORNL, 2009). Six well-instrumented tractors and six trailers all with a mix of standard dual tires and NGWBSTs accumulated more than 688,000 miles at an aggregated average speed of 58 mph traveling through 39 eastern and midwestern states and Canada. Every combination of tires that included some mix of NGWBSTs provided statistically significant fuel-consumption reductions at all three of the selected gross weight groupings. Those applications that replaced both tractor and trailer duals with NGWBSTs achieved a 9.3 to 9.6 percent reduction in fuel consumed when carrying medium and high payloads.

A combined tractor-trailer C_{rr} of 0.0045 is achievable in the next 5 years (TIAX, 2009, p. 4-53) and could yield a fuel-consumption reduction of about 15 percent. These reductions are compared to the supposed EPA non-low-rolling-resistance tractor-trailer 0.0077 C_{rr} weighted baseline noted above, and result from the corresponding 40 percent reduction in C_{rr}.

Potential pavement damage due to changes in tire shape and load is a high concern for regulators. Road wear and damage are known to be directly controlled by contact (hertz) stress. This stress increases with the 4th power of unit load. As a result, vehicle weight and the distribution of that weight over the tire contact patches are critical factors in determining the durability of road surfaces. Studies show that NGWBSTs do not increase contact stress in the pavement compared to the stress generated by standard dual tires. As a

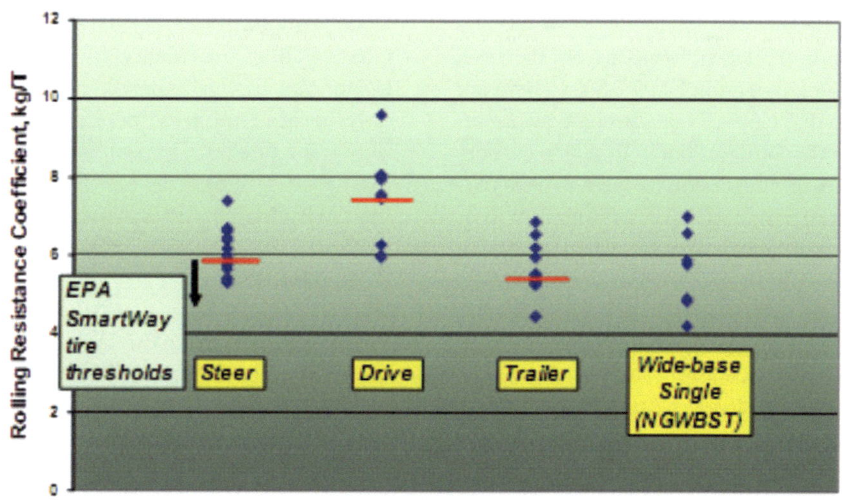

FIGURE 5-4 Example rolling resistance coefficients for heavy-duty truck tires. (Data in this figure are $C_{rr} \times 1{,}000$). Acronyms are defined in Appendix I. SOURCE: Courtesy of Michelin Tire North America. C. Bradley and S. Nelson, Michelin Tire North America, "Truck Tires and Rolling Resistance," presentation to the National Research Council Committee to Assess Fuel Economy Technologies for Medium- and Heavy-Duty Vehicles, February 4, 2009, Washington, D.C.

result, pavement damage is expected to be similar for these two types of tires for primary roads.

Some of the concerns regarding pavement damage can be attributed to past experiences and studies of the original 65-series on-off-road and regional single tires (e.g., 385/65R22.5 and 425/65R22.5). These tires are normally operated at elevated service inflation pressures, with consequently reduced tire footprint size and peak tire-road contact stresses that surpass those of traditional dual tires. Studies of the original "65-series singles" concluded that increased road damage could be attributed to these tires, and this naturally generated concern for state and federal departments of transportation when NGWBSTs were developed and put in use (Al-Qadi, 2007a,b).

Other issues still concerning users of NGWBSTs are tread life, truck stability in blowout, running on a flat, availability of nearby repair facilities, and integrity of drive traction. Tire manufacturers believe that these issues have been resolved (Al-Qadi, 2007a,b). Many carriers have been replacing duals on existing equipment with NGWBSTs at an accelerated pace, seeking the anticipated fuel savings contribution of NGWBSTs. A recent trade press article identified eight medium and large carriers that have been converting, some beginning many years ago (TT, 2011). It has been observed that when replacement of dual with wide single tires are made to existing axles, manufacturers caution that improper wheel selection can reduce load-carrying capability (AM, 2010). Carriers need to be made aware of the importance of such caveats. Potential safety issues associated with blowouts of NGWBSTs do not appear to have become a remarkable problem. Nevertheless, the Department of Transportation (DOT) should thoroughly review the real-world ramifications of documented NGWBST blowouts. Many carriers are routinely using NGWBSTs because they are satisfied that these issues are minimal, but many other fleet owners will need to be convinced. Most Class 8 tractor-trailer combination trucks can benefit from the fuel-consumption reductions contained in NGWBSTs. There are some application issues that carriers must become aware of, such as trailer spread-axle configurations. In short, carriers must monitor the details of conversion as well as new purchases. Another benefit of using NGWBSTs is an 800 lb or greater weight reduction for an 18-wheeler, when used with aluminum wheels.

Maintenance of tire pressure is important both to tire life and rolling resistance. A 20 percent pressure reduction in all tires causes a fuel-consumption increase of 2 to 3 percent, associated with an increase in rolling resistance (NRC, 2010, p. 114). Tire pressure monitoring systems with in-wheel sensors have become available in recent years and are gaining acceptance in fleet application. As usual, the cost/benefit trade-off is an important element in the acceptance of such new technologies.

At least two standard laboratory test procedures are in common use for determining truck (and light-duty vehicle) rolling resistance characteristics: they are the SAE J1269 multi- and single-point, and the draft International Organization for Standardization (ISO) 28580 single-point with interlaboratory correlation procedure. Unfortunately, tire manufacturers have not voluntarily collaborated in efforts to facilitate the comparison of rolling resistance and other characteristics across company boundaries. Furthermore, given that C_{rr} is not constant over the tread life of tires, it would be a great improvement toward tire selection if such laboratory procedures added a requirement to measure C_{rr} at the tire mid-life (circa 150,000 miles). This is an area needing DOT or DOE attention, because fuel-consumption benefits are being lost because of limited carrier acceptance of the NGWBSTs.

Response to Recommendations from NRC Phase 1 Review

The 21CTP agreed that it is appropriate for the DOT, EPA, and DOE to report tire performance data to buyers and sellers. It noted that the EPA SmartWay Transport Partnership publicly reports the fuel-savings capability of low-rolling-resistance tires, including the wide-base single families of tires (see 21CTP response in Appendix C to Recommendations 5-9 on Parasitics from [NRC, 2008]). The DOE also recently reported the favorable results of its testing with NGWBSTs.

Findings and Recommendations

Finding 5-5. Next-generation wide-base single tires (NGWBSTs) can provide a combination tractor-trailer with an immediate 10.5 percent fuel-consumption reduction and up to a 15 percent reduction in the next 5 years, but many fleets do not yet embrace the technology.

Recommendation 5-2. The DOE should set the goal for reduced rolling resistance for the tires of the combination tractor-van trailer, rather than for the tractor drive wheels only, because improved-performance trailer tires are equally important to realizing the full benefit of reduced rolling resistance designs. This benefit can be achieved by combining the EPA base values for steer and drive tires in the EPA/NHTSA GHG rule, with an assumed trailer tire C_{rr} value of about 0.0072.

Finding 5-6. Carriers need to follow carefully the recommendations of axle manufacturers for replacing dual tires with single-wide tires to ensure that the integrity of the load system is not compromised.

Recommendation 5-3. The 21CTP should consider producing a comprehensive summary that can be updated giving the prescriptions and precautions that carriers should consider when *retrofitting* NGWBSTs onto original equipment axles fitted with dual wheels and tires. This effect might best be managed in conjunction with the American Trucking Association's (ATA's) Technology and Maintenance Council, which has drafted such a Recommended Practice and is a specialist in creating such directives for ATA membership (ATA, 2007).

Finding 5-7. There is no rolling resistance test procedure with interlaboratory correlation universally employed as an industry standard.

Recommendation 5-4. The 21CTP, strongly supported by DOT and EPA (the latter through its SmartWay program), should conduct an authoritative study of the several barriers (e.g., related to tread life, truck stability in blowouts, run-flat tires, and other topics) to the widespread carrier adoption of NGWBSTs. The DOT should specifically support reduction of barriers to NGWBST acceptance by requiring the universal use by tire manufacturers of a rolling resistance test procedure like that in ISO (International Organization for Standardization) 28580, to ensure that comparative interlaboratory data exist.

AUXILIARY LOADS

Auxiliary loads are typically driven by a mechanical drive from the propulsion engine. Examples include compressed air for brake systems; air conditioning compressors; electrical energy for lighting, hotel loads, heating, ventilation, and air-conditioning (HVAC) fans and battery-charging systems; power-steering pumps; and power take-off (PTO) drives, where used. Trailer refrigeration loads are normally satisfied with a second, dedicated engine. Power demand on the propulsion engine for these functions might range up to 9 percent, depending on the truck duty cycle and environment.

Seemingly missing from this list of "auxiliaries" is the engine cooling fan. However, this is not a significant omission when the principal truck operation perspective is to be a line-haul tractor-trailer at high road speed. The line-haul fan power consumption has been summarized for a 1,800 engine rpm fan-on condition as 40 to 60 hp. However, for high road speed, the fan-on time was estimated at 5 percent (NRC, 2010, p. 110). Together these figures suggest a time average consumption of 2 to 3 hp, sufficiently small not to warrant attention. However, the committee notes that there are operating conditions, for example, at lower road speeds, climbing hills, and with the air conditioner on, when the fan on will be consuming 20 to 50 hp for appreciable periods. A variable-speed fan drive can reduce the necessary fan speed and accompanying power consumption to only that necessary to achieve both radiator and air-conditioner-condenser temperatures and pressure.

Goal 3 above requires a reduction in essential auxiliary loads by 50 percent. In Table 5-1, auxiliary load is identified as 9 percent of the demand on the engine. A 50 percent reduction, to 4.5 percent of the demand, would provide a corresponding 4.5 percent reduction in fuel consumption on a level road at 65 mph, assuming unchanged brake thermal efficiency (BTE). Auxiliary power requirements are derived from many vehicle functions, as described above (DOE, 2011).

The most often imagined solution for removing these loads from their current drive is the electrification of the various drives, which was called the More Electric Truck project in the 21CTP. Electrically driving many of these systems can achieve power on demand, using energy only when and where it is needed. In that way, nonpropulsion energy like that from waste heat recovery (WHR) systems, hybrid system regenerative braking systems, or auxiliary idle reduction systems might supply the needed electricity. A 21CTP goal identified for hybrid drive units in Chapter 4 (see revised Goal 1) provides for "electrification of power accessories with goals of reduced weight and cost, operation in engine-off conditions and a service life of 1 million miles/15 years." Truck electrification will have to be achieved without a degradation of current subsystem durability and reliability, and advanced electrical systems will have to be cost-effective. Multiple projects were initiated and successfully completed in the 21CTP program by the end of 2007. Average power savings of up to 2 percent of fuel are the largest estimated in the NRC Phase 1 report (NRC, 2008, p. 74).

Two of the SuperTruck platforms will investigate electrically driven auxiliaries, as those projects integrate WHR, hybrid-electric, and idle-reduction systems. It is expected that those projects may be better able to quantify the actual savings of electrically driven auxiliaries in real-world duty. Chapters 4 and 6 of this report are devoted to hybrid systems and idle reduction, respectively.

Response to Recommendations from NRC Phase 1 Review

The 21CTP reported that it is exploring the cost and benefits of continuing R&D in the auxiliaries area compared to the potential for efficiency improvements by other heavy-truck technologies (see Appendix C, response to Recommendation 5-1 on parasitics from NRC, 2008).

Finding and Recommendation

Finding 5-8. The More Electric Truck may achieve about one-third of the auxiliaries' reduction goal for a loaded tractor-trailer. Better quantification is expected to result through two of the SuperTruck projects.

Recommendation 5-5. The Partnership should renew R&D efforts to further reduce fuel consumption related to auxiliary power demands.

WEIGHT REDUCTION

Truck weight reduction affects fuel consumption by reducing tire rolling resistance and unrecovered energy used when accelerating or grade climbing. The energy required to overcome resistances is approximately linearly dependent on the weight of the vehicles. Generally for Class 8 trucks, there is more incentive for reducing the vehicle weight for weight-limited trucks, which can replace the reduced tare weight with cargo. For a 35,000 lb tractor-trailer fully loaded with 45,000 lb of cargo, reducing weight by 1,000 lb increases load-carrying capacity by the same amount and therefore reduces payload-specific fuel consumption, expressed in gallons per 1,000 ton-miles, by a little more than 2 percent. Unfortunately, for a volume full (cubed-out) trailer application, the fuel-consumption reduction is only about 0.5 percent per 1,000 lb of tare weight reduction, because payload remains constant (NESCCAF/ICCT, 2009, p. 50). Approximately 60 percent of tractor-trailer loads are cubed-out before grossed out (weight full) (NRC, 2010). Opportunities for fuel efficiency impact vary considerably by truck type and class and duty cycle. Vehicle weight effects are more important for duty cycles with frequent starts and stops (NRC, 2010, Table 5-16). For urban delivery vehicles, a 10 percent reduction in weight can reduce fuel consumption by as much as 7 percent.

Because of budget reductions for the 21CTP, no funding was provided for lightweight materials (other than for propulsion materials) from FY 2007 through FY 2009. However, prior to the FY 2007 budget reduction, there had been numerous projects aimed at vehicle weight reduction. Several projects involving the national laboratories and industry led to useful examples of weight reduction. A complete list of 21CTP projects up to 2007 can be found in the 21CTP Quad Sheets (DOE, 2007).

New Incentives for Vehicle Weight Reduction

The incentive for reducing the weight of weight-limited Class 8 trucks has taken a new urgency as trucks have been adding weight particularly with emissions-compliance devices. Emissions control components have added as much as 400 lb to the typical tractor. Aerodynamic devices are slowly growing in popularity, adding several hundred pounds in some cases, especially trailer devices (the trade-off of adding weight to improve aerodynamics is a good one, because aerodynamic drag is a major contributor to fuel consumption at highway speeds). Weight reduction will be needed to offset other components such as auxiliary power units (~400 lb) added to reduce fuel consumption normally expended during idle. For city buses and urban delivery vehicles, weight reduction becomes important to offset the hybrid systems that many of these vehicles are adding.

Furthermore, the new EPA/NHTSA standards to reduce fuel consumption and greenhouse gas emissions from medium- and heavy-duty vehicles are expected to achieve reductions in CO_2 and fuel-consumption from 7 to 20 percent reduction from current, 2010, baselines (EPA/NHTSA, 2011). As selected truck tractor technologies are expected to build on the EPA SmartWay configurations, some weight increase issues will be encountered—for example, with the use of tractor aerodynamic components and idle reduction components.

For these reasons, the 21CTP reinstated funding of the lightweight materials research beginning late in 2010 ($500,000) and into 2011 ($2.12 million) (see Chapter 1, Table 1-2). The FY 2011 budget of $2.12 million will be directed to lightweight materials initiatives for the SuperTruck projects. Several weight-reduction initiatives are funded under the SuperTruck projects themselves.

Weight-Reduction Goals

Weight-reduction goals vary by vehicle; the targets for all vehicle classes range from 10 to 33 percent. For Class 8 tractor-trailer combinations the goal had been from 15 to 20 percent, with the specific goal of 5,000 lb. Recently the goal has been modified. The new goal is a 10 percent reduction in weight for a baseline tractor-trailer tare weight of 34,000 lb (specific goal of 3,400 lb) along with a longer-term stretch goal of 20 percent reduction in weight.

Opportunities and New Initiatives

For a Class 8 truck there are numerous opportunities for reducing vehicle weight by introducing lighter-weight materials, albeit often at a cost premium. The more obvious opportunities lie in the body structure (~19 percent of total tractor weight), the chassis/frame components (~12 percent), and wheels and tires (~10 percent). Truck manufacturers have been substituting lightweight materials for a number of components in the cab, chassis, and wheels. Examples include composite structure in the cab, aluminum panels, aluminum wheels, and aluminum fuel tanks. There are also weight-reduction opportunities afforded by extensive use of aluminum for both tractor and trailer (NRC, 2010, see Figure 5-38).

The Pacific Northwest National Laboratory (PNNL), in cooperation with PACCAR, Novelis, and Magna, is developing an aerodynamic lightweight cab structure.[6] The objective is to develop a low-cost forming method using aluminum alloy that allows the use of aluminum instead of steel or other materials. A target of 40 percent weight savings is the objective. PNNL forming equipment will be used, and its impact on corrosion and ability to be painted will be evaluated. The forming technology will be transferred to Magna for the fabrication of a full-scale cab. The project will begin in FY 2011.

As part of the SuperTruck projects, additional lightweight materials initiatives are planned. For example, Navistar, which of the 21CTP industry partners provided the most

[6] Jerry Gibbs, DOE, "Materials Support for the 21st Century Truck: Lightweighting and Propulsion Materials," presentation to the committee, November 15, 2010, Washington, D.C.

detail for its weight-reduction effort, plans to explore the following.[7]

- Use of composites in cab-in-white,
- Use of composites in trailer floor,
- Carbon composite drums,
- Lightweight rotors and aluminum calipers,
- Plastic fuel tanks,
- Single lightweight driveshaft, and
- Aluminum cross-members.

It may also be possible to transfer and expand technical work that had been carried out on light-duty vehicle activities to heavy-truck applications. Two examples were presented to the committee (see footnote 6). The first example involves the potential application of magnesium. Studies at the PNNL and ORNL will explore the possibility of improving the ductility of magnesium and its low-temperature formability. The researchers will also explore the development of new joining methods (magnesium to steel), as conventional automotive joining methods are not applicable to magnesium. Galvanic corrosion limits applications of magnesium currently—the team will explore the application of low-cost ceramic coatings as a solution to this problem.

The second example involves the processing of carbon fiber for composite applications. Although glass-reinforced composites, often in the form of sheet molding compound, have been used in high-volume automotive and in heavy-truck applications for years, higher-performance carbon-reinforced composites have not. The primary obstacle has been the cost of the carbon fiber itself. However, there are other issues associated with the application of high-performance carbon composites. The fabrication process is slow compared with steel stamping processes. Furthermore, assembly is complicated by the fact that joining methods typical of automotive assembly processes are not applicable. Other issues needing resolution include coloration, recycling methods, and the ability to repair damaged parts.

Progress is being made, however, in adapting carbon composites for automotive applications. Indeed, carbon composites have been used for years in the motor-sports industry, and they are more recently finding applications in low-volume specialty automotive products. And, in fact, certain characteristics of carbon composites are superior to conventional steel. Clearly the higher strength and stiffness of carbon composites enable part-to-part weight reduction. Composites offer opportunities for part consolidation as well as the ability to form complex shapes that could not be achieved by steel stamping. Composites are corrosion-resistant. For low volumes, tooling costs can be lower than those typical of metal stamping.

The DOE recently issued a Funding Opportunity Announcement (FOA) that includes opportunities for additional cost-shared research in the area of lightweight materials. It includes additional research in magnesium and carbon-fiber composites.[8]

Response to Recommendations from NRC Phase 1 Review

Because of severe reductions in the 21CTP budget, no budget was allocated to lightweight materials research from 2007 through 2009. In the NRC Phase 1 report (NRC, 2008), it was recommended (Recommendation 5-2) that scarce budget resources be allocated to projects with higher potential payoff for fuel-consumption reduction (e.g., engine efficiency) than payoff through the application of lightweight materials. Furthermore, the NRC Phase 1 report suggested that it should be the responsibility of the truck manufacturers to take the next steps of system integration, product validation, and production application of lightweight materials.

Finding

Finding 5-9. Several projects that were carried out prior to 2007 have shown the potential for the reduction in weight of individual components and subsystems. However, to date there has been no integrated full-vehicle project to show that the goal of reducing the weight of a Class 8 tractor-trailer by 3,400 lb can be achieved. Moreover, the NRC Phase 1 report had recommended that such a project, using prototype components, vehicle integration, and full-vehicle system analysis, should be carried out by industrial partners—led by original equipment manufacturers. The new SuperTruck program appears to be a response to this suggestion.

THERMAL MANAGEMENT

Thermal management projects have been coordinated with the DOE Heavy Vehicle Systems Optimization Program. The primary focus has been on increasing cooling radiator capacity without increasing size. Heat rejection and radiator size are issues that have several effects on truck design. Incurring higher heat rejection normally mandates a larger cooling package, which adds weight and cost. The larger cooling package also requires more airflow, and the air passes through the turbulent and restrictive areas under the hood, increasing the vehicle aerodynamic drag. A larger cooling package also requires changes to the shape of the truck's front, which limits aerodynamic design options and results in increased drag. Because of these factors, lower heat rejection and/or a more efficient cooling system contribute to lower aerodynamic drag, lower truck weight, and potentially lower cost.

[7] Anthony Cook, Navistar, Inc., "Navistar's SuperTruck Program," presentation to the committee, September 8, 2010, Washington, D.C.

[8] Financial Assistance Funding Opportunity Announcement, Number: DE_FOA-0000239, December 16, 2010.

Good progress has been achieved by the application of nanoparticle-containing coolants. Nanofluids contain highly conductive particles (1 to 100 nanometers) suspended in liquids to provide significantly increased heat transfer from cooled systems, compared to conventional glycol-water systems. These nanofluid coolant projects were planned for 7-year duration, with completion anticipated in 2012. The nanofluid cooling system work has been managed at the Argonne National Laboratory (ANL), evidently with limited trucking company involvement. Nevertheless, a heavy-duty truck demonstration is planned for the nanofluid coolant project(s). Research status forecasts a radiator size reduction of more than 10 percent (ANL, 2009a, 2010b).

Modeling of underhood cooling airflow has also been pursued. Both technologies were driven by the engine manufacturers' addition of cooled exhaust gas recirculation (EGR) beginning in 2002 to lower NO_x by combustion modification. EGR rates have continued to increase in some of those engines to enable compliance with the latest NO_x standard reduction in calendar year 2010. The committee has not found that high-technology cooling radiator designs, coordinated with increased EGR, have been incorporated in heavy truck systems to date. It is noted that there is an ANL project associated with the nanofluid projects, called Efficient Cooling in Engines with Nucleated Boiling. This project is attempting to exploit the fact that the boiling heat-transfer coefficient is more than five times the liquid heat-transfer coefficient. A two-phase boiling heat-transfer regime is sought, then altering the cooling system to manage a continuous gas-liquid condition. Such a two-phase heat-transfer system with a net higher heat-transfer than a conventional nonboiling cooling system further supports the notion to reduce radiator size (ANL, 2009b).

The committee believes that the removal of barriers to production implementation is a key next-step action. The project leaders have listed barriers for nanofluids: adequate heat-transfer coefficient increase, high viscosity of nanofluids, pumping power increase, and fluid production costs (ANL, 2009a); and for the nucleate boiling radiators: coolant boiling heat-transfer coefficients, coolant two-phase pressure-drop data, and acceptable boiling limits of both critical heat flux and flow instability (ANL, 2009b).

Furthermore, current efforts to achieve a major fuel-consumption reduction by means of the addition of WHR and hybrid electric systems will add yet higher cooling demands on the radiator system. Given the fuel-consumption reduction goal, aerodynamics needs to be improved substantially, and smaller radiators would facilitate changes to the frontal tractor shape for reduced air pressure and aero-friction. Cummins noted in its SuperTruck presentation to the committee that it would address this issue of tractor frontal shape as affected by cooling system capacity/efficiency needs. And the DOE has initiated a new CRADA with Cummins with this intention, reported at the June 2010 DOE Merit Review presentations. There is an associated CRADA with Peterbilt, to optimize coolant boiling to reduce coolant system size and power consumption (ANL, 2010a). This second project could support these same themes but was not reported by the SuperTruck participants. These projects are illustrative of areas in which technology development cost and risk can be managed under the DOE VSST protocol (DOE, 2010a). Further, the Partnership has identified a number of potential topics that might make a big contribution to the breakthrough sought, such as advanced window glazing, concepts for direct heating and cooling of the vehicle occupants, heat-generated cooling techniques, and thermoelectrics (DOE, 2011).

Other next steps appear to be to provide consistent funding to permit nanofluid performance-enhancing discoveries to be tested in truck-cooling demonstrations in aerodynamically efficient trucks, thereby integrating several technologies.

Response to Recommendations from NRC Phase 1 Review

The DOE has continued to emphasize the cooling system nanofluid developments through a series of ongoing projects at the ANL. It has also initiated the two CRADAs noted above: Cooling Boiling in Head Region-PACCAR, Integrated Underhood Thermal and External Aerodynamics-Cummins, VSS004 (ANL, 2010a). The Cummins SuperTruck project will evaluate tractor frontal shape as affected by cooling system capacity/efficiency needs.

The DOE also agreed to find methods to provide status reporting as part of its overall assessment of the Partnership's processes and progress. The DOE believes that the SuperTruck projects provide a unique opportunity to bring together manufacturers and suppliers for coordinated systems R&D.

Finding and Recommendation

Finding 5-10. Heavy-duty truck thermal management objectives are growing in importance as new systems to improve both engine and truck efficiency, particularly waste heat recovery systems, become reality. These are accompanied by new heat management issues and are expected to be added to trucks in the current decade.

Recommendation 5-6. The Partnership should continue priority support of nanofluid and high-efficiency underhood cooling systems, as well as review other potential technical concepts, and validate them as an integrated system.

FRICTION AND WEAR

The DOE initiated a study focused on developing a comprehensive friction model for engine and drivetrain systems. The project purports to improve existing computational models, which are based only on fluid film lubrication. Noting that most engine and driveline components operate under both mixed and boundary lubrication regimes, a comprehensive model that can adequately predict friction in both sliding and rolling contact regimes is the objective.

Completed work identified that friction at a lubricated interface (especially under the boundary regime) is a complex phenomenon involving the shearing of the lubricant fluid film, the tribochemical boundary films, and the near-surface material. Next steps will attempt to quantify and model all three regimes. The expected 6-year project was begun at the ANL in 2010 (ANL, 2010c). This project is also illustrative of one in which technology development cost and risk can be managed under the DOE VSST protocol. It is noted that no truck OEM companies are yet collaborating with the contractor in this project.

As described in the subsection titled "Lubricants" in Chapter 3, the Engine Manufacturers Association initiated a lubricant project beginning in April 2010 as the EPA/NHTSA GHG/fuel efficiency rule was anticipated. The primary objective is an improved fuel efficiency contribution through implementation of a High Temperature High Shear (HTHS) viscosity property. Early industry testing with HTHS controlled oils has shown promise for fuel-consumption reduction while other parameters continue to provide traditional performance. The notion of this HTHS property is believed to be applicable to drivetrain units as well as engines. This potential should be considered within the projects that the 21CTP considers for friction and wear reduction.

In addressing the committee, two of the SuperTruck presenters indicated that friction reduction was included in their projects. However, it is believed that these efforts are being directed toward engine friction specifically and not to driveline or auxiliaries.

The DOE presented fuel and lubricants development plans to the committee.[9] The DOE suggested that it might play a valuable role in developing an understanding of microfluidic transport and tribological film formation. The first project noted above appears to be a good beginning. The DOE cited the long-standing engine and drivetrain friction-reduction needs, anticipating that half of the target improvement might be achieved with engine oil enhancements. The DOE surely can make a helpful contribution, but it will need to become engaged with driveline and auxiliary system developers and manufacturers as well as becoming involved in the heavy-duty-engine-industry oil development process. The 21CTP certainly provides an appropriate forum.

The DOE (2011) white paper identified a list of friction and wear activities that were proposed to begin in 2011 and to be completed in 2018. These activities may well make a needed contribution in this difficult area. A project plan needs to be developed within the Partnership to solidify this list (see below) and begin the work.

[9] Kevin Stork, DOE, "DOE Fuel & Lubricant Technology R&D," presentation to the committee, November 15, 2010, Washington, D.C.

- Integration of friction and wear models to reduce powertrain loss,
- Development of advanced coatings for reduced friction and wear, and
- Down-selection and demonstration of engineered surface applications for reduced wear and friction.

Revision of Goal 5 within the new Vehicle Power Demand Orientation

The "friction and wear" component of Goal 5 requires the following: "Develop and demonstrate technologies that reduce powertrain and driveline losses" (DOE, 2011, p. 3). Note that this deals with both powertrain and drivetrain friction.

In Table 5-1, drivetrain load is identified as 6 percent of the demand (50 percent reduction of the load will reduce fuel consumption by 3 percent, assuming unchanged engine BTE). The drivetrain loads are those resulting from engine torque transmitted through units like a transmission(s) and rear axle(s) gearing, and not including auxiliaries or tires, as partitioned in the DOE (2011) white paper. Drivetrain loads derive from windage of gears running in lubricant and friction within components of the geared units.

Goal 5 has been carried over from the DOE (2006) white paper "Parasitics," which included powertrain components, often referred to as engine accessories, that are needed to operate an engine on a dynamometer and on the EPA Federal Test Procedure (FTP). These are the oil pump, fuel pump, water pump, and any other engine fluid pump (like an EGR pump). Those loads are integral to engine operation contributing to the brake thermal efficiency and not to vehicle power demand. To represent only the power consumptions of vehicle power demand, it is appropriate that the term "powertrain" be removed from the Goal 5.b statement.

There is also a need for an updated study of the current driveline power demand of 12 hp. It is noted that this 12 hp, reported in the most recent white paper (DOE, 2011), is the same value as that in the previous white paper (DOE, 2006) which represented both powertrain and driveline "parasitic" (friction) losses together. The committee expects that the driveline-only power demand (friction) will be significantly revised downward.

Response to Recommendations from NRC Phase 1 Review

The NRC Phase 1 review report (NRC, 2008) estimated that a 50 percent reduction in drivetrain losses would reduce fuel consumption by 1.2 percent. Similarly, that review suggested that a 50 percent reduction in powertrain (engine) friction losses would reduce fuel consumption by 3.7 percent. The two losses sum to about 5 percent, somewhat lower than the stated goal. Further, Recommendation 5-8 from that review suggested that the DOE reassess the technical feasibility of these 50 percent reductions while retaining adequate durability and reliability (NRC, 2008, p. 80).

The Partnership responded that it agreed with the recommendation (see Appendix C), and that each of the three SuperTruck teams will continue the research in these areas of friction, wear, and lubrication in the engine and drivetrain. Both Daimler and Navistar have friction-reduction phases in their projects. The DOE believes that the 50 percent reduction goal feasibility (combined powertrain and driveline losses) will receive reasonably thorough evaluation through the SuperTruck projects. The committee suggests that the DOE will need to be proactive with the SuperTruck contractors to ensure that they allocate adequate resources in order to achieve the expected thorough evaluation.

Findings and Recommendation

Finding 5-11. There is a need for an updated study of the current driveline power demand of 12 hp. Furthermore, to represent vehicle power demand power consumptions only, it is appropriate that the term "powertrain" be removed from the 21CTP Goal 5.b. statement.

Recommendation 5-7. The term "powertrain" should be removed from the 21 CTP Goal 5.b statement. In addition, the Partnership should update its study on the driveline power demand of 12 hp.

Finding 5-12. There has been no apparent collaboration on lubricant projects between the DOE and OEM partners.

OVERALL FINDING AND RECOMMENDATION

Finding 5-13. Summarizing the committee's findings on vehicle power demands: Project prioritization by the 21CTP roughly follows the consumption ranking of the several heavy-duty truck operating loads listed in Table 5-1 and technology risk. However, sometimes market forces provide considerable impetus for quite good development and implementation—for example, in tire rolling resistance and, to a lesser extent, trailer aerodynamic components. The DOE has identified a strong role in which technology development costs and risks are high, as in its vehicle systems simulation and testing activities for heavy-duty trucks. It has generally followed these principles, to address high cost and risk in the vehicle power demand projects. The SuperTruck projects will provide a unique Partnership opportunity to provide both further high-risk technology results for certain vehicle power demand reductions and real-world validation of numerous integrated systems.

Recommendation 5-8. Although it is tempting to assume that the SuperTruck projects will address all of the technologies required to reduce tractor-trailer fuel consumption, in practice many technologies may be left behind, particularly those that are not yet very mature. The Partnership should carefully review the technologies that have been identified and determine whether any technologies to reduce vehicle power demand are not being adequately addressed by the SuperTruck program. The DOE should define projects and find funding to support the development of technologies beyond the scope of SuperTruck.

REFERENCES

Al-Qadi, I. 2007a. New Generation of Wide-Base Tire and Its Impact on Trucking Operations, Environment, and Pavements. Transportation Research Board 86th Annual Meeting. Paper No. 07-2432. January 21-25. Washington, D.C.

Al-Qadi, I. 2007b. Impact of Wide-Based Tires on Pavement and Trucking Operation. Presented at the International Workshop on the Use of Wide-Base Tires, Federal Highway Administration, Turner-Fairbank Highway Research Center. October 25-26. McLean, Virginia.

AM (ArvinMeritor, Inc.). 2010. Product Information Letter #515, Rear Drive Axle Application Guidance and Recommendations for Single Wide-Base Tires Mounted on Outset Wheels. January. Troy, Michigan.

ANL (Argonne National Laboratory). 2009a. Nanofluid Development for Engine Cooling Systems. Available at http://www1.eere.energy.gov/vehiclesandfuels/pdfs/merit_review_2009/vehicles_and_systems_simulation/vssp_21_timofeeva.pdf.

ANL. 2009b. Efficient Cooling in Engines with Nucleated Boiling. Yu. Project VSSP_14_Yu. May 19.

ANL. 2010a. CRADAs: Cooling Boiling in Head Region—PACCAR, Integrated Underhood Thermal and External Aerodynamics-Cummins. Routbort. VSS004, June, 8.

ANL. 2010b. Erosion of Radiator Fluids by Nanofluids. Department of Energy. Available at http://www1.eere.energy.gov/vehiclesandfuels/pdfs/merit_review_2010/propulsion_materials/pm008_routbort_2010_o.pdf. Accessed May 5, 2011.

ANL. 2010c. Friction Modeling for Lubricated Engine and Drivetrain Components. Ajayi. June. Available at http://www1.eere.energy.gov/vehiclesandfuels/pdfs/merit_review_2010/propulsion_materials/pm026_ajayi_2010_p.pdf. Accessed June 1, 2010.

ARB (Air Resources Board). 2008. Final Regulation Order to Reduce Greenhouse Gas Emissions from Heavy-Duty Vehicles. Available at http://www.arb.ca.gov/regact/2008/ghghdv08/ghgfro.pdf. Accessed May 5, 2011.

ATA (American Trucking Association). 2007. New Generation Wide Based Single Tires. Position Paper, Brian Routhier, Revision 9, December 21.

ATDynamics. 2010. Four East Coast Trucking Fleets Put TrailerTails® on Tractor-Trailers to Reduce Fuel Consumption in Advance of the Holiday Season. Arlington, Virginia. Available at http://www.atdynamics.com/trailertail_faq.htm. Accessed January 1, 2011.

DOE (U.S. Department of Energy). 2006. 21st Century Truck Partnership Roadmap and Technical White Papers. Doc. No. 21CTP-003. December. Washington, D.C.

DOE. 2007. 21st Century Truck Partnership. Project Quad Sheets. Doc. No. 21CTP-004. February. Washington, D.C.

DOE. 2010a. Effort to Reduce Truck Aerodynamic Drag Through Joint Experiments and Computations. Project No. V5506 (K. Salari, Principal Investigator). November. Washington, D.C.

DOE. 2010b. Vehicle System Simulation and Testing. Multi-Year Program Plan, 2011-2015. December. Washington, D.C.: Vehicle Technologies Program.

DOE. 2011. White Paper Draft on Vehicle Power Demands. Washington, D.C.: Office of Vehicle Technologies.

EPA (Environmental Protection Agency). 2011. SmartWay Transport Verified Technologies. Available at http://www.epa.gov/smartwaylogistics/transport/what-smartway/verified-technologies.htm. Accessed April 20, 2011.

EPA/NHTSA (Environmental Protection Agency/National Highway Traffic Safety Administration). 2010. Greenhouse Gas Emissions Standards and Fuel Efficiency Standards for Medium- and Heavy-Duty Engines and Vehicles. EPA-HQ-OAR-2010-0162. NHTSA-2010-0079. FRL-9219-4. November 30.

EPA/NHTSA. 2011. Greenhouse Gas Emissions Standards and Fuel Efficiency Standards for Medium- and Heavy-Duty Engines and Vehicles. Federal Register, Vol. 76, No. 179, September 15. Available at http://www.nhtsa.gov/staticfiles/rulemaking/pdf/cafe/2011-20740.pdf.

Navistar. 2008. Navistar Awarded Department of Energy Funding to Develop Aerodynamic Trailers. Available at http://ir.navistar.com/releasedetail.cfm?ReleaseID=354505.

NESCCAF/ICCT (Northeastern States Center for a Clean Air Future/International Council on Clean Transportation). 2009. Reducing Heavy-Duty Long Haul Combination Truck Fuel Consumption and CO_2 Emissions. Miller, ed. October.

NRC (National Research Council). 2008. Review of the 21st Century Truck Partnership. Washington, D.C.: The National Academies Press.

NRC. 2010. Technologies and Approaches to Reducing the Fuel Consumption of Medium- and Heavy-Duty Vehicles. Washington, D.C.: The National Academies Press.

ORNL (Oak Ridge National Laboratory). 2009. Effect of Wide-Based Single Tires on Class-8 Combination Fuel Efficiency, Oscar Franzese, Helmut Knee, and Lee Slezak. October 28.

TIAX. 2009. Assessment of Fuel Economy Technologies for Medium- and Heavy-Duty Vehicles. Report prepared for the National Academy of Sciences by TIAX LLC. July 31. Cupertino, Calif.

TT (Transport Topics). 2011. Low Rolling Resistance Tires Gain Traction with Carriers Seeking to Save Fuel, Money. S. Bennett. June 20.

6

Engine Idle Reduction

INTRODUCTION

Overnight and workday idling of trucks is estimated to consume well over 2 billion gallons of fuel annually in the United States (DOE, 2011). Extended idling by commercial trucks costs truck owners about $6 billion annually and wastes more than 1 percent of the petroleum used in the United States. Much of this petroleum use could be avoided by installing idle reduction technologies, adopting more efficient freight-scheduling policies, or in some cases simply turning off the engines. In addition to the fuel consumed, idling produces emissions and noise. Overnight idling is used to keep a truck's cab and/or sleeper heated or cooled, to keep the fuel and engine warm in winter for easier starting, to provide power to operate electrical appliances, and to keep the batteries charged. A long-haul truck idles an estimated 1,800 to 2,400 hours per year when parked overnight (DOE, 2011). Workday idling includes creeping along in queues at ports and depots. Every hour that a truck idles unnecessarily is equivalent in fuel consumption to about 4 to 5 miles of driving and adds an estimated $0.15 in maintenance costs.[1]

Solutions to eliminate overnight idling are shown in Table 6-1. They include engine controls (for automatic shutdown/start-up systems), fuel-operated heaters (FOH), auxiliary power units (APUs), battery-powered heating and cooling systems, and shore power or truck stop electrification (also called Electrified Parking Spaces [EPS]) (NRC, 2010). The attributes provided by each of these idle reduction technology solutions are shown in the table. The cells in Table 6-1 are shaded light green to indicate favorable attributes, yellow shading indicates mild drawbacks, and dark orange indicates major drawbacks. The U.S. Environmental Protection Agency (EPA), through its SmartWay program's contacts with truck manufacturers and fleets, estimates that approximately 30 percent of the existing fleet has some type of idle reduction technology.[2] The most prevalent onboard technology, determined by a survey conducted by the American Transportation Research Institute (ATRI) of 55,000 truckers, was direct-fuel-fired heaters (32 percent), followed by battery-powered air conditioners (24 percent), while auxiliary power units APUs were used by 12 percent of the respondents (American Transportation Research Institute, 2006).

The overall goal of the engine idle reduction portion of the 21st Century Truck Partnership (21CTP) is to reduce fuel use and emissions produced by idling engines. The metric for this goal, which was provided to the committee in a November 2010 presentation and in previous versions of 21CTP white papers since 2007, was an 85 percent reduction in idling fuel consumption in the period 2002 to 2017.[3] An August 2010 white paper draft revised the goal to a two-thirds reduction, based on discussions with industrial partners on the most appropriate and achievable goals using a variety of factors (DOE, 2010). However, in the February 2011 "21CTP White Paper on Idle Reduction," the quantification of this goal was deleted (DOE, 2011). To date, the 21CTP has not been able to carry out surveys to measure quantitatively the progress being made toward the previously stated goal owing to the absence of funding for such studies. Only qualitative observations can indicate the increased adoption rate of these devices for which the primary drivers have been (1) the high cost of diesel fuel and (2) the regulatory measures adopted in some states and cities to reduce idling.

There are restrictions on engine idling in 46 states and jurisdictions. Many states have strict regulations in more than one city, whereas the regulations of other states are statewide. Sometimes the regulations for a city are different from those of the state. Some of the localities have started enforcing

[1] Glenn Keller, ANL, "Idle Reduction Accomplishments," presentation to the committee, November 15, 2010, Washington, D.C.

[2] Answers provided by Ken Howden, DOE Office of Vehicle Technologies, to committee questions 5(b).

[3] Glenn Keller, ANL, "Idle Reduction Accomplishments," presentation to the committee November 15, 2010, Washington, D.C. This overall goal has been used to formulate the 21CTP technical goals.

TABLE 6-1 Comparison of Attributes of Idle Reduction Systems

Idle Reduction Technology	Heating	Cooling	Electric	Requires Recharge Infrastructure	Service Fee	Emissions Control Needed?	Idle Time Avoided per Year	Fuel Use (gal/hr)	% Benefit	Cost
Engine Control	Yes	Yes	Yes	No	No	No	1,500 to 2,400	~0.5	3%	$1,000 to $4,000
Heater	Yes	No	No	No	No	No	500 to 800	0.2 to 0.3	1.3 to 2.3%	$1,000 to $3,000
Auxiliary Power Unit	Yes	Yes	Yes	No	No	In California	1,500 to 2,400	0.2 to 0.3	4 to 7%	$6,000 to $8,000
Battery	Yes	Yes	Some	Yes	No	No	1,500 to 2,400	—	5 to 9%	$3,000 to $8,000[a]
Shore power	Yes	Yes	Yes	Yes	Yes	Yes	1,500 to 2,400	—	5 to 9%	~$100

NOTE:
Green: Favorable Attribute
Yellow: Mild Drawback
Dark Orange: Major Drawback
[a] May require a diesel particulate filter, at an additional cost of $3,000.
SOURCE: NRC (2010), Table 5-24, p. 125.

anti-idling regulations more aggressively (DOE, 2011; Delphi, 2010). The California Air Resources Board (CARB) adopted a rule that since 2007 has not only limited idling to 5 minutes, but also requires automatic shutoff devices. Philadelphia bans the idling of heavy-duty diesel-powered motor vehicles, with exceptions made during cold weather.

FUNDING

The NRC Phase 1 report did not contain a breakdown of the 21CTP budget for idle reduction through 2008 (NRC, 2008). Likewise, the 21CTP budget for idle reduction efforts was not available for FY 2009 and FY 2010 and the FY 2011 President's Congressional Budget Request (see Table 1-2 in chapter 1 of this report). Similarly, a budget forecast for meeting the idle reduction goals that extend through 2017 to reduce fuel use and emissions produced by idling engines was not provided to the committee. Therefore, an assessment of the probability of achieving the goals for idle reduction technologies cannot be made at this time. However, as noted in the section, "Goals," the American Recovery and Reinvestment Act (ARRA) of 2009 did provide funds for idle-reduction related projects.

GOALS

In the NRC Phase 1 report, seven 21CTP goals for engine idle reduction were addressed (NRC, 2008). In its November 15, 2010, presentation to the committee, the 21CTP slightly modified these goals for 2010 and added one new goal. Those goals are presented in bold type in this section. The action items addressing these goals provide a path toward accomplishing the overall goal of the idle reduction portion of the 21CTP. The status of action items addressing each of these goals is discussed in this section.

21CTP Engine Idle Reduction Goal 1. Continue industry/government collaboration to promote the development and deployment of cost-effective technologies for reducing fuel use and emissions due to idling of heavy-duty engines.

For more than a decade, the Department of Energy (DOE) has carried out cooperative research and development (R&D) to characterize and address the reduction of fuel use and emissions during the idling of heavy-duty engines. The NRC (2008) Phase 1 report discussed the R&D work focused on idling reduction technologies. All of the 21CTP partners, both government and industry, have ongoing roles in developing and implementing a coherent program of idling reduction, as described below:

- The DOE analyzes technology needs and performs the appropriate R&D to help make cost-effective technology available for implementation. The results of the analyses enable a systematic comparison of potential strategies, including emission credits, positive incentives, and regulations to install appropriate idle reduction technology.
- The Environmental Protection Agency (EPA) and the Department of Transportation (DOT) have been named to lead the effort in implementation.

- A major goal of the DOD is to reduce the logistical footprint of deployed forces, primarily though savings in fuel consumption.
- The 21CTP industrial partners and their suppliers need to work together to make idle reduction technologies an affordable and cost-effective part of their vehicles' design, seamlessly integrating their choice of technologies into their products.
- Local, state, and regional air quality agencies have teamed up with the EPA and DOE's Clean Cities coalitions to form regional collaboratives to address diesel engine emissions, with idling reduction as a major component of their efforts.

21CTP Engine Idle Reduction Goal 2. Expand the current educational programs for truck and bus owners and operators to implement enabling technologies and operational procedures to eliminate unnecessary idling.

The DOE has established or encouraged the following initiatives to educate stakeholders on the benefits of idle reduction and the opportunities to implement technologies and procedures to eliminate unnecessary idling:

- The EPA, through the SmartWay Transport Partnership, has sponsored numerous idle reduction outreach efforts and events, including technical papers, articles, and presentations.
- The DOE Clean Cities Program has sponsored outreach activities to educate Clean Cities' coordinators and fleet managers about the benefits of idling reductions and the technologies available, through white papers, webcasts, and presentations at various professional meetings. The DOE has produced idle reduction fact sheets and other educational materials.
- Through the Clean Cities Program, the DOE has broadened its involvement in idling reduction to include light- and medium-duty vehicles in addition to heavy-duty vehicles.
- The "National Idling Reduction Network News" is a DOE-sponsored electronic newsletter whose primary distribution each month reaches almost 1,500 readers.
- The DOE has produced idling reduction fact sheets and other educational materials to make drivers and fleet owners aware of reasons not to idle.
- The following DOE and industry publications address idling reduction: Argonne National Laboratory's *Idling: Cruising the Fuel* In*efficiency Expressway* (ANL, 2009) and Cummins' *Idle Talk: How the Regulations Affect You* (Cummins, 2008).

21CTP Engine Idle Reduction Goal 3. Investigate a mix of incentives and regulations to encourage trucks and buses to find other more fuel-efficient and environmentally friendly ways to provide for their power needs at rest.

The ARRA of 2009 has provided the following funding for idle-reduction-related projects (DOE, 2011):

- $65 million for the purchase and installation of idle reduction equipment for on-road diesel vehicles and educational outreach about the benefits of idling reduction (see Goal 2). This project includes APUs, fuel-operated heaters, battery-powered air conditioners, engine block heaters, and engine start-up/shutoff idle control systems and other emission reduction projects, such as engine re-powers (the replacement of an in-use, existing engine with a remanufactured engine or a new engine with lower emissions), replacements, or installation of diesel oxidation catalysts in cases where these projects were bundled with idling reduction projects. Examples of projects funded include the following:[4]
  —Installing 163 diesel-fired heaters in the city of Chicago fleets and 155 units in the city of Portland, Oregon, fleets;
  —Augmenting state funding for the installation of 562 idle reduction technologies by the Wisconsin Department of Commerce program that competitively awards money for APUs to truckers;
  —Providing funding in Nebraska to equip approximately 187 vehicles with EPA-verified idling reduction equipment;
  —Adding fuel-fired heaters to school buses in Mississippi, Wisconsin, Minnesota, Michigan, Maryland, South Dakota, and North Dakota; and
  —Retrofitting of 180 long-haul trucks with APUs by the Colorado Department of Public Health and Environment [5,6]
- $32 million for truck stop electrification (TSE). The funds will provide for the purchase and installation of wayside single-system (no onboard equipment required) and dual-system EPS (DOE, 2011).
  —A single-system EPS supplies all needed services through a duct inserted into the cab window. Single-system electrification requires no retrofit on the truck, and therefore minimal upfront cost by the user;
  —A dual-system EPS is simply a plug at a parking spot that enables the trucker to tap into the electric power grid to power onboard electrical devices. Dual-system electrification involves installing some combination of an inverter/charger, electric engine block heater, electric fuel heater, and electric heating/cooling device for the cab and sleeper conditioning, and electric idle control on the truck.
  Currently, the single system is more widespread.

[4] 21CTP response to committee questions from its March 31-April 1, 2011 meeting.

[5] Glenn Keller, ANL, "Idle Reduction Accomplishments," presentation to the committee, November 15, 2010, Washington, D.C.

[6] Answers provided by Ken Howden, DOE Office of Vehicle Technologies, to committee question 20(a).

Approximately 165 electrified parking spaces have been completed to date. Examples of TSE projects that are being funded include the following:
—30 AirDock units on the Maine Turnpike ($1.2 million from EPA ARRA);
—90 Shorepower™ units off Interstate 10 in Arizona;
—14 CabAire units in New Haven, Connecticut; and
—CabAire units on the Pennsylvania Turnpike.[7]

- Cascade Sierra Solutions has received a $22.2 million grant for a 3-year program known as Shorepower Truck Electrification Project (STEP) for the construction of approximately 1,200 electrified parking spaces at 50 truck stops across the United States. These DOE ARRA funds are matched with private-sector funds. In addition, approximately $10 million is being provided in purchase rebates of up to 20 percent of the cost of idle reduction equipment for users of the STEP network of electrified parking spaces.[8]
- Grants from the Diesel Emissions Reduction Act (DERA) of 2005 funds have been used at the DOE and EPA to fund a variety of idling reduction projects, such as the following:
—$1.13 million each to Cascade Sierra Solutions, Community Development Transportation Lending Services, and Owner-Operator Independent Drivers Association for revolving loans and low-cost financing for emission- and idling-reduction equipment for trucks (EPA DERA).

The EPA, working with the DOT, states, and private lenders, is developing innovative, market-based, and sustainable funding opportunities, such as low-interest loans through EPA's SmartWay Program, to replace traditional grants to allow the truck and rail industries to purchase and use idle reduction technologies. Low-interest loans are expected to be a more sustainable incentive than grants, which typically expire after a period of time. Low-interest loans allow truck owners who are unable to make initial investments because of limited capital to pay over time with their fuel savings.

While all of these developments were under way, a major provider of EPSs, IdleAire, filed for bankruptcy in June 2008 and shut down operations. However, by the summer of 2010, Convoy Solutions, LLC, dba IdleAire, began reopening EPS sites. The DOE indicated that all new efforts directed toward EPS are focused on locating the EPSs along major freight corridors.[9]

[7] Glenn Keller, ANL, "Idle Reduction Accomplishments," presentation to the committee, November 15, 2010, Washington, D.C.; and answers provided by Ken Howden, DOE Office of Vehicle Technologies, to committee question 20(b).

[8] 21CTP response to committee questions from its March 31-April 1, 2011, meeting.

[9] Glenn Keller, ANL, "Idle Reduction Accomplishments," presentation to the committee, November 15, 2010, Washington, D.C.

The patchwork of anti-idling regulations nationally has been an impediment to broader use of anti-idling measures. The EPA has no legal authority to promulgate anti-idling laws, or any driving time or behavior limits on truck drivers. The EPA's legal authority rests with promulgating emissions standards to vehicles and engines. (21CTP Response to NRC [2008] Recommendation 6-4; see Appendix C in this report.) However, the regulatory environment is currently changing. Specifically, as noted later in this chapter, the proposed EPA and NHTSA, "Greenhouse Gas Emissions Standards and Fuel Efficiency Standards for Medium- and Heavy-Duty Engines and Vehicles" (EPA/NHTSA, 2010a) contain a provision indicating that, if a manufacturer chooses to use idle reduction technology to meet the standard, then it would require an automatic main engine shutoff after 5 minutes to help ensure that the idle reductions are realized in-use.

Finding 6-1. The DOE, EPA, and DOT have funded a wide variety of idle reduction projects focused on implementation. A consolidated list of the funding provided for these projects was not provided to the committee, however, and the effectiveness of these projects could not be evaluated. The national patchwork of anti-idling regulations is an impediment to broader use of anti-idling measures.

Recommendation 6-1. The DOE, EPA, and DOT should develop a consolidated list of the funding provided for the idle reduction projects, review the effectiveness of these projects, and formulate a coordinated and consistent plan to encourage the adoption of idle reduction technologies to meet the goal of reducing fuel use and emissions produced by idling engines by at least two-thirds by 2017. The EPA and DOT should work to find incentives for states to promulgate uniform anti-idling regulations.

21CTP Engine Idle Reduction Goal 4. Facilitate the establishment of consistent electrical codes and standards that apply to both on-board and stationary electrification technologies.

The NRC Phase 1 report on review of the 21CTP described the status of the relevant electrical codes at that time (NRC, 2008). The report focused on changes or additions that were needed in two areas: (1) onboard wiring for the truck and (2) Electrified Parking Spaces. The onboard wiring codes were established in Society of Automotive Engineers (SAE) Standard J2698, finalized and published in 2008 and titled "Primary Single Phase Nominal 120 VAC Wiring Distribution Assembly Design-Truck and Bus." This SAE standard has addressed the known issues with onboard wiring for the truck identified in the NRC Phase 1 report (NRC, 2008).

The National Electric Code (NEC) Part 626, titled "Electrified Parking Spaces," was approved in 2008 and addressed these topics: how to plug in, the voltages to supply, and a suggested common plug style. Part 626 clarifies that automobile parking areas are not subject to Part 626, so that the SAE J1772 coupler to power a truck is allowed but not required. Part 626 states that the receptacle needs to be a three-wire grounded type and that each truck stop parking spot needs both 208 vac and 120 vac receptacles.[10] This NEC standard has addressed the known issues with stationary electrification technologies identified in the NRC Phase 1 report (NRC, 2008).

21CTP Engine Idle Reduction Goal 5. Promote the development and demonstration of cost-effective add-on idling-reduction equipment that meets driver cab comfort needs, has a payback of 2 years or less, and produces fewer emissions of NO_x and PM than a truck meeting 2010 emission standards.

The NRC Phase 1 report noted that Webasto Airtop 2000 diesel-fueled cab heaters tested by Schneider National provided a 2.4 percent improvement in fuel economy and a payback of less than 2 years for a fuel price of $2.40/gal. The current list price of the Webasto cab heater is $1,745. Although the DOE has not worked with Schneider since the initial testing program, Schneider has indicated that all 6,000 trucks purchased since 2003 had been retrofitted or factory installed with cab heaters, and 80 percent of its fleet, or 8,000 trucks, were to be retrofitted by the winter of 2007/2008 (Maronde and Slezak, 2006). Webasto and Bergstrom battery-powered cooling systems based on phase-change medium that is charged during normal operation of the truck's air conditioning system were also evaluated by Schneider National. It was concluded at that time that these cooling systems needed further work, which was not specified, before they could be widely deployed.

DOE's earlier development of phase-change materials for stand-alone cab cooling had identified deficiencies with this concept. Subsequently, the resolution of these deficiencies has led to the commercial release of the Webasto Blue Cool product. The Blue Cool product was reported to be the first thermal storage APU with shore power connectivity. Cab comfort and electrification are provided without idling. Tests confirmed that Blue Cool provided sufficient cooling under most ambient conditions. The in-cab-mounted air handler delivers chilled air to the bunk for up to 10 hours without consuming any fuel. Blue Cool charges itself while the vehicle is in motion and does not require additional batteries. The electrical load of the circulation pump and fans during cooling is 3.5 to 10 A. Webasto claims that Blue Cool has the shortest return on investment among idle reduction products, although the DOE did not perform testing or analysis to confirm this claim, and sufficient information was not available to determine if this system met the 2-year payback objective (Webasto, 2010). The current price of the Webesto Blue Cool system ranges from $5,295 to $6,595, depending on whether an Airtop 2000 heating unit is also provided.

A team consisting of Espar, Navistar, and Walmart evaluated 20 trucks with Espar (2010) Airtronic bunk heaters, Espar engine preheaters, and Bergstrom Nite battery-powered, electric A/C units, and 5 trucks with ThermoKing Tripac APU systems for heating, cooling, and accessory power. Both configurations provided acceptable performance. Following this evaluation, Walmart retrofitted its entire fleet of 7,000 trucks with Thermo King TriPac units as a result of Walmart's settlement in 2006 with the EPA for clean air violations related to idling trucks at stores in Connecticut and Massachusetts. The TriPac unit includes a Tier IV-compliant 2-cylinder diesel engine with a diesel particulate filter (DPF) for state of California operation, an alternator for truck battery charging, an A/C unit, and a fuel-fired heater (Thermo King, 2010). The TriPac unit is designed to meet anti-idling and emissions regulations nationwide, including CARB requirements and is claimed to be the sales leader of APU systems in the industry (Thermo King, 2010).

With the tightening of diesel emissions regulations in 2010, some of the diesel APUs are no longer available; others, like the truck diesel engines, have had to be equipped with particulate filters and NO_x traps, thereby increasing their costs and making the achievement of the 2-year payback goal more difficult (DOE, 2011).

The military needs APUs to reduce in-field fuel consumption and related logistical costs and to reduce thermal and audible identification signatures during silent watch. APUs are quieter than idling primary engines, and they have a reduced thermal signature, making them less detectable on the battlefield. The U.S. Army Communications-Electronics Research, Development, and Engineering Center (CERDEC) Laboratory is working to demonstrate the feasibility of a diesel engine APU on the M915A5 long-haul tractor. Diesel APUs are being considered since "silent watch" is not a requirement for these trucks. In FY 2008, contracts were awarded to Dewey Electronics and Cummins Power Generation to develop stand-alone APU/environmental control unit (ECU) prototypes. A contract to Red Dot Corporation is expected to conclude in FY 2011 with the demonstration of two APU/ECU prototypes integrated onto two M915A5 tractor trucks. The resultant system design is projected to save up to 870 gal/yr and to achieve a simple payback period of less than 5 years (assuming fully burdened fuel cost of $15/gal). Reducing fuel use is key because approximately two-thirds of the ground fleet is used to deliver fuel to the other third in the battlefield.

With $500,000 funding from the EPA, the North Carolina State University (NCSU) conducted a 34-month Truck OEM APU Prep Kit Design and Installation Project that was concluded in August 2008 (Tazewell et al., 2008). In this project, Volvo was awarded a contract to develop a Prep Kit to facilitate idle reduction technology installations and demonstrate APUs in at least 20 trucks in the field study and to track idle reduction usage, truck idling, and driver acceptance. The

[10] Answers provided by Ken Howden, DOE Office of Vehicle Technologies, to committee question 19(d), March 1, 2011.

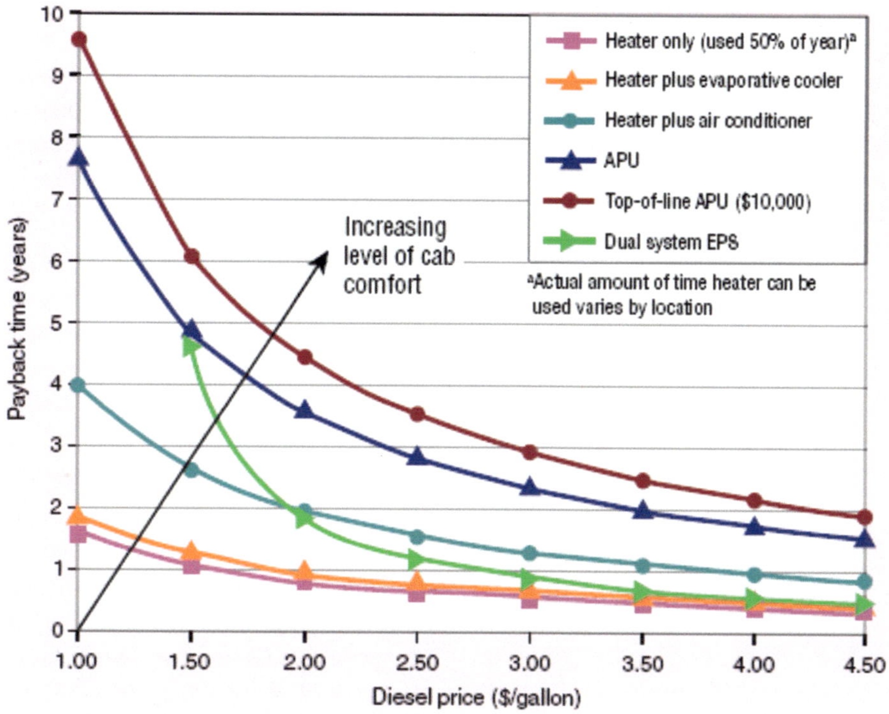

FIGURE 6-1 Payback time versus fuel price, by device, used 2,000 hours per year. Acronyms are defined in Appendix I. SOURCE: L. Gaines and D. Santini, Economic Analysis of Commercial Idling Reduction Technologies, Argonne National Laboratory. Available at http://www.transportation.anl.gov/pdfs/TA/372.pdf.

APUs consisted of three components, a Kubota Z482 2-cylinder water-cooled diesel engine, a generator, and a heating, ventilation, and air conditioning (HVAC) system.

The field study was divided into two fleets: the first, Fleet A, had a self-reported annual idling rate of 2,500 hours using Volvo's largest cabs, with predominately single drivers; the second, Fleet B, had a self-reported annual idling rate of 800 hours using Volvo's midsize cab with predominately team drivers. The key results from this study were as follows:

- Annual fuel use was reduced for all stops by 22 percent and 5 percent for Fleets A and B, respectively.
- NO_x emissions were reduced for all stops by 46 percent and 14 percent for Fleets A and B, respectively.
- Research concluded that 100 percent usage of the APU instead of the base engine could result in a 36 to 47 percent reduction in fuel use, an 80 to 90 percent reduction in NO_x emissions, and a 10 to 25 percent reduction in particulate matter (PM) emissions.

The study concluded that driver behavior plays a significant role in determining APU benefits. The data showed that APUs were used by single drivers for an average of 59 percent of the idling time and by team drivers for an average of only 25 percent of the idling time.

The economics of APUs are sensitive to initial APU costs, idling time, actual APU usage, and fuel costs. The NCSU study found engine idle fuel flow rates of approximately 0.6 gal/h instead of the 0.8 gal/h that has been quoted by the EPA and NHTSA, and APU fuel flow rates of approximately 0.32 gal/h instead of the 0.2 gal/h that has been quoted by the EPA and NHTSA (EPA/NHTSA, 2010b, 2011b). Using an initial APU cost of $8,400 and an annual idle time of 2,130 hours and $4.00/gal fuel cost, the simple payback period is 3.5 years.

The Argonne National Laboratory (ANL) has studied payback periods for several idle reduction systems and provided the graph in Figure 6-1 showing the payback time versus fuel price.[11] For a $4.00/gal fuel price, the top-of-the-line APU that has an initial cost of $10,000 and is used 2,000 hours per year has a projected payback period of 2.2 years. Other APUs with lower initial costs have a projected payback period of less than 2 years (approximately 1.8 years).

These projections by ANL show shorter payback periods than the projections made by NCSU, because NCSU found that measured base engine idle fuel flow rates were lower than generally assumed and that measured APU fuel

[11] See Gaines and Santini at http://www.transportation.anl.gov/pdfs/TA/372.pdf. 2010.

flow rates were higher than generally assumed. The NCSU report (Tazewell et al., 2008) concluded that APUs generally used more fuel than the published amounts, whereas the 20 newer trucks in this study used less fuel than the commonly published 1 gal/h at idle. The DOE cites ranges for the base engine fuel flow rates between 0.64 and 1.15 gal/h, depending on the idle speed and use of air conditioning. The 21CTP has not as yet determined the representative values for the payback-period calculation, because this would require more study with a wider range of truck models, ages, and fleets.[12]

The EPA and NHTSA, in the regulatory impact analysis for the "Greenhouse Gas Emissions Standards and Fuel Efficiency Standards for Medium- and Heavy-Duty Engines and Vehicles," have provided the following values for an APU that can be used to calculate payback period (EPA/NHTSA, 2010b, 2011b):

- Annual idling hours: 1,800 hours;
- Base engine fuel usage: 0.8 gal/h; and
- APU fuel usage: 0.2 gal/h.

The annual fuel savings for an APU were calculated as follows:

$$\text{Annual Fuel Savings} = 1{,}800\text{ h} \times (0.8 - 0.2\text{ gal/h}) \times \$4.00/\text{gal} = \$4{,}320/\text{yr}.$$

Therefore, assuming an APU capital cost of $8,400, as before, a simple payback period of approximately 2 years results. At a lower fuel cost of $3.00/gal, the payback period is extended to 2.6 years.

An annual maintenance cost for an APU was not provided by the DOE, so net maintenance cost savings was not included in the calculation of the simple payback period. Net maintenance cost savings, consisting of maintenance cost savings for the main engine (1,800 h/yr × $0.15/h = $270) offset by maintenance cost for the APU, would affect the payback period by approximately 6 percent, depending on the APU maintenance cost and fuel cost.

The DOE informed the committee in March 2011 that it does not currently have plans to address production-level systems not meeting the 2-year payback period, because market forces will likely drive improvements in these systems. For new technology development, the DOE is working with the SuperTruck program participants on various idle reduction concepts and will encourage the participants to consider the potential costs and paybacks of these concepts wherever possible.[13]

Finding 6-2. A variety of add-on idle reduction systems are commercially available. In earlier studies, a diesel-fuel-fired heater met the 2-year payback goal, but full-function systems (with heating and cooling) had payback periods extending beyond 2 years, owing primarily to high initial cost and less than 100 percent usage during idling. Recent studies of the payback period by the EPA and NHTSA, ANL, and NCSU have provided a range of results related to different assumptions for initial costs, truck engine idle time and APU fuel flow rates, and actual usage times. These studies have projected simple payback periods ranging from 2 to 3.5 years.

Recommendation 6-2. The DOE should conduct a study that includes wide ranges of truck models, ages, and fleets to determine payback periods for the range of commercially available add-on idle reduction systems. The DOE should continue to encourage the deployment of add-on idle reduction systems through communications to manufacturers and end users.

21CTP Engine Idle Reduction Goal (New). Reduce the thermal load of the truck heating, ventilating, and air conditioning (HVAC) system during driver rest periods through implementation of efficient cab insulation systems and low thermal transmission glazing.

A reduction of cabin energy load, through the addition of insulation and window glazing, coupled with controls to reduce peak energy loads, could enable the downsizing of APUs and battery-powered systems to reduce cost and weight while enhancing their performance. To assess the HVAC load reduction potential in truck sleeper cabins, the DOE funded the development of CoolCalc, an analysis tool that allows users to create sleeper cabin models and predict cabin temperatures in different environmental conditions. The main objective of the project was to identify and evaluate design opportunities to reduce the thermal load inside the truck tractor cabs and to enable advanced idle reduction technologies. The DOE has released the software to industry partners.

The DOE also funded the National Renewable Energy Laboratory (NREL) CoolCab project that investigated insulation and reflective glazing to reduce the thermal load and improve the cab's climate-control efficiency. This project included the thermal testing of Volvo 770 and Kenworth T660 cabs. The results of the Kenworth T660 cab thermal soak test were used to validate the CoolCalc model of the vehicle. Predicted peak soak average air temperature on sunny days was within 0.4°C of the measured value. Development of a CoolCalc model of the Volvo 770 cab is under way. Thermal test results show the heating load in a cab sleeper could be reduced up to 20 percent with high R-value insulation. For FY 2011, the CoolCalc models will be used to quantify the impact of thermal load reduction technologies, such as insulation and reflective glazing, on cooling and heating thermal loads. These results will be used to determine

[12] Answers provided by Ken Howden, DOE Office of Vehicle Technologies, to committee question (25).

[13] Answers provided by Ken Howden, DOE Office of Vehicle Technologies, to committee questions, March 1, 2011.

TABLE 6-2 Status of Navistar Auxiliary Power Unit System Versus Program Goals

Parameter	Unit	Goal	Status
Temperature	TMC RP-432[a]	Maintain temperature in range for 10 hours	Met
Truck idle time	Hours per year	<200	Met
Idle fuel consumption	Gallons per hour	<0.25	Met
Emissions	Tier 4	Comply	Met
Particulate matter	Grams per hour	<0.2	Met
NO_x	Grams per hour	<25	Met
Price	18 month payback	\$5,000	Not met
Maintenance	Service interval	500 hours	Met
5-year life	B10	10,000 hours	Met

[a] Specifies minimum 68°F/maximum 78°F sleeper temperatures at 0°F/100°F ambient temperatures.
SOURCE: Data from Casey (2008).

the reduction in sizing of the APU or other idling reduction technology.[14]

21CTP Engine Idle Reduction Goal 6. Produce a truck with a fully integrated electrically powered truck cab HVAC system to reduce idling-reduction system component duplication, weight, and cost, by 2012.[15]

The DOE recognizes that costs could be reduced through the complete, nonduplicative integration of idle reduction equipment into the original truck design. Effectiveness in reducing workday idling could be improved by hybridization and by development of systems that reduce idling during creep modes.

To address this goal initially, the DOE funding helped Navistar to complete engineering development to provide the option of ordering factory-installed APUs as original factory-installed equipment (Casey, 2008). Navistar's idle reduction system had four elements:

- Auxiliary power unit: 2-cylinder water-cooled diesel generator-set generating 6 kW of power at 120 V A/C, purchased from Mechron;
- Electric air conditioner: A stand-alone system mounted in the sleeper compartment;
- Cab and engine heater: Fuel-fired coolant heater purchased from Espar and integrated into the truck's existing coolant loop; and
- Improved cab insulation.

[14] Answers provided by Ken Howden, DOE Office of Vehicle Technologies, to committee question 19(d), March 1, 2011.

[15] "Produce a truck" was interpreted by the committee to mean to design, engineer, and manufacture a truck for sale.

In addition, Navistar also developed an aftermarket APU wiring accommodation kit as an alternative to its factory-installed APU system.

Navistar had a goal at the beginning of the program to sell 2,000 factory-installed APU idle reduction systems under this program. The goal was not met with the factory-installed APU systems; only 65 units were sold. However, Navistar did sell 2,628 trucks with wiring accommodation kits for aftermarket APUs, indicating that the goal of 2,000 trucks with APUs installed by the end of 2007 was achieved. The goal of 2,000 factory-installed idle reduction systems was exceeded by the 4,325 factory-installed fuel-fired heater systems. However, these factory-installed idle reduction systems did not address the nonduplicative aspect of this goal.

The status of the Navistar APU system versus the goals for the program are shown in Table 6-2. Fuel savings were tracked on five fleet vehicles, and fuel economy increased from 6.38 mpg to 6.99 mpg, providing a 9.6 percent improvement. The Navistar factory-installed APU system did not meet the 18-month payback goal. The cost advantage of aftermarket APU units was likely due to the addition of federal excise tax applied to factory-installed APUs as well as margin added by Navistar for purchasing and installing the unit. However, following this project, the Energy Improvement and Extension Act of 2008, which is a part of the Emergency Economic Stabilization Act of 2008 (Public Law 110-343), modified the Internal Revenue Service (IRS) code and allows for APU units to be exempt from paying the 12 percent federal excise tax. At the conclusion of this project, Navistar announced in March 2008 that it was developing the MaxxPower APU, a 1-cylinder APU to provide a more cost competitive APU offering while meeting California's 2008 APU emissions requirements.

To address the goal of a fully integrated, electrically powered truck cab HVAC, the DOE, through the ANL, established the Caterpillar More Electric Truck (MET) program

(Lane et al., 2004), which was initiated in 2000 and ended in 2007. The objective of the program was to reduce loads on the engine by using electrically powered accessories including the HVAC, water pump, brake air compressor, oil pump, and cooling fan with revised cooling system (Stone et al., 2004). The results showed a 1.3 percent reduction in fuel consumption on the road due primarily to the electric water pump and electric brake air compressor, and a 2.7 percent reduction in fuel consumption during steady state conditions due to the electric cooling fan with a revised cooling system.

The DOE concurred with Recommendation 6-7 from the NRC (2008) Phase 1 report (see Appendix C)—to continue R&D of the system components used in the More Electric Truck program in order to provide further improvements in idle reduction. In 2007, additional work was anticipated to reduce fuel consumption further in the following areas:

- Mild hybrid energy storage using nickel metal hybrid batteries (NiMH);
- Advanced cooling system components (electric thermostat valve and cooling fan, high-efficiency aftercooler); and
- Decoupling the air compressor from the engine.

The DOE was not able to apply any funding to this program, so as a result, no significant activity toward achieving the 2012 goal of a fully integrated, electrically powered truck cab HVAC system to reduce idling reduction component duplication, weight, and cost has been conducted.[16] However, the SuperTruck program is expected to pursue the concept of integrated systems similar to the More Electric Truck program. All three of the SuperTruck program teams, Cummins-Peterbilt, Detroit Diesel, and Navistar, will be addressing idle reduction, as discussed in Chapter 8 of this report.[17,18,19]

21CTP Engine Idle Reduction Goal 7. Develop and demonstrate viable fuel cell APU systems for military and other users, in the 5-30 kW range, capable of operating on JP-8 fuel with 35 percent efficiency (based on the fuel's heating value) by 2015.

Delphi's solid oxide fuel cell (SOFC) APU converts chemical energy in conventional fuels directly into useful electrical power without combustion, resulting in minimal criteria emissions. Delphi has been developing the SOFC since 2000, and is currently working on the fourth generation.[20] Cummins plans to use Delphi's SOFC APU for hotel loads for idle reduction as part of its SuperTruck program. This fuel cell will use ultralow-sulfur diesel fuel. Similarly, Detroit Diesel has also shown a fuel cell APU as a component of its SuperTruck program. Navistar is not planning to use an APU for the SuperTruck program, because a hybrid drive system will be used to charge the large-capacity batteries to provide power for idle reduction functions.

Delphi Corporation and Peterbilt Motors recently announced the demonstration of a Delphi SOFC APU powering a Peterbilt Model 386 truck's hotel loads.[21] The Delphi SOFC APU provided power for the vehicle's electrical system and air conditioning and maintained the truck's batteries while the diesel engine was turned off. Recently, Delphi demonstrated the SOFC APU to the public during the November 2010 Hybrid Truck Users Forum (HTUF) conference in Dearborn, Michigan.[22]

The key subsystems of the SOFC are the SOFC stacks, the fuel reformer, the system controller, and the output power conditioner (Shaffer, 2004). The SOFC stacks operate at a temperature of 750°C, which results in long warm-up times, currently ranging from 2 to 5 hours, with a goal of 1 hour. With the high operating temperature, the exhaust energy is expected to be sufficient to heat the sleeper compartment at close to no-load idle, and possibly the entire passenger cabin when it is used for a break. The electrolyte of the SOFC is yttria-stabilized zirconia, a zirconium-oxide based ceramic.[23] The SOFC contains no precious metals. Delphi is currently developing the fourth-generation SOFC stack. The A-Level design APU, currently operating, contains the Gen 3 stack, while the latest B-Level design APU, which contains the Gen 4 stack, was being assembled as of May 2011. The reformer, which uses a proprietary, automotive formulation catalyst containing precious metals, was developed to produce carbon monoxide (CO) and hydrogen (H_2) under non-carbon-forming conditions. The output power conditioner converts stack voltage (22 volts for a 30-cell stack module) to the requested output voltage.[24]

The fuel flow rates and specific fuel consumption values for the SOFC APU in its typical operating mode are shown in Table 6-3 and compared to a two-cylinder diesel APU and to the truck's diesel engine continuously idling. Previously, the NRC (2010, p. 122) reported that, for carbon-based fuels,

[16] Answers provided by Ken Howden, DOE Office of Vehicle Technologies, to committee questions, March 1, 2011.

[17] Donald Stanton, Cummins, "Cummins-Peterbilt SuperTruck Program," presentation to the committee, November 9, 2010, Washington, D.C.

[18] David Kayes, "Detroit Diesel's Super Truck," presentation to the committee, September 9, 2010, Washington, D.C..

[19] Anthony Cook, Navistar, Inc., "Navistar's Super Truck Program," presentation to the committee, September 9, 2010, Washington, D.C.

[20] Dan Hennessy, Delphi, "Solid Oxide Fuel Cell Development at Delphi," presentation to the committee, January 31, 2011, Washington, D.C.

[21] Delphi, Peterbilt Test Solid Oxide Fuel Cell APU. Available at http://www.ccjdigital.com/Delphi-peterbilt-test-solid-oxide-fuel-cell. Accessed December 7, 2010.

[22] Delphi Demonstrates Solid Oxide Fuel Cell, Showcases Capability to Save Fuel and Cut Emissions During Truck Stops. Available at http://www.Delphi.com/news/pressRelesaes/pr_11_11_001/. Accessed December 7, 2010.

[23] Personal communication from Thomas Peffley, Delphi, to committee member W.R. Wade, July 7-8, 2011.

[24] Answers provided by Ken Howden, DOE Office of Vehicle Technologies, to committee questions 8(a), 6(a), and 7(a).

TABLE 6-3 Comparison of Fuel Consumption Rates for Various Types of Idle Operation

Type of Operation at Idle	Power Output (kW)	Fuel Flow Rate (gal/h)	Specific Fuel Consumption (gal/kW-h)	References
Engine idle		0.8		EPA/NHTSA (2010b)
Two cylinder diesel APU	5 kW	0.20 to 0.33	0.04 to 0.066	Table 6-4 in this chapter
SOFC APU typical A-Level design	1.5 kW	0.156	0.11	

NOTE: Acronyms are defined in Appendix I.
SOURCE: Personal communication, Thomas Peffley, Delphi, July 7-8, 2011.

the fuel-cell-powered APU can achieve the same fuel consumption improvement as that of conventional APUs. More importantly, because of the lower power output of the SOFC APU relative to the diesel APU, the specific fuel consumption (gal/kW-h) of the SOFC APU is approximately twice that of the diesel APU.

Relative to diesel APUs, the SOFC APU provides the following advantages:

- Projected to meet 2010 EPA emissions regulations (even though the reformer produces emissions),
- Very quiet (<60 dBA), and
- Projected longer maintenance intervals and better durability.

The SOFC APU also has a number of issues, including the following:

- Warm-up time of 2 to 5 hours to reach an operating temperature 750°C. Delphi has a goal of 1.5 hours.
- The SOFC APU to be kept operating at idle throughout the workday to maintain temperature and requires an idle fuel flow of approximately 50 percent of the typical operating condition fuel flow. Delphi is evaluating the use of the SOFC APU to power part of the truck electrical loads when the truck is being driven.
- Output of 1.5 kW for an A-Level build design, which is significantly below the DOE's goal and competitive 5 kW diesel APUs. Delphi is forecasting that a B-Level design will provide 3.0 kW output. Delphi has stated that its SOFC could provide up to 5 kW of power, but it believes that 3.0 kW output is sufficient, based on discussions with truck manufacturers.[25]
- A 25 percent efficiency (using diesel fuel), which is significantly below the DOE's goal of 35 percent.
- The continuing need for a desulfurizer bed with a cartridge that requires maintenance every 6 months, even when operated with ultralow-sulfur diesel (ULSD) fuel.
- Weight of 500 lb, which exceeds Delphi's target of 400 lb and equals that of a diesel APU. On a specific pounds per kilowatt/kW basis, the Delphi SOFC APU is 3.3 times heavier than a diesel APU. Delphi is working on weight reduction of all major subsystems of the SOFC APU and the truck interface and mounting structure.

The 25 percent efficiency of the A-Level design of Delphi's SOFC APU was obtained at a reported fuel flow rate of 0.156 gal/h of ULSD fuel and 1.39 kW output. The B-Level design APU is expected to improve efficiency to 30 percent, while further improvements are expected to achieve the DOE's goal of 35 percent.[26] Delphi indicated that changing from diesel fuel to JP-8 would lower efficiency because of changes in fuel processing and the fuel's high sulfur content, although tests on JP-8 have not been conducted.[27] Although JP-8 is the standard military fuel, diesel fuel is expected to be used by the APU in commercial truck applications.

Delphi told the committee that it is trying to move the SOFC APU out of the laboratory, but it did not provide a production date. The first B-Level design APU is expected to be installed on a truck that will be used in fleet service in the fall of 2011. Delphi is now focused on the commercial viability of the SOFC APU and is emphasizing the following areas:

- Manufacturability and cost reduction (with the objective of being competitive with a diesel APU; significant cost reductions of the SOFC stack are needed).

[25] Answers provided by Ken Howden, DOE Office of Vehicle Technologies, to committee question 15.

[26] The status of the technology and efficiency for SOFCs was based not only on presentations to the committee and answers supplied by the 21CTP to the committee, but also through personal communications between Wallace Wade, committee member, and representatives of Delphi (Dan Hennessy on May 18 and June 23, 2011, and Thomas Peffley on July 7 and 8, 2011).

[27] Answers provided by Ken Howden, DOE Office of Vehicle Technologies, to committee question 17(c).

- System-level durability on a test bench for the Gen 4 stack (currently 60 thermal cycles completed; issues are being addressed, particularly seal degradation, prior to continuing to the goal of 400 thermal cycles).[28]
- System-level durability and validation (440 hours and 2,200 miles of operation on a truck have been obtained as of January 2011; extended durability operation on a truck is the next step). Accelerated tests will be run to simulate the durability goals of 28,800 and 1 million miles for heavy-duty truck applications.

Delphi did not provide an estimated cost of the SOFC APU when it met with the committee. However, Delphi later indicated that the life–cycle cost of the SOFC APU (including initial cost, fuel cost, and maintenance costs) is expected to be competitive with "midrange" diesel APUs that are compliant with 2010 emissions standards (Delphi, 2010).

The DOE has provided funding for the development of the SOFC program at Delphi through the Solid State Energy Conversion Alliance (SECA) (Office of Fossil Energy) and through the Office of Energy Efficiency and Renewable Energy (EERE) Fuel Cell Technologies Program. The SOFC program began at Delphi in 2000. After initial studies, the DOE, through the National Energy Technology Laboratory (NETL), entered into a 10-year, $138 million cost-sharing program with Delphi and its partner Battelle to develop and test an SOFC APU that can be mass-produced at low cost for commercial and military applications.[29] Early development focused on the use of gasoline, natural gas, and synthetic coal gas before switching to diesel fuel. In addition, Delphi has received government funding that has been used for general system development as well as component development, including the SOFC stack. Most recently, Delphi has received the following SOFC APU-specific awards:[30]

- Tank-Automotive Research, Development and Engineering Center (TARDEC) Fuel Cell Based Ground Vehicle Auxiliary Power Units ($2.9 million; project completed February 2009);
- DOE Cummins/PACCAR SuperTruck program ($1.0 million, current program); and
- DOE R&D Demonstration of Fuel Cells (Delphi partnered with PACCAR) ($2.4 million, current program).

A battery-operated air-conditioning system is a lower-priced competitor to the Delphi SOFC APU. However, since these systems typically produce 3,000 British thermal units (Btu) to a little over 6,000 Btu cooling, they can maintain comfort in a cabin or sleeper compartment only if it is already cooled by the vehicle's engine-powered air conditioning. With only 20 to 40 percent of the performance of an APU, they generally are not capable of initial cool-down. If they are used in areas where winter heating is needed, a diesel-fuel-fired heater is a necessary additional cost.

As discussed in the NRC Phase 1 report, the DOD was supporting a variety of companies with various (1) fuel reformers, (2) SOFC, and (3) polymer electrolyte membrane (PEM) fuel cells. At that time it was reported that the DOD had two fuel cell APU programs under way:

- The U.S. Army CERDEC fuel cell APU program was focused on testing and evaluating prototypes. During FY 2007 through FY 2010, several JP-8, ULSD, and DF-2 compatible desulfurization/reformation subsystems were evaluated from the following contractors:[31]
  —IdaTech: Stand-alone desulfurizer and steam reformer integrated with low-temperature PEM fuel cell;
  —Altex Technologies: Stand-alone organic sulfur trap with steam reformer and coupled with a high-temperature PEM fuel cell;
  —Precision Combustion Inc.: Stand-alone autothermal reformer designed for use with an SOFC; and
  —Aspen Products Group: Second-generation desulfurizer integrated with an autothermal reformer for use with an SOFC.

Although feasibility and modest fuel efficiency benefits were demonstrated, the long-term reliability of components and catalyst durability remain challenges. A system development contract with Altex Technologies will conclude in FY 2011, resulting in the delivery of a 5 kWe/10 kWt co-generation system (e = electrical, t = thermal) that is compatible with field kitchen applications.

- The U.S. Army TARDEC fuel cell program had two contracts awarded under a Broad Agency Announcement to Altex Technologies Corporation for a high-temperature PEM fuel cell and United Technologies Research Center for a SOFC.[32] The contracts are for a 3-year effort to deliver a fuel cell APU that operates with JP-8 that fits under armor on the Abrams tank in FY 2013.

The U.S. Army TARDEC's National Automotive Center (NAC) demonstrated a fuel cell APU system in a Peterbilt 385 tractor. SunLine Services Group was the prime contractor, and Southwest Research Institute was the technical integrating contractor (Montemayor, 2006; DOE, 2011).

[28] Answers provided by Ken Howden, DOE Office of Vehicle Technologies, to committee questions 8(a), 6(a), and 7(a).

[29] Answers provided by Ken Howden, DOE Office of Vehicle Technologies, to committee questions, March 1, 2011.

[30] Answers provided by Ken Howden, DOE Office of Vehicle Technologies, to committee question 5(a).

[31] Answers provided by Ken Howden, DOE Office of Vehicle Technologies, to committee question 27(a).

[32] Answers provided by Ken Howden, DOE Office of Vehicle Technologies, to committee question 27(b).

Three different configurations of fuel cells were alternatively installed in the truck:

- A 5 kW solid oxide fuel cell from General Dynamics/ Acumentrics (which failed after 40 hours),
- Two 1.2 kW Ballard Nexa PEM fuel cells to provide power for the air-conditioning system and the coolant pump, and
- A 20 kW Hydrogenics PEM fuel cell to provide power for the air-conditioning system and radiator cooling fan.

The fuel cells used onboard compressed hydrogen, because liquid fuel reformer systems were not available when this program began in 2000. With the Hydrogenics system installed on the truck, a 13 percent improvement in diesel fuel economy was measured, but the amount of hydrogen used was not available.

The final report for the Sun Transit Agency program stated that a diesel reformer fuel cell hybrid electric truck remains an elusive goal. However, SOFCo, a company specializing in the development of SOFC and fuel processor technology, and Delphi were identified as leading the effort to develop a diesel reformer/fuel cell unit.

Finding 6-3. The Delphi SOFC APU provides several advantages over diesel APUs, but it has significant issues in its current development status, including the following: low efficiency of 25 percent versus the DOE's goal of 35 percent, a low demonstrated output power of 1.5 kW versus 3.0 kW believed to be sufficient by Delphi, although typical diesel APUs provide 5 kW output, limited demonstrated durability, 2- to 5-hour warm-up time to the 750°C operating temperature, and the need to keep it operating at idle throughout the workday to maintain temperature. The 10-year funding for this program expires in 2011.

Recommendation 6-3. The DOE should reassess the viability of the SOFC APU, particularly for application to the SuperTruck program, considering the following: (1) SOFC APU is still in the laboratory, (2) the low efficiency of 25 percent versus the DOE goal of 35 percent, (3) the low 1.5 kW output compared to the typical 5 kW diesel APUs, (4) the disadvantages associated with the requirement for continuous operation at 750°C, and (5) the expiration of funding from the DOE Office of Fossil Energy and EERE Fuel Cell Technologies Program of the DOE Office of Energy Efficiency and Renewable Energy after 10 years of development. The DOE should coordinate more closely with the DOD in its fuel cell APU developments to ensure that the best technology is being pursued for the 21CTP's Goal 7 in the engine idle reduction focus area; that goal relates to the development and demonstration of viable fuel cell APU systems for military and other users. (This recommendation is a follow-on to Recommendation 6-8 in the NRC Phase 1 report.)

Following the reassessment called for in Recommendation 6-3, the DOE will need to determine, if the viability of the SOFC APU is reconfirmed, whether the additional development work required to meet the SOFC APU goals can be contained within the Super Truck program, because the funding for the SOFC APU over the past 10 years of development expires in 2011. Delphi also expects funding from SECA and other contracts with the DOE, DOD, and state sponsors to continue.

SUMMARY OF IDLE REDUCTION TECHNOLOGIES

The EPA's SmartWay program has evaluated the fuel-saving benefits of various devices through grants, cooperative agreements, emissions and fuel economy testing, demonstration projects, and technical literature review. As a result, the EPA has determined that the following idle reduction technologies provide fuel-saving and/or emissions reducing benefits when used properly in their designed applications:

- Electrified Parking Spaces (truck stop electrification),
- Auxiliary power units and generator sets,
- Fuel operated heaters,
- Battery air-conditioning systems,
- Thermal storage systems, and
- Automatic shut-down/start-up systems.

A listing of specific products that the EPA has verified for each of these categories can be found on the EPA website.[33] The listing is quite extensive and illustrates that the commercialization of idle reduction technologies is well under way and has accelerated since the NRC Phase 1 report was published in 2008. The 21CTP has not conducted any detailed analysis of the individual idle reduction products, and so it is not able to comment on the performance of these products.

The functionalities and costs of the idle reduction technologies discussed in this chapter that are under development or in production are summarized in Table 6-4.

EFFECT OF GREENHOUSE GAS EMISSIONS STANDARDS AND FUEL EFFICIENCY STANDARDS

The EPA and NHTSA addressed idle reduction technologies in their final rules for "Greenhouse Gas Emissions Standards and Fuel Efficiency Standards for Medium- and Heavy-Duty Engines and Vehicles" issued on September 15, 2011 (EPA/NHTSA, 2011a). The final rules recognize the following idle reduction technologies (with EPA and NHTSA considering that the baseline Class 8 vehicle consumes 0.8 gal/h of diesel fuel) (EPA/NHTSA, 2010b):

[33]Available at http://www.epa.gov/smartway/transport/what-smartway/verified-technologies.htm.

TABLE 6-4 Cab Comfort Technology Summary

System	Elements of System	Cooling	Heating	Charge Batteries	Fuel Consumption Rate	Installed Cost	Maintenance Cost/Yr
None[a]	Vehicle engine idling (2001 truck)	Yes	Yes	Yes	0.77 gal/h cooling 0.98 gal/h heating	$0	$150/30,000 miles for oil change[b]
None[a]	Vehicle engine idling (2007 truck)	Yes	Yes	Yes	0.53 gal/h heating 0.72 gal/h cooling	$0	$150/30,000 miles for oil change[b]
Automatic start/stop	Vehicle engine idling	Yes	Yes	Yes	0.8 gal/h if on 0.0 gal/h if off	$1,200	$150/30,000 miles for oil change[b]
Cab bunk heater	Diesel fuel burner, heat exchanger, and fan	No	Yes	No	0.04 to 0.06 gal/h	$1,300	$110
Evaporative cooler	Thermal storage using graphite matrix	Yes	No	No	0.015 gal/h (3.5 to 10 amps from vehicle batteries)	$1,800	$100
Battery- powered air-conditioning systems	Battery, motor, vapor compression air conditioning components	Yes	No	No	0.15 gal/h	$4,000 without battery upgrade	$200
Diesel APU[a]	Diesel engine, generator, particulate filter, NO_x trap	Yes	Yes	Yes	0.20 to 0.33 gal/h	$8,000 (add $1,000 for DPF)	$400
Electrified parking space (single system)	Heating, cooling module on pedestal connected to window-mounted module (includes communications entertainment)	Yes	Yes	Yes	0	$10 ($9,000 to $16,700 infra-structure cost)	$1.00/h -$2.45/h usage cost
Electrified parking space (dual system)	Onboard equipment (e.g., inverter/charger, electric heating/cooling device) powered by extension cord	Yes	Yes	Yes	0	($2,500 to $6,000 infra-structure cost)	$1/h usage cost
SOFC APU	Solid oxide fuel cell, reformer, output power conditioner	Yes	Yes	Yes	0.2 gal/h	$8,000 to $9,000	N/A

NOTE: Acronyms are defined in Appendix I.
[a] May not be available due to local regulations and/or non-compliance with new 2010 emission regulations. [b] Possible reduction in overhaul time.
SOURCE: Based on L. Gaines and D. Santini, Economic Analysis of Commercial Idling Reduction Technologies. Available at http://www.transportation.anl.gov/pdfs/TA/372.pdf, and L. Gaines, Which Idling Technologies Are the Best? See references.

- Auxiliary power unit, which powers the truck's heating, cooling, and electrical system and typically uses 0.2 gal/h of diesel fuel;
- Fuel operated heater, which provides heating services to the truck and typically uses 0.04 gal/h of diesel fuel;
- Battery air-conditioning systems, which provide cooling to the truck; and
- Thermal storage systems, which provide cooling to the truck.

Another alternative involves Electrified Parking Spaces with or without modification to the truck.

The final rules include extended idle reduction technology as an input to the Greenhouse Gas Emission Model (GEM) for Class 8 sleeper cabs. The manufacturer would input a value (see below) based on the idle reduction technology installed in the truck. If a manufacturer chooses to use idle reduction technology to meet the standard, then it would require an automatic main engine shutoff after 5 minutes to help ensure that the idle reductions are realized in-use. However, the agencies are not mandating the use of idle reductions or idle shutdown but rather are allowing their use as one part of a suite of technologies feasible for reducing fuel consumption and meeting the proposed standards.

The EPA's and NHTSA's value (0.5 gal/1,000 ton-mile saved) for the idle reduction technologies was determined using an assumption of 1,800 idling hours per year; 125,000 miles traveled; a baseline fuel consumption of 0.8 gal/h; and an APU fuel consumption of 0.2 gal/h ((0.8 – 0.2) gal/h × 1,800 h/(19 tons × 125,000 miles × (1,000 tons)/1,000 tons) = 0.5 gal/1,000 ton-miles saved). Relative to the 2,500 idling hours for single drivers and 800 idling hours for team drivers found in the NCSU (Tazewell et al., 2008) study mentioned earlier, the EPA and NHTSA used 1,800 idling hours per year for Class 8 trucks with sleeper cabs, which, the committee assumes, may have been a blending of idling hours for single and team drivers. As an example, for a Class 8 mid-roof, sleeper cab with a 2017 proposed standard of 7.2 gal/1,000 ton-miles, idle reduction technology could provide nearly 30 percent of the reduction required to achieve the standard (assuming a total reduction of 1.8 gal/1,000 ton-miles to meet the 7.2 gal/1,000 ton-miles standard, by assuming the standard is a 20 percent reduction [which is within the EPA/NHTSA range of 9 to 23 percent] from the 2010 status, subsequently calculated to be 9.0 gal/1,000 ton-mile). The 0.5 gal/1,000 ton-mile reduction in fuel consumption amounts to a 6 percent reduction in overall fuel consumption (0.5 gal/1,000 ton-mile/9.0 gal/1,000 ton-mile × 100 = 6 percent).

Finding 6-4. Idle reduction technologies could provide 6 percent reduction in overall fuel consumption for Class 8 long-haul trucks with sleeper cabs, which is nearly 30 percent of the 20 percent reduction in the fuel consumption required to meet the EPA/NHTSA proposed 2017 fuel consumption standards.

Recommendation 6-4. The 21CTP should review and potentially revise its idle reduction plans and goals in view of the fact that the proposed 2017 fuel efficiency standards provide an incentive for the adoption of idle reduction technologies as a means for achieving these standards for Class 8 long-haul trucks with sleeper cabs.

RESPONSE TO RECOMMENDATIONS FROM NRC PHASE 1 REPORT

Seven findings and recommendations were made regarding idle reduction technologies in the NRC (2008) Phase 1 report (Findings and Recommendations 6-1 to 6-4 and 6-6 to 6-8) (Finding and Recommendation 6-5 was omitted in the Phase 1 report). The DOE concurred with all of the recommendations except Recommendation 6-4 (see Appendix C in this report), thereby reconfirming the 21CTP engine idle reduction goals that are directed toward substantially reducing energy consumption and exhaust emissions due to heavy-duty-vehicle idling.

Recommendation 6-4 suggested that the EPA renew its efforts to promulgate national anti-idling regulations. The 21CTP commented that the EPA has no legal authority to promulgate anti-idling laws, or any time or behavior limits on truck owners. However, as noted above with respect to Goal 3, the patchwork of anti-idling regulations nationally have been an impediment to the broader use of anti-idling measures and efforts. Finding 6-1 and Recommendation 6-1 in this chapter address this issue by recommending that the EPA and the DOT should work to find incentives for states to promulgate uniform anti-idling regulations.

GOALS FOR FY 2012

In the February 2011 "21CTP Draft White Paper on Idle Reduction" (DOE, 2011), the 21CTP no longer recognizes the previously reviewed goals that extended from the NRC Phase 1 review through 2010. Instead, the 21CTP is recommending the following five goals for FY 2012.

- *21CTP Goal 1 Recommended for FY 2012:* Work with OEM truck manufacturers to obtain data on the number of new trucks being ordered with idle reduction options.
- *21CTP Goal 2 Recommended for FY 2012:* Conduct a fleet survey to gather data on the amount of in-use idling hours that are accumulated by type of heavy-duty vehicle.
- *21CTP Goal 3 Recommended for FY 2012:* Acquire data from the EPA SmartWay Program to measure fuel savings and emissions reductions associated with the various types of idle reduction equipment available.
- *21CTP Goal 4 Recommended for FY 2012:* Establish a nationwide multi-mode idle reduction education program.
- *21CTP Goal 5 Recommended for FY 2012:* Promote the incorporation of idle reduction equipment on new trucks as fuel saving devices as they are identified through the DOE SuperTruck program.

The 21CTP stated in the February 2011 idle reduction white paper: "Without funding dedicated to this effort [i.e., the above goals], it is quite difficult, if not impossible, for the 21st Century Truck Partnership to accomplish these goals" (DOE, 2011). The white paper states: "Assuming there is funding, the action items [previously identified as goals through 2010] . . . lay out a path to accomplishing the stated objective." In contrast, the committee finds that the new goals, which focus on measuring the usage and benefits of idle reduction and the incorporation of idle reduction technologies on new trucks, are generally not supported by the "action items," which focus on cost-effective add-on idle reduction technologies; the development of some specific technologies such as electrically powered HVAC systems, cab insulation, and fuel cell APUs; and education programs and incentives to encourage the deployment of cost-effective technologies to reduce fuel use and emissions due to idling.

Finding 6-5. In February 2011, the 21CTP deleted the quantification of the overall goal to reduce fuel use and emissions produced by idling engines. The 21CTP issued five new goals for idle reduction and designated the goals that had been in place through 2010 as "action items." The new goals are generally not supported by the "action items." A separate budget for idle reduction for FY 2012 has not been proposed, although idle reduction will be addressed by the SuperTruck program. The 21CTP has stated that, "without funding dedicated to this effort [the idle reduction goals], it is quite difficult, if not impossible, to accomplish these goals" (DOE, 2011).

Recommendation 6-5. The 21CTP should revise its new idle reduction goals to include metrics, funding, and timing for the overall goal of reducing fuel use and emissions produced by idling engines. The associated "action items" should be supportive of these goals.

REFERENCES

American Transportation Research Institute. 2006. Idle Reduction Technology—Fleet Preferences Survey. February.

ANL (Argonne National Laboratory). 2009. Idling, Cruising the Fuel *In*efficiency Expressway. Center for Transportation Research. September. Available at http://www1.eere.energy.gov/cleancities/pdfs/idling_reduction_primer.pdf. Accessed October 23.

Casey, C. 2008. Release of Factory-Installed Idle Reduction Systems for International Sleeper Trucks. Final Scientific/Technical Report DE-PS26-05NT42485. International Truck and Engine Corporation. September 5.

Cummins. 2008. Idle Talk: How the New Regulations Affect You. February. Available at http://www.Cumminsnorthwest.com/PDF/IdleTalk.pdf. Accessed October 23, 2011.

Delphi. 2010. Delphi truck fuel-cell APU to hit road in 2010. Automotive Engineering Online. Available at http://sae.org/mags/aei/SEEWC/8222. Accessed December 7, 2010.

DOE (U.S. Department of Energy). 2010. 21CTP Draft White Paper on Idle Reduction. August. Washington, D.C.: Office of Vehicle Technologies.

DOE. 2011. 21CTP Draft White Paper on Idle Reduction. February. Washington, D.C.: Office of Vehicle Technologies.

Espar. 2010. Worldwide Leader in Marine and Vehicle Heaters. Available at http://www.espar.com. Accessed December 7, 2010.

EPA/NHTSA (U.S. Environmental Protection Agency/National Highway Traffic Safety Administration). 2010a. Greenhouse Gas Emissions Standards and Fuel Efficiency Standards for Medium- and Heavy-Duty Engines and Vehicles. Dockets No. EPA-HQ-OAR-2010-0162 and No. NHTSA-2010-0079. October 25. Available at http://www.regulations.gov.

EPA/NHTSA. 2010b. Draft Regulatory Impact Analysis, Proposed Rulemaking to Establish Greenhouse Gas Emissions Standards and Fuel Efficiency Standards for Medium- and Heavy-Duty Engines and Vehicles. EPA-420-D-10-901. October.

EPA/NHTSA. 2011a. Greenhouse Gas Emissions Standards and Fuel Efficiency Standards for Medium- and Heavy-Duty Engines and Vehicles. Federal Register, Vol. 76, No. 179, September 15. Available at http://www.nhtsa.gov/staticfiles/rulemaking/pdf/cafe/2011-20740.pdf.

EPA/NHTSA. 2011b. Final Rulemaking to Establish Greenhouse Gas Emissions Standards and Fuel Efficiency Standards for Medium- and Heavy-Duty Engines and Vehicles. Regulatory Impact Analysis. EPA-420-R-11-901. August.

Gaines, L. Which Idling Reduction Technologies Are the Best? Argonne National Laboratories. Available at http://www.transportation.anl.gov/pdfs/EE/533.pdf.

Gaines, L., and D. Santini. 2006. Economic Analysis of Commercial Idling Reduction Technologies. Argonne National Laboratory. Available at http://www.transportation.anl.gov/pdfs/TA/372.pdf.

Lane, W., S. Diamond, and J. Routbort. 2004. More Electric Truck. FY2004 Annual Report.

Maronde, C., and L. Slezak. 2006. Idle Reduction Demonstration Projects. SAE Commercial Vehicle Engineering Congress and Exhibition. November 1.

Montemayor, A.F. 2006. Phased Introduction of Fuel Cells into a Class 8 Tractor. SWRI Project No. 03983. January 31.

NRC (National Research Council). 2008. Review of the 21st Century Truck Partnership. Washington, D.C.: The National Academies Press.

NRC. 2010. Technologies and Approaches to Reducing the Fuel Consumption of Medium- and Heavy-Duty Vehicles. Washington, D.C.: The National Academies Press.

Shaffer, S. 2004. Development Update on Delphi's Solid Oxide Fuel Cell System. SECA Review Meeting. Boston, Mass.

Stone, L.M., J.F. Birkel, R.D. Nine, and L.A. Slezak. 2004. Advanced Electric Systems and Aerodynamics for Efficiency Improvements in Heavy Duty Trucks. Final Report. Contractor: Caterpillar, Inc. Contract No. DE-FC26-04NT42189.

Tazewell, A., H.C. Frey, P. Kuo, and J.R. Stone. 2008. EPA Truck OEM APU Prep Kit Design and Installation Project. Final Report. August. North Carolina State University.

Thermo King. 2010. Thermo King Website. Available at http://www.thermoking.com/tripac/. Accessed December 6, 2010.

Webasto. 2010. Webasto Website. Available at http://www.webastoshowroom.com/bluecooltruck/product_infor.html. Accessed December 6, 2010.

7

Safety

INTRODUCTION

Safety is an important goal of the 21st Century Truck Partnership (21CTP), with an overall objective to "promote the development and early adoption of technologies and processes to improve truck safety, resulting in the reduction of fatalities and injuries in truck-involved crashes, thus enabling benefits related to congestion mitigation, emission reduction, reduced fuel consumption, and improved productivity" (DOE, 2010). While working in concert with the U.S. Department of Energy (DOE) and truck manufacturers, the U.S. Department of Transportation (DOT), which has primary responsibility for safety, provides leadership for truck safety. Participating agencies within the DOT include the National Highway Traffic Safety Administration (NHTSA), with responsibility for safety regulations for all motor vehicles; the Federal Highway Administration (FHWA), with responsibility for highways, freight management, and operations; and the Federal Motor Carrier Safety Administration (FMCSA), with responsibility for developing federal regulations that promote commercial carrier safety and industry productivity. Presentations from all three agencies were given to the committee (see Appendix B) and are discussed throughout this chapter.

OVERVIEW OF GOALS AND TIMETABLES

The 21CTP works collaboratively with the DOT to enhance vehicle safety. The overall goals of this collaboration are as follows (DOE, 2010):

1. "To ensure that advancements in truck design and technology to improve fuel efficiency do not have any negative impacts on safety."
2. "Conversely, to ensure that efforts to improve safety do not reduce efficiency and, where possible actually contribute to improvements in overall motor carrier industry system efficiency."

As discussed in the following sections of this chapter, the committee was unable to find any convincing case that technologies applied to reduce fuel consumption would degrade heavy-duty-truck safety. However, the wording of Goals 1 and 2 above could be interpreted by some that improving fuel consumption could be accomplished at the expense of sacrificing safety. The committee does not believe that this would happen, because the DOT and NHTSA have a primary focus on improving vehicle safety. Nevertheless, the Partnership might consider rewording these goals to be less ambiguous.

The DOT agencies all contribute to safety through their various areas of responsibility. The role of the NHTSA is to monitor highway safety in total (not just for heavy-duty trucks) and to conduct analysis to determine the potential benefit of vehicle safety systems, and to issue Federal Motor Vehicle Safety Standards (FMVSS) when appropriate.[1] The Office of Freight Management and Operations of the FHWA strives to promote the efficient movement of freight and oversees the enforcement of federal regulations on the size and weight limits of trucks.[2] The Federal Motor Carrier Safety Administration has as its mission to reduce crashes, injuries, and fatalities involving large trucks and commercial buses. The FMCSA is responsible for establishing safe operating requirements for commercial vehicle, drivers, carriers, vehicles, and equipment in interstate commerce.[3]

The FMCSA has set specific goals for truck and bus safety. These goals are set at fatality per 100 million vehicle miles traveled (VMT). In particular, the goals for 2007 through 2011 are as follows:

[1] R. Kreeb, DOT, NHTSA, "Safety," presentation to the committee, November 15, 2010, Washington, D.C.

[2] R. Schmitt, DOT, FHWA, "Overview of DOT Truck Safety and Productivity Activities," presentation to the committee, September 8, 2010, Washington, D.C.

[3] L. Loy, DOT, FMCSA, "Overview of DOT Truck Safety and Productivity Activities," presentation to the committee, September 8, 2010, Washington, D.C.

- 2007: 0.175 per 100 million VMT;
- 2008: 0.171 per 100 million VMT;
- 2009: 0.167 per 100 million VMT;
- 2010: 0.164 per 100 million VMT; and
- 2011: 0.160 per 100 million VMT.

(For comparison, it is noted that the fatality rate for *all* vehicle accidents in the United States in 2009 was 1.13 fatalities per 100 million VMT.[4] The extent to which these and other goals have been met is discussed later in this chapter.

Finding 7-1. The wording of 21CTP Safety Goals 1 and 2 as now written might be subject to misinterpretation by some as allowing the compromise of safety in the effort to improve fuel consumption.

Recommendation 7-1. The Partnership should review the wording of its safety goals and consider rewording them so as to unambiguously state that safety will not be compromised in reducing fuel consumption.

NATURE OF LARGE-TRUCK ACCIDENTS—A BRIEF OVERVIEW

Combination trucks (defined as tractor-trailer and single-unit trucks towing trailers) are involved in about 75 percent of the fatalities resulting from medium- and heavy-duty truck and bus accidents.[5] In 2009, a total of 3,380 fatalities were due to large-truck crashes—this was a reduction of 20 percent from 4,245, the number of fatalities in 2008. Of the 3,380 fatalities in 2009, 2,551 were occupants in the other vehicle, and 503 were occupants of the truck (DOT, 2010b). It is typically the case that in truck accidents involving two vehicles, 75 percent or more of the fatalities involve the occupants of the other, usually smaller, vehicle. In accidents involving both a light vehicle and a large truck, the driver of the light vehicle is cited as being at fault most of the time, with some studies showing the driver of the light vehicle at fault as much as 78 percent of the time (see NRC [2008] for more detail and additional references). Most of the fatal crashes involving trucks occurred in rural areas (64 percent), during the daytime (67 percent) and on weekdays (80 percent) (DOE, 2010).

Only about 300 fatalities occur each year in accidents involving truck Classes 5 and 6 combined, primarily because of their typically lower speed in urban daylight settings and many fewer miles traveled compared to Class 8 trucks (DOE, 2010).

[4] See http://www.NHTSA.gov/PR/NHTSA-05-11. Accessed June 22, 2011.

[5] Unless otherwise noted, accident statistics cited in this chapter are for the United States.

Total fatalities for bus-related accidents in 2008 were 307, of which 41 were occupants in motor coaches.[6] Commercial buses represent a very small percentage of fatal crashes, only 0.5 percent of the total. Very few fatalities occur due to school bus accidents. In 2009 occupants in a school bus had 3 fatalities, although 91 pedestrian fatalities were associated with school bus accidents.[7]

More details regarding the nature of heavy-duty-truck and bus accidents can be found in the NRC Phase 1 report (NRC, 2008, Chapter 7), on the NHTSA website, or in the University of Michigan Transportation Research Institute's Trucks Involved in Fatal Accidents Database.[8] Because the vast majority of fatalities and injuries associated with truck and bus accidents are due to combination-truck accidents, most of this chapter focuses on technologies that might reduce combination truck accidents.

CRASH-AVOIDANCE STRATEGIES

Vehicle design and performance characteristics play an important role in truck crashes. The 21CTP places emphasis on technology that can enhance truck roll stability, improve braking performance, and reduce jackknifing. Additional crash avoidance technologies include driver warning, driver assist, and driver monitoring as well as onboard safety system monitoring (DOE, 2010). In addition, the DOT is exploring technologies to improve the frequency and thoroughness of in-service truck inspections. Many crash-avoidance technologies such as electronic stability control (ESC) and roll stability control are commercially available.

For research on heavy-truck safety, most of it devoted to crash-avoidance study, the NHTSA has an annual budget of about $2.1 million. The FMCSA budget is approximately $17.4 million, including analysis and research.

Several crash-avoidance technologies are addressed in this chapter: (1) braking and stability control, (2) collision warning, (3) safety system diagnostics, (4) driver behavior and performance, (5) smart roadside, and (6) intelligent transportation systems.

Braking and Stability Control

Material prepared by the NHTSA suggests that improved braking performance could reduce heavy-duty-truck accidents, particularly those for which the truck would rear-end another vehicle.[9] In 2009, the NHTSA published a final rule on amend-

[6] L. Loy, DOT, FMCSA, "Overview of DOT Truck Safety and Productivity Activities," presentation to the committee, September 8, 2010, Washington, D.C.

[7] See the Fatality Analysis Reporting System (FARS) data tables, School Bus Related. Available at http://www-fars.nhtsa.dot.gov.

[8] See http://www-fars.nhtsa.dot.gov, and www.umtri.umich.edu/expertiseSub.php?esID=29. Accessed June 22, 2011.

[9] R. Kreeb, DOT, NHTSA, "Overview of DOT Truck Safety and Productivity Activities," presentation to the committee, September 8, 2010, Washington, D.C.

ing FMVSS No. 121 to improve the stopping distance of trucks. By 2011, most new trucks will be required to have the capability to reduce their stopping distance 30 percent more than had previously been required. (This is an important improvement, but the stopping distance is still much longer than that of light-duty vehicles.) The improved braking performance can be accomplished by the use of larger drum brakes or air disc brakes. In time, the NHTSA would be expected to conduct field tests to assess the effects of this new braking requirement.

It should be noted that the field performance of antilock braking systems (ABSs) required by FMVSS No.121 on all air-braked vehicles of 10,000 lb or greater manufactured after March 1, 1997, has shown mixed results. In a comprehensive study published by the NHTSA in July 2010 (Kirk, 2010), it was found that there was a statistically significant 6 percent reduction in the number of crashes where ABS is assumed to be influential, and a large reduction in jackknives and off-road overturns; yet it was found that there was not a statistically significant reduction in fatal crash involvement. Although improved braking was influential in reducing the number of accidents as noted above, it is possible that accidents that are so severe as to cause a fatality cannot be avoided simply by improved braking. In addition, drivers need to be trained not to push the ABS technology to its limits.

Over the past 5 years, truck manufacturers have been offering electronic stability control on several truck models, and ESC has become standard on some truck models. DOE (2010) provides a detailed explanation of how ESC works. Because the application of stability control systems is fairly recent, there are insufficient real-world data to assess its effectiveness. However, studies have shown that the systems do offer potential for accident and fatality reduction. In Woodrooffe et al. (2009), crash scenarios were selected from national databases and examined to assess the potential benefit of stability systems on 5-axle tractor semitrailers. Assuming that all 5-axle tractor semitrailers were equipped with ESC systems, the expected annual safety benefit related to combined rollover and directional (yaw) instability is a reduction of 4,659 crashes, 126 fatalities, and 5,909 injuries.

Anticipatory automatic braking and speed control systems may also be used for accident prevention. However, these systems were not included in the materials prepared by or presented by the DOE or DOT, and therefore were not evaluated by the committee.

Collision Warning

Advancements in collision warning systems for heavy-duty trucks have continued over the past several years. The 21CTP supports this area, because it may have potential for significant benefit in improving highway safety. Warning systems currently available include the following:

- Lane departure warning (LDW),
- Forward collision warning (FCW),
- Side object detection, and
- Rear object detection.

These systems use radar, video detection, ultrasonic, and other sensor systems combined with sensor input analysis algorithms to determine if a crash situation is developing, and then they warn the driver (DOE, 2010). Some systems not only warn the driver but also take control of the vehicle by de-throttling or braking.

In the NRC Phase 1 report (NRC, 2008, Chapter 7), it was reported that, based on field operational tests (FOTs) that had been completed at that time, LDW systems could potentially provide a reduction in accidents for single-vehicle roadway departure of a little more than 20 percent. In a more recent study, estimates were made of the cost-benefit potential of LDW systems (Houser et al., 2009). General Estimates System (GES) data were used to estimate outcomes from different lane departure crashes.[10] Then, using information from the aforementioned field operational test, efficacy rates were determined in order to estimate the types of crashes that could be prevented using LDW systems. Assuming that the systems had been in place from 2001 to 2005, and recognizing that certain types of accidents could not have been prevented by LDW (e.g., loss of steering control from brake lock-up), it was estimated that the mean average annual preventable fatalities could be 147 and preventable injuries could be 2,642.

The DOT has taken an approach of integrating forward collision, rear-end impact, road departure, and lane changing warning systems into what it calls Integrated Vehicle-Based Safety Systems (IVBSS). This program also involves the University of Michigan Transportation Research Institute, Battelle, Eaton, PACCAR, Conway, Navistar, Takata, and the Michigan Department of Transportation. A field operational test was recently completed—it was a 10-month test involving 10 trucks and 20 drivers. Some key findings of the FOT were encouraging (DOT, 2010a):

- Drivers stated that the system made them more aware of the traffic environment;
- Most of the drivers would recommend the purchase of such a system, would prefer to drive a truck with such a system, and thought that such systems would increase driving safety; and
- Seven drivers said that the system potentially prevented them from having a crash.

In an independent evaluation of the FOT results, the John A. Volpe National Transportation Center estimated that the integrated system would be 11 percent effective in preventing accidents of the type targeted by IVBSS, and therefore could prevent, annually, 13,000 crashes involving trucks.[11]

[10] See http://www.nhtsa.gov/people/ncsa/nass_ges.html.

[11] See http://www.umtri.umich.edu/public/ivbss/IVBSS_Final_Public_Meeting_Presentations.pdf.

Safety System Diagnostics

The DOT has two initiatives in the area of safety system diagnostics: tire pressure monitoring and brake systems diagnostics. Tire pressure monitoring systems have become common on light-duty vehicles and could be of particular importance for both safety and life–cycle costs if used on heavy line-haul trucks, particularly as the industry moves toward single wide-base tires. Properly inflated tires not only enhance safety and durability, but also reduce fuel consumption. In a recent study it was found that 1 in 14 tires was as much as 20 psi underinflated (approximately 20 percent of recommended pressure). This can lead to higher tire procurement costs, and it is estimated that underinflated tires on heavy-duty trucks results in fuel consumption increases of about 0.6 percent.[12] Yet to date, only about 5 percent of the heavy-duty truck fleet has tire pressure monitoring systems.

The industry has been moving aggressively with more powerful braking systems. However, with 10 wheel ends on a typical tractor-trailer, brake maintenance is a challenge for most fleets and one of the highest-cost maintenance components. In general, there is no feedback to the driver until an emergency stop is necessary. A road-check study conducted in 2002 found that of the 49,032 vehicles checked at random, 22 percent were pulled out of service for noncompliance, and more than half of those pulled were because of brake-related issues (Lang, 2005).

In a study of the effectiveness of brake monitoring systems, the FMCSA conducted a field trial to evaluate several systems to measure brake stroke, shoe lining wear, and temperature on a fleet of buses: in-city buses were selected to provide a harsh braking protocol and because the fundamental brake design on transit buses is similar to that of heavy-duty Class 8 trucks. In general, the systems tested performed well (see Order et al., 2009, for more detail).

To date the market penetration of onboard brake monitoring systems on trucks is near zero, although there is some application (10 to 15 percent) of the systems on transit buses. Trucking and bus companies that have rigorous preventive maintenance inspection programs would not benefit sufficiently to justify the added cost of onboard systems. Currently there is no plan to introduce regulations requiring onboard systems. Instead, the FMCSA has added the use of Performance Based Brake Testers (Performance Based Brake Tester [PBBT] Test and Procedure Guidelines, Commercial Vehicle Safety Alliance Training Course, Revised January 2010) to the standard North America Vehicle inspection procedure (FMCSA, 2002).[13]

[12] L. Loy, DOT, FMCSA, "Overview of DOT Truck Safety and Productivity Activities," presentation to the committee, September 8, 2011, Washington, D.C.

[13] Personal communication regarding onboard brake testing from Bob Kreeb, NHTSA, and Luke Loy, FMCSA, to committee member Larry Howell, December 8, 2010.

Driver Behavior and Performance

The FMCSA, with the help of the NHTSA, conducted a study to determine the causes of truck crashes (Craft, 2007). In 963 crashes involving trucks, from April 2001 to December 2003, there were 249 fatalities. Of the crashes studied, it was concluded that 87 percent were caused by driver behavior, that of either the truck driver or the driver of the other vehicle; 10 percent were caused by vehicle failure; and 3 percent were caused by the environment. (Of the vehicle failures, brake problems were most often cited.) Critical reasons for driver faulty behavior, in descending order of frequency, included the following: interruption of traffic flow, unfamiliar roadway, inadequate surveillance, driving too fast, illegal maneuver, inattention, fatigue, illness, false assumption about the other driver's action, and distraction inside the vehicle. Impaired driver behavior due to alcohol and drugs also contributes to truck crashes, but in most cases involving a truck and a light vehicle, the driver of the light vehicle is the one who is impaired (DOT/NHTSA, 2006). However, in a study of driver distraction in commercial vehicles, it was found that drivers were engaged in non-driving-related tasks in 71 percent of crashes (Olsen et al., 2009). (The apparent difference between the two studies as to the frequency of "distraction" as a causal factor is due to the fact that in the Olsen et al. study, "distraction" included inattentiveness, drowsiness, and secondary driving tasks such as checking the rearview mirror.)

Clearly, driver behavior and performance are important factors in highway safety, and the DOT has for this reason put significant emphasis on the subject. It is beyond the scope of this report to cover in detail all the studies and programs in place on driver behavior, but a few highlights are addressed.

The NHTSA has engaged in a number of driver-distraction research studies, including an observational study of driver cellular telephone use; driver distraction with wireless communication systems and route guidance systems; and the impact of inattention on crash risk (DOE, 2010). The FMCSA has initiated a program to determine the effectiveness of onboard monitoring in reducing accident risk, including the observation of driver behavior, fatigue monitoring, lane departure warning, forward collision warning, and hours of service monitoring. Onboard monitoring systems will be installed in 270 trucks across three motor carrier fleets, to be deployed by 2011, with results from the field operational tests in 2013.[14] In the meantime, President Obama, on October 1, 2009, issued an Executive Order, *Federal Leadership on Reducing Text Messaging While Driving*, stating that federal employees shall not engage in text messaging when driving government vehicles or in personal vehicles while

[14] L. Loy, DOT, FMCSA,,"Overview of DOT Truck Safety and Productivity Activities," presentation to the committee, September 8, 2010, Washington, D.C.

on government business.[15] Moreover, in 2009, the FMCSA issued a rule, effective October 27, 2010, prohibiting texting by commercial drivers while operating in interstate commerce.[16] Many states have banned texting and/or cell phone usage while a person is driving (GHSA, 2011).

Smart Roadside

The FMCSA is conducting research to improve the manner in which state, local, and federal officials interact with commercial vehicle operators and drivers at the roadside. The objective is both to improve the efficiency and comprehensiveness of operations and at the same time to ensure that operators are adhering to applicable regulations. As an example, Level 1 inspections include the examination of the driver's license, medical examiner's certificate and waiver, hours of service, seat belt, brake system, fuel system, lighting, and many other vehicle systems. A Level 1 inspection usually takes about 40 minutes (DOE, 2010). With about 4 million trucks in service, it is likely that many will go more than a year without being inspected because of the time required. Smart roadside will use wireless technology to transmit driver, vehicle, and carrier information, including an electronic hours-of-service log to an inspection station.[17]

Smart roadside Phase 1, concept development, was completed in 2008, and Phase 2, prototype testing, in 2009. Phase 3, field operations testing, is in progress and expected to be completed in 2011. In Phase 3, real-time and automated safety checks are being demonstrated. The data include driver identification, license status, and log information, as well as vehicle lights, brakes, and tires.[18] This technology looks promising for ensuring that more trucks and drivers are operating safely.

Intelligent Transportation Systems

The vision of intelligent transportation systems is that every vehicle operating on the nation's highways will be a sensor probe with the capability to communicate with all other vehicles (vehicle-to-vehicle, V2V) and with the infrastructure (vehicle-to-infrastructure). The objectives are to enhance traffic management, reduce congestion, enable on-road vehicle inspection by means of wireless transfer of data, and prevent crashes. Realization of this capability will require the installation of dedicated short-range communication devices at intersections, on roadsides, and within the vehicles (DOE, 2010). Specific truck-related applications would include the following:

- Electronic No-Zone, a V2V communication system that will allow the truck driver to be aware of nearby vehicles, including any that might be in blind-spot areas, and vehicles close to the truck will be made aware of the truck;
- Technology that will alert drivers that they are approaching a slowed or stopped vehicle; and
- Curve speed information to warn the driver if the truck that he or she is driving needs to slow down as it approaches a curve.

More information is available from the Intelligent Transportation Systems Joint Program Office of the U.S. DOT Research and Innovation Technology Administration.[19]

Finding 7-2. Vehicle crashworthiness and occupant protection systems have seen extensive deployment, have contributed greatly to improved highway safety, and have achieved extensive North American fleet penetration. The next important step is to prevent crashes altogether.

Recommendation 7-2. The committee supports the emphasis that the DOT and the 21CTP are giving to crash-avoidance technologies and recommends that crash-avoidance technologies continue to be given high priority and technical support.

COMMERCIAL VEHICLE WEIGHT AND SIZE

The DOT recognizes that it may be possible to increase motor carrier efficiencies by allowing increases in commercial vehicle weight or trailer size. Consideration must be given to the potential of greater highway damage. Although heavy-duty trucks weighing more than 40,000 lb account for only 5 percent of total highway traffic, they account for more than 50 percent of highway damage.[20] (Damage can be mitigated by distributing the load over more axles.) Safety could be an issue, too, although the use of heavier vehicles could be offset by a reduction in the total number of heavy vehicles on the road. The Transportation Research Board (TRB) has recommended additional study to assess the impact on highway safety of the use of heavier commercial vehicles and of longer or multiple trailers (TRB, 2010a). In any case, the National Research Council report *Technologies and Approaches to Reducing the Fuel Consumption of Medium- and Heavy-Duty Vehicles* found that when allowed over the entire fleet, increasing vehicle size and weight could yield fuel savings of 15 percent or more (NRC, 2010). Fur-

[15] See http://www.whitehouse.gov/the-press-office/executive-order-federal-leadership-reducing-text-messaging-while-driving.

[16] See Docket No. FMCSA-2009-0370, Limiting the Use of Wireless Communication Devices, also, in the *Federal Register*, September 27, 2010 (Vol. 75, No. 186, p. 59118).

[17] See http://www.fmcsa.dot.gov/facts-research/presentations/6_wireless_roadside_inspections_Loftus_vid_508.pdf. Dated March 4, 2008.

[18] L. Loy, DOT, FMCSA, "Overview of DOT Truck Safety and Productivity Activities," presentation to the committee, September 8, 2010, Washington, D.C.

[19] See http://www.its.dot.gov/index.htm, Accessed April 6, 2011.

[20] J. Nicholas, DOT, FHWA, "Safety," presentation to the committee, November 15, 2010, Washington, D.C.

ther, in that report, it is recommended that Congress give serious consideration to liberalizing vehicle weight and size (NRC, 2010, Recommendation 7-2, p. 177). A similar suggestion has been made by the TRB in the aforementioned study. To acquire real-world data, with the support of the FHWA, commercial truck weight pilot studies are being conducted in Maine and Vermont to assess the benefits as well as potential safety issues with road and bridge infrastructure as combination-vehicle weights up to 99,000 lb travel on the interstate system (DOE, 2010).

PROGRESS TOWARD GOALS

In 2009, there were 33,808 highway fatalities in the United States, the lowest number of deaths since 1950 (DOT, 2010b). Fatalities declined in 2009 from 2008 in all vehicle categories, including motorcycles, for which the number of fatalities had been increasing for the previous several years. For truck accidents in which there was a fatality, the total number of fatalities decreased from 4,250 in 2008 to 3,380 in 2009, a decrease of 20 percent. The goals of the FMCSA for reducing truck and bus fatalities per 100 million vehicle miles traveled were met in 2006 (target, 0.179; actual, 0.176); 2007 (target, 0.175; actual, 0.169); 2008 (target, 0.171; actual, 0.152); and 2009 (target, 0.167, actual, 0.121).

The significant decline in highway fatalities is certainly good news. There have been a number of studies aimed at identifying the contributions to the reduction. One potential contributing factor could be the recession that began in 2008. People typically travel less during a recession; in particular, nonessential travel is reduced. Data have confirmed a month-to-month reduction in fatalities during every recession going back several decades (DOT, 2010c). However, there is evidence that many other factors are contributing as well. In fact, there is a long-term trend of highway fatality reduction in spite of the up-and-down cycles associated with recessions (DOT, 2010c). During the past 10 years, there has been an increase in seat belt usage and continuing improvements in occupant protection systems in most vehicles, including the application of frontal and side air bags. Better occupant protection in light-duty vehicles could be contributing to the reduction of light-duty vehicle occupant fatalities in truck V2V accidents. Unfortunately, however, even with seat belt usage at about 85 percent, more than half of the passenger vehicle occupant fatalities in 2009 were unbelted (DOT, 2010b). Thanks to advocacy groups and stricter laws, the number of alcohol-related highway fatalities had declined from the 1980s into the 1990s, but that number has leveled off for the past decade at approximately 37 percent of all fatalities.[21] Certainly, more rapid response of emergency vehicles plays an important role.

[21] See http://www-fars.nhtsa.dot.gov/Trends/TrendsAlcohol.aspx.

ADDITIONAL OPPORTUNITIES

Clearly, significant progress in highway safety has been made in the United States, but when it comes to highway safety, there should always be efforts to strive to do even better. And there are additional opportunities. In a comprehensive study comparing U.S. progress in highway safety with that of other developed countries, the TRB found a number of areas that deserve further attention (TRB, 2010b). The TRB found that although the United States achieved a 19 percent reduction in fatalities from 1995 to 2009, other nations have done better. Annual traffic fatalities have declined in France by 52 percent and in the United Kingdom by 39 percent, for example. Traffic fatalities have declined by about 50 percent during that time span in 15 high-income countries. The reader is referred to the TRB (2010b) 188-page report for more detail, but a few highlights of the report's conclusions are noted. First, at a general level, the TRB found that successful national safety programs are characterized by the *overall management* rather than by particular interventions. The elements of the management program include a systems perspective that integrates engineering design, traffic control, regulatory control, and public health methods to identify and reduce risk; specific goals and milestones, and accountability to meet those goals; and regular monitoring to measure progress and to identify problems. The report notes that the U.S. programs are typically deficient with respect to this ideal management model.

The Phase 1 NRC review of the 21CTP recommended (NRC, 2008, Recommendation 7-1) that the 21CTP and the DOT should develop a prioritized list of all heavy-truck safety projects. The 21 CTP's response to this recommendation (see Appendix C in this report) was that it could not be done because of the various independent federal agencies that are involved. Yet, the TRB has recommended an approach that would require a higher level of management integration by the DOT and the states than currently exists (TRB, 2010b). Although it is beyond the scope of this committee's charge to make recommendations covering highway safety overall, it is clear that a more integrated management approach to highway safety would also be beneficial to truck safety and should be given serious consideration.

The TRB (2010b) report also notes specific suggestions that are worthy of consideration. In the area of alcohol-impaired driving, the TRB report notes that the legal blood alcohol content (BAC) limit is 0.08 BAC in the United States, whereas it is only 0.05 BAC or lower in Australia, Canada, Japan, and nearly every country in Europe. Fatalities in accidents in which alcohol was a factor have remained nearly constant, at about 37 percent of all fatalities during the past decade in the United States.[22] The TRB report suggests that it may be possible to reduce alcohol-related fatalities by reducing the legal BAC limit and by enforcing it more strictly.

[22] See http://www-fars.nhsta.dot.gov/trends/trendsalcohol.aspx. Accessed April 6, 2011.

Successful speed management programs in other countries target major road systems and use intensive enforcement. This has led to reduction in top speeds by from 3 to 4 miles per hour and is credited with an estimated fatality reduction of from 15 percent to 20 percent. Seat belt usage is another area assessed by the TRB study. Although the average seat belt usage in the United States is near 85 percent, it is more than 90 percent in most of the other nations studied in the TRB report. The TRB suggests that an increase in seat belt usage in the United States by 5 percent might save an additional 1,200 lives annually (TRB, 2010b). The TRB report offers suggestions for actions that state and federal agencies should consider in response to the report's recommendations.

As noted earlier, the majority of accidents involving tractor-trailer combinations occur as a result of a smaller vehicle striking the tractor or trailer. A significant number of such crashes occur because a light-duty vehicle runs into the back of the trailer. In some cases, the smaller vehicle underrides the trailer, causing intrusion into the passenger compartment. The NHTSA issued a rule requiring the installation of structural guards on the back of trailers for the purpose of preventing underride. These guards were to have been installed on all trailers with a gross vehicle weight (GVW) of 10,000 lb or more, manufactured on or after January 24, 1998 (DOT, 2010d). The DOT (2010d) study found that the structural guards have had little success in reducing the number of fatalities that occur as a result of accidents in which a smaller vehicle rear-ends a trailer. More recent testing by the Insurance Institute for Highway Safety has shown that midsize cars impacting certain underride guards at closing speeds of 35 miles per hour can result in significant passenger compartment intrusion and that certain guards failed at speeds as low as 25 miles per hour.[23] In the spirit of looking for additional opportunities, the committee suggests that the DOT explore the potential benefit of modifying the requirements for the structural guards going forward.

Finding 7-3. The DOT has met its heavy-truck safety goals for the past 4 years. However, the committee observes that the TRB's 2010 study *Achieving Traffic Saftey Goals in the United States: Lessons from Other Nations* has shown that other nations have established more aggressive initiatives and goals with impressive results, and those results suggest that even greater improvement in highway safety is possible in the United States. The committee also notes that overall improvements in highway safety also yield improvements in heavy-duty truck safety, because most heavy-duty truck fatal accidents involve a light-duty vehicle.

Recommendation 7-3. The DOT should evaluate the conclusions and recommendations of the TRB study *Achieving Traffic Saftey Goals in the United States: Lessons from Other Nations* of highway safety in other nations, and consider the possibility of establishing more aggressive initiatives and goals for highway safety in general. The DOT should also consider establishing more aggressive goals for heavy-duty truck safety.

HEAVY-DUTY TRUCK SAFETY AND FUEL CONSUMPTION

As noted in the section above titled "Overview of Goasl and Timetables," an overarching goal of the 21CTP is to ensure that new systems and technologies to reduce truck fuel consumption do not degrade safety and that improvements in heavy-duty truck safety do not increase fuel consumption. As shown in *Technologies and Approaches to Reducing the Fuel Consumption of Medium- and Heavy-Duty Vehicles*, (NRC, 2010, p. 29), the primary energy losses of a heavy-truck-and-trailer combination are due (in decreasing order) to the power plant, vehicle aerodynamics, tire rolling resistance, auxiliary loads, and drivetrain. Modifications to the engine for reduced fuel consumption generally have no direct impact on vehicle safety.

Modifications for reduced aerodynamics drag typically include efforts to reduce the drag coefficient. The committee is aware of no negative impact on safety due to improved aerodynamic performance of heavy-duty trucks except for the potential of side panels on trailers or other devices falling off.

The primary goals for reducing rolling resistance are to ensure proper inflation pressures on existing truck tires and to eventually replace dual tires with single wide tires. Although it is worthwhile to further explore the potential changes in stopping distance in going to single wide tires, any changes most likely would be offset by new stopping distance requirements for heavy-duty trucks.

Highway accidents are often caused by excessive speed, which also increases aerodynamic drag and therefore fuel consumption. Thus, in this example, adhering to posted speed limits should both improve highway safety and reduce fuel consumption.

A reduction in highway accidents in general will reduce congestion due to the slowdown at the crash site. As noted in NRC (2008, Chapter 7), congestion is an important cause of increased and unnecessary fuel consumption; thus, reducing congestion-causing accidents will also reduce fuel consumption.

There is some possibility for interaction between fuel consumption and safety. For example, side panels on trailers and other devices for improving aerodynamics should be adequately secured to ensure that they do not fall off and present a road hazard. And the rear structural guards on trailers add weight to the trailer, albeit a very small percentage relative to the weight of the tractor-trailer combination. Should a next-generation wide-base single tire fail, it is

[23] See http://iihs.org/externaldata/srdata/docs/sr4602.pdf. Accessed June 15, 2011.

possible that wheel damage could occur as the vehicle pulls off the roadway. These interactive effects are expected to be negligibly small. However, as vehicle manufacturers adopt new components and systems to reduce fuel consumption, it will be important for the DOT to monitor these and other vehicle modifications to ensure that safety issues do not emerge.

Finding 7-4. Some of the potential safety improvements considered by the committee may have negligible impact on fuel consumption and, in some cases, appear to have positive implications. However, further study of the potential highway safety impact of high productivity vehicles is warranted.

REFERENCES

Craft, R. 2007. The Large Truck Crash Causation Study (LTCCS), Analysis Brief, LTCCS Summary. FMCSA-RRA-07-017. July. Washington, D.C.: Federal Motor Carrier Safety Administration.

DOE (U.S. Department of Energy). 2010. White Paper on Safety. Working draft, September 1, 2010. Updated 21st Century Truck Partnership Roadmap and Technical White Papers. Washington, D.C.: Office of Vehicle Technologies.

DOT (U.S. Department of Transportation). 2010a. Integrated Vehicle-Based Safety Systems: Heavy-Truck Field Operational Test, Key Findings Report. DOT HS 811 362. August. Washington, D.C.

DOT. 2010b. Highlights of 2009 Motor Vehicle Crashes. Traffic Safety Notes. DOT HS 811 363. August. Washington, D.C.

DOT. 2010c. An Analysis of the Significant Decline in Motor Vehicle Traffic Fatalities in 2008. DOT HS 611 346. June. Washington, D.C.

DOT. 2010d. The Effectiveness of Underride Guards for Heavy Trailers. DOT HS 811 375. October. Available at http://www-nrd.nhtsa.dot.gov/Pubs/811375.pdf.

DOT/NHTSA (DOT/National Highway Traffic Safety Administration). 2006. Large Truck Crash Causation Study: An Initial Overview. DOT HS 810 646. August. Washington D.C.

FMCSA (Federal Motor Carrier Safety Administration). 2002. U.S. Transportation Secretary Mineta Announces FMCSA Rule Permitting Performance Brake Testing Technology. DOT FMCSA 15-02. August 9. Washington, D.C. Available at http://www.fmcsa.dot.gov/about/news/news-releases/2002/080902.htm. Accessed September 8, 2010.

GHSA (Governors Highway Safety Association). 2011. Cell Phone and Texting Laws. June. Available at http://www.ghsa.org/html/stateinfo/laws/cellphone_laws.html. Accessed June 21, 2011.

Houser, Amy, D. Murray, S. Shackelford, R. Kreeb, and T. Dunn. 2009. Analysis of Benefits and Costs of Lane Departure Warning Systems for the Trucking Industry. FMCSA-RRT-09-022. February. Washington, D.C.: Federal Motor Carrier Safety Administration.

Kirk, A. 2010. The Effectiveness of ABS in Heavy Truck Tractors and Trailers. HS 811 339. July. Washington, D.C.: U.S. Department of Transportation.

Lang, M. 2005. Technical Brief: On-Board Sensors for Determining Brake System Performance. FMCSA-PSV-05-001. July. Washington, D.C.: Federal Motor Carrier Safety Administration.

NRC (National Research Council). 2008. Review of the 21st Century Truck Partnership. Washington, D.C.: The National Academies Press.

NRC. 2010. Technologies and Approaches to Reducing the Fuel Consumption of Medium- and Heavy-Duty Vehicles. Washington, D.C.: The National Academies Press.

Olsen, R., R. Hanowski, J. Hickman, and J. Bocanegra. 2009. Driver Distraction in Commercial Vehicle Operations. FMCSA-RRR-09-042. September. Washington, D.C.: Federal Motor Carrier Safety Administration.

Order, D., D. Skorupski, R. Stinebiser, and R. Kreeb. 2009. Fleet Study of Brake Performance and Tire Pressure Sensors. FMCSA-PSV- 09-001, FMCSA-PSV-05-001. July. Washington, D.C.: Federal Motor Carrier Safety Administration.

TRB (Transportation Research Board). 2010a. Regulation of Weights, Lengths, and Widths of Commercial Motor Vehicles. Special Report 267. Washington, D.C.

TRB. 2010b. Achieving Traffic Safety Goals in the United States, Lessons from Other Nations. Special Report 300. Washington, D.C.

Woodrooffe, J., D. Blower, G. Timothy, P. Green, B. Liu, and P. Sweatman. 2009. Safety Benefits of Stability Control Systems for Tractor-Semitrailers. DOT HS 811-205. October. Washington, D.C.: U.S. Department of Transportation.

8

SuperTruck Program

INTRODUCTION AND BACKGROUND

This chapter provides an overview and the committee's evaluation of the SuperTruck projects that began in 2010. The SuperTruck projects were designed to provide complete long-haul Class 8 trucks that incorporate a wide range of technologies that have been developed under the 21st Century Truck Partnership (21CTP) over the past decade. This program offers the opportunity to demonstrate which technologies can actually be implemented in a truck, as well as the opportunity to perform vehicle-level testing of the technologies. The chapter introduction describes the basic design and intent of the SuperTruck projects. The second section describes the goals, timetables, tasks, and deliverables of the SuperTruck projects. The third section examines in detail the budgets available for these projects, and the fourth section looks at the SuperTruck project teams and the technologies that they plan to evaluate. The chapter concludes with the committee's evaluation of the SuperTruck project plans, goals, and overall approach.

Nature of the SuperTruck Projects

On January 11, 2010, the Department of Energy (DOE) announced the selection of nine projects, totaling more than \$187 million, to improve the fuel efficiency of heavy-duty trucks and passenger vehicles. The funding includes more than \$100 million from the American Recovery and Reinvestment Act of 2009 (ARRA).[1] The projects require that private industry contribute at least 50 percent of the project cost, and so a total of more than \$375 million will be provided for research, development, and demonstration projects (DOE, 2010a).

Of the funding recipients announced by DOE, three projects will focus on measures to improve the overall efficiency of long-haul Class 8 trucks. These projects will receive \$115 million in DOE funding to develop and demonstrate full-vehicle system-level technologies by 2015. The total cost of the full-vehicle projects will be more than \$230 million, including contractor contributions to the funding. Three projects have been selected for awards under the SuperTruck program:[2]

1. *Cummins, Inc., Columbus, Indiana*—Develop and demonstrate a highly efficient and clean diesel engine, an advanced Waste Heat Recovery (WHR) system, an aerodynamic Peterbilt tractor-trailer combination, and a solid oxide fuel cell auxiliary power unit (APU) to reduce engine idling.
2. *Daimler Trucks North America, LLC, Portland, Oregon*—Develop and demonstrate technologies including optimized combustion, engine downsizing, electrification of auxiliary systems such as oil and water pumps, waste heat recovery, improved aerodynamics, hybridization, and possibly a fuel cell auxiliary power unit to reduce engine idling.
3. *Navistar, Inc., Melrose Park, Illinois*—Develop and demonstrate technologies to improve truck and trailer aerodynamics, combustion efficiency, waste heat recovery, hybridization, idle reduction, and reduced rolling resistance tires.

The objective of the three SuperTruck projects is to develop and apply technologies leading to a system-level demonstration of highly efficient and clean diesel-powered Class 8 trucks that (DOE, 2010b):

- Achieve a 50 percent increase in vehicle freight efficiency measured in ton-miles per gallon, which

[1] Funding Opportunity Announcement (DOE, 2010b, pp. 6-8).

[2] Volvo Technology of America was awarded a fourth SuperTruck project in August 2011. Details of the technology development plan were not available to the committee during the preparation of this report.

translates to a 33 percent reduction in load-specific fuel consumption (gallons per 1,000 ton-miles);
- Achieve at least a 20 percent improvement through engine efficiency development, and achieve 50 percent brake thermal efficiency (BTE) under highway cruise conditions, which translates to a 16.7 percent reduction in fuel consumption due to engine improvements; and
- Evaluate potential approaches to 55 percent BTE in an engine via modeling, analysis, and potentially also laboratory tests.

Deliverables include computer simulation and hardware testing, as well as full-vehicle demonstrations using realistic drive cycles. An additional deliverable called for in the Funding Opportunity Announcement (FOA) is that "the systems developed shall be validated as cost effective via a business case analysis and will be reviewed for commercialization potential in later project phases as part of the phase gate review process" (DOE, 2010b, p. 7).

Each of the three teams is composed of a number of partners, including engine and truck original equipment manufacturers (OEMs), suppliers, fleet owners, universities, and DOE laboratories. Although each team has the same objectives, different technologies have been selected by the different teams to meet these objectives. For example, Navistar and Daimler plan to use hybridization in their approach to meeting the 50 percent vehicle freight efficiency target, whereas Cummins does not. In addition, the Navistar and Daimler teams plan to use different types of hybrid systems. Later in this section, the teams and the technical approach used by each team are identified. In general, each team will seek to improve vehicle freight efficiency through improved powertrain efficiency, idle reduction, reduced aerodynamic drag, and reduced tire rolling resistance, among other technologies.

Background and Relationship to Previous 21st Century Truck Projects

The SuperTruck projects can be considered a logical extension of prior research and development (R&D) activities of the 21CTP, in the sense that many of the technologies that will be applied in a system-level demonstration began as R&D initiatives and component-level demonstrations. Indeed, the National Research Council's (NRC's) Phase 1 report on the 21CTP had several recommendations for system-level demonstrations and a recommendation that industry partners assess cost objectives required to achieve commercial viability (NRC, 2008). The SuperTruck projects should result in more accurate estimates of the commercial viability of the various technologies. In Table 3-9 of the NRC (2008) Phase 1 report, it is noted that a shortcoming of component testing is that such hardware demonstrations are not subject to the realistic packaging constraints typical of commercialization. Prototype vehicle demonstrations should address this concern. Cummins intends to demonstrate 50 percent BTE under highway cruise conditions,[3] as requested in the program objectives and as recommended by the NRC Phase 1 report. The other teams are expected to provide similar demonstrations. Recommendation 4-6 in the NRC Phase 1 report suggests continued development and demonstration of heavy-duty hybrid truck technology, as will be addressed by the Navistar and possibly also by the Daimler SuperTruck teams. The SuperTruck projects plan to address Recommendation 5-1 of the NRC Phase 1 report, which suggests continued evaluation of systems that can improve idle reduction, along with the study of the cost-effectiveness of such systems. The Cummins and Daimler teams plan to evaluate fuel cell APUs, and the Navistar team plans to use the hybrid system battery to provide idle reduction. With regard to lightweight materials research, the NRC Phase 1 report's Recommendation 5-3 notes that it should be the responsibility of truck manufacturers to take the next steps of system integration, product validation, and production of a lightweight truck—an opportunity afforded by the SuperTruck program. Many of the fuel-saving technologies that will be implemented by the SuperTruck teams add significant weight to the vehicle, so all the teams have plans to implement offsetting weight reductions. In short, the SuperTruck program appears to address several of the shortcomings noted in the NRC (2008) Phase 1 report.

GOALS, TIMETABLES, TASKS (ACTIVITIES), AND DELIVERABLES

Project Goals

The project goals were listed by DOE (2010b) in the Funding Opportunity Announcement (FOA). The overall goal for Class 8 tractor-trailers is to develop and demonstrate a 50 percent total increase in vehicle freight efficiency measured in ton-miles per gallon (equivalent to a 33 percent load-specific fuel consumption reduction [gal/1,000 ton-mile]). This will be achieved through efficiency improvement in advanced vehicle systems technologies and advanced engine technologies. The project duration will be up to 5 years. At least 20 percent of this 50 percent improvement will be through the development of a heavy-duty diesel engine capable of achieving 50 percent BTE on a dynamometer under a load representative of a level road at 65 mph (see Chapter 3 in this report). Specific technology developments mentioned in the FOA include ancillary systems, waste heat recovery, materials, and electrification in addition to advanced combustion techniques.

The project efficiency goals must be met while adhering to prevailing (Environmental Protection Agency [EPA]) 2010

[3] Donald Stanton, Cummins, "Cummins-Peterbilt SuperTruck Program," presentation to the committee, September 8, 2010, Washington, D.C.

emissions standards as well as the vehicle safety and regulatory requirements that apply to Class 8 tractor-trailers. The FOA stipulates that the vehicle efficiency improvement will require an integrated team that includes an engine manufacturer, a truck OEM, and a trailer manufacturer, along with suppliers, national laboratories, universities, fleet operators, and other stakeholders, to ensure maximum benefit.

Additional fuel-saving technologies listed in the FOA include electrical or mechanical drivetrain hybridization with energy storage and regeneration, along with reductions in aerodynamic drag, rolling resistance, weight, and main engine idle. The FOA specifies that any demonstration of achievement must utilize a test cycle proposed by the team that is representative of a typical long-haul Class 8 truck, and including a minimum of 75 percent of the distance traveled under highway conditions, with a vehicle weight of 65,000 lb. The level of improvement is based on a comparison to a similarly configured "best-in-class" 2009 commercially available Class 8 vehicle. Comparable vehicle performance, such as acceleration times and grade capability, is to be maintained by the SuperTruck vehicle.

A second project goal is to identify key pathways to achieving a 55 percent BTE heavy-duty diesel engine, through modeling and analysis conducted in parallel with work toward the overall goal of a 50 percent improvement in vehicle freight efficiency. Critical components and/or systems needing specific additional development to achieve this 55 percent BTE goal should also be identified. This engine must be capable of meeting 2010 emissions standards, and it must be commercially viable, which implies a requirement for cost-effectiveness.

Relationship Between SuperTruck Goals and Previous (2006) 21CTP Goals

The most outstanding difference between the SuperTruck goals given above and the 21CTP goals established in 2006 is that the 2006 goals included no requirement for a vehicle measurement of efficiency improvement. This omission in the 2006 goals was noted in the NRC Phase 1 report in Overall Report Recommendation 1-1, stating in part: "more (major truck) manufacturers should be participants" and goals should be "strategic to reducing fuel consumption of heavy-duty vehicles" (NRC, 2008, p. 2). These omissions have been corrected in the current project goals, which require a vehicle demonstration of fuel-consumption reduction. It is particularly refreshing that the SuperTruck demonstration is required to be conducted under real-world operation and gross vehicle weight (GVW) conditions.

Although the 2006 21CTP goal provided for achieving a 20 percent increase in (peak) BTE to 50 percent, the 2008 NRC Phase 1 report, in Recommendation 3-1, clarified that "objective and consistent criteria (were not) used to assess the success or failure of achieving that key goal" (NRC, 2008, p. 3). This shortcoming has been corrected in the current project goals, in which an engine dynamometer demonstration is required over a simulated real-world duty cycle, and the 50 percent BTE goal is to be achieved at the "highway cruise" condition.

It is worth noting that the target for 50 percent BTE under "highway cruise" conditions is significantly more stringent than the original 21CTP goal of 50 percent peak BTE. The original goal was for the engine's best operating point, which is typically at or near full load at relatively low engine speed. Cruise speed is typically somewhat higher than the best BTE speed, and cruise load is lower than the typical best BTE load point. In addition, the vehicle aerodynamic and rolling resistance goals of SuperTruck will lead to lower vehicle power demand on the engine at cruise, which makes the target of 50 percent BTE at cruise even more challenging. The SuperTruck contractors may find that this change in goals introduces a significant technical challenge.

The 2006 21CTP goals provided for an effort to develop component technologies for reaching a 55 percent (peak) BTE, with a particular focus on low-temperature combustion (LTC) (NRC, 2008). The current project reduces the 55 percent BTE focus to one providing for modeling and analysis, including a requirement to comply with 2010 criteria emissions. The FOA calls for identifying critical components and/or systems needing specific development, and finally evaluating the overall system for commercial viability. No test demonstration appears to be required, although two SuperTruck teams (Cummins and Navistar) showed the committee plans for a test cell demonstration of the 55 percent target.

In summary, comparing the goals for these SuperTruck projects to the previous (2006) 21CTP goals, it appears that the DOE has implemented more robust requirements for demonstrations under near-real-world conditions and has required the contracting teams to include a wide range of technical specialties. Now that a number of technologies have been demonstrated on a laboratory scale under previous 21CTP projects, this new approach should serve the trucking industry and the public more favorably over the next several years.

BUDGETS

The $115 million in DOE funding awarded to the SuperTruck program was divided as follows among the three project teams:[4,5]

[4] Patrick Davis, DOE, "U.S. Department of Energy Vehicle Technologies Program Overview," presentation to the committee, September 8, 2010, Washington, D.C.

[5] Volvo Technology of America was awarded $19 million in SuperTruck funding in August 2011. Volvo Technology of Sweden was awarded a similar amount by the Swedish government under a separate contract. The two contracts will combine to provide a SuperTruck program similar in scope to the other three contracts.

- Cummins team: $38,831,115;
- Daimler team: $39,559,868; and
- Navistar team: $37,328,933.

According to the aforementioned FOA (DOE, 2010b), a 50 percent or greater cost sharing is required of the project teams, but in many cases the requirement for a 50 percent cost sharing will be exceeded. For example, Navistar noted that it was providing $34 million of "in kind funding," and its partners will provide $16.6 million. When combined with the $37.3 from the DOE, the total Navistar project funding is about $88 million over a 5-year span. It should also be noted that while the Cummins and Daimler teams are directly supported by ARRA funds, the Navistar team is funded from internal DOE resources. This resulted in a delayed start for the Navistar team and will also necessitate annual funding of the contract as DOE funds permit.

In addition to the funding provided for the SuperTruck program directly, there are also projects in the 21CTP budget request that provide support at the component and subsystem level for many of the technologies that will be applied in the SuperTruck projects. Table 1-2 in Chapter 1 shows details of funding for non-SuperTruck-related 21CTP projects. However, that table also includes funding for one of the three SuperTruck projects, namely Navistar, which was $4.35 million in FY 2010 and $7.3 million in FY 2011, and so these amounts would have to be subtracted to arrive at the non-SuperTruck funding level. There is no clear definition in Table 1-2 of which line items are SuperTruck-related and which are not. These non-SuperTruck projects are in areas of technology that are also being explored by the SuperTruck teams, so the effort in the non-SuperTruck projects should complement the SuperTruck project efforts and help fill the technology pipeline for the SuperTruck projects. The SuperTruck projects funding of $115 million is over 5 years, or an average of $23 million per year.

In summary, the DOE has made a significant investment in this effort to improve the overall fuel efficiency of heavy-duty trucks. The SuperTruck projects will form the backbone of the 21CTP work over the next several years, and they will consume most of the budget.

TECHNOLOGIES AND TEAMS

Summarizing the Three SuperTruck Team Plans

In addition to describing the two high-level goals, the DOE in its FOA document (DOE, 2010b) also suggested numerous subtopics that the teams might evaluate for contribution to those goals. The committee finds the topic list substantially inclusive. Each of the three SuperTruck teams made a presentation to the committee providing an overview of its development plans and identifying its team partners, to the extent they were firmed up. Further, both Cummins and Daimler have made presentations at the DEER 2010 (Directions in Energy-Efficiency and Emissions Research) conference, where some additional details of their SuperTruck plans were revealed.

A summary of the three SuperTruck Team plans is shown in Table 8-1. The first column lists technologies that the teams plan to evaluate. This list includes technologies suggested by the DOE, as well as suggestions from the contracting companies. For example, all three companies added optimization opportunities for catalytic exhaust systems that may provide measurable fuel-consumption reductions. Entries described as "(Implied)" indicate that the companies had listed the topic but without any elaboration to describe the approach that they intend to take. It is not clear whether the "implied" technologies will be included in the final project, or what form the technologies might take. Two of the teams made confidential presentations to the committee regarding their plans, including details that were not publicly disclosed. Technologies and approaches that were described only in these confidential presentations are not shown in Table 8-1.

A review of Table 8-1 shows that the three studies will take different approaches to the program goals, although there is overlap. Many of the technology topics are sufficiently broad so as to provide considerable leeway for unique paths and multiple solutions within a specific technology field.

The program targets specify a 50 percent increase in vehicle freight efficiency (ton-miles per gallon) and specify that 20 percent of the improvement must come from improved engine efficiency in terms of BTE. Because fuel economy works in a multiplicative rather than an additive fashion, a 20 percent improvement in engine efficiency combines with a 20 percent reduction in vehicle power demand to produce the required 50 percent overall fuel economy target.[6] Items 1 through 6, as well as 8 and 9 in Table 8-1 would directly contribute to reduced load on the engine. The SuperTruck teams have not predicted efficiency performance values for any of these technologies in their public materials.

In vehicle aerodynamics, with the exception of Navistar, it is not clear which technologies are being developed beyond those found in the 2010 NRC report on medium- and heavy-duty vehicle fuel consumption reduction (NRC, 2010). Navistar has identified both an "active fifth wheel"

[6] A 20 percent increase in BTE leads to a 20 percent increase in fuel economy, and a 16.7 percent decrease in fuel consumption. A 20 percent reduction in vehicle power demand yields a 20 percent reduction in fuel consumption (assuming constant BTE), and a 25 percent increase in fuel economy. These two improvements combine as follows to meet the 50 percent FE (or 33 percent FC) target:

New Fuel Economy = Old FE × ((1 + % increase in FE from BTE) × (1 + % increase in FE from Reduction in Load)).

New FE = Old FE × ((1 + 20%) × (1 + 25%)) = Old FE × (1.2 × 1.25) = 1.5 × Old FE.

Thus, New Fuel Economy is 50 percent improved, as a result of improving BTE by 20 percent and reducing vehicle power demand by 20 percent.

TABLE 8-1 Comparison of SuperTruck Projects and Technologies to Be Explored by Each of Three Teams

Item No.	Technologies That the Teams Plan to Evaluate[a]	Industry/Government Lab/Academic/Trucking Fleet Teams		
		Engine mfg.: Cummins	Engine mfg.: Detroit Diesel[c]	Engine mfg.: Navistar
		Truck OEM: Peterbilt	Truck OEM: Freightliner (Daimler Trucks)	Truck OEM: Navistar
		Trailer mfg.: Utility	Others TBD	Trailer mfg.: Wabash Suppliers: ATDynamics, Alcoa, Behr, Bosch, Federal Mogul, Michelin, ArvinMeritor. National labs and universities: TPI /LLNL, ANL
		Suppliers: Eaton, Delphi, Modine, Dana, Bridgestone, Van Dyne		Truck operators: Safeway, Swift[d]
		National labs and universities: ORNL, Purdue		
		Truck operator: U.S. Express[b]		
1	Vehicle aerodynamics	SmartWay tractor, trailer gap closure, full trailer skirts, aft body plates, optimized cooling to reduce aerodynamic impact, other features TBD[e]	Smartway tractor and trailer, other technologies implied	SmartWay+ tractor, trailer gap device, full trailer skirts, aft body plates, active fifth wheel, height lowered at highway speed
2	Transmission	Advanced	(Implied)	*See* Hybrid powertrain
3	Hybrid powertrain	Infrequent start/stop favors waste heat recovery—no hybrid planned	Hybrid system type not specified[f]	Diesel-electric dual mode (series/parallel)
4	Road load management	GPS, adaptive cruise, driver feedback	Predictive cruise control using 3D digital map database (shown as an example in presentation)	
5	Rolling resistance	Reduced rolling resistance	(Implied)	Super single tires/wheels
6	Axles	Smart axle (2-wheel/4-wheel drive)	Long-haul tandem (possibly 6 × 2)	Smart 6 × 2 tandem
7	Idle management	Solid oxide fuel cell APU	Fuel cell APU	Hotel-loads from hybrid
8	Weight reduction	Features TBD	Cross-members[f] other features TBD	Cab and trailer composites, plastic fuel tanks, aluminum wheels/brake rotors, aluminum cross-members and driveshafts, carbon composite brake drums
9	Solar panels		(Implied)	
	Engine			
10	Base engine (PCP, friction/parasitics)	Increased PCP, changes to combustion modes implied	In-cylinder pressure sensor	Combustion feedback
11	Fuel system	(Implied)	Bosch APCRS	Increased injection pressure, parallel gasoline injection option
12	Advanced LTC	Increased PCCI regime, lifted flame diffusion burn	PCCI studies	Diesel and gasoline injection (dual fuel)
13	Controls	Powertrain router as network coordinator[g]	Features TBD	Features TBD
14	Electrically driven components	(Implied)	Some accessories	Some accessories
15	Waste heat recovery	Rankine cycle, mechanical drive to engine	Rankine cycle, drives electric generator and/or turbocompound[h]	Rankine cycle, drives electric generator

continues

TABLE 8-1 Continued

Item No.	Technologies That the Teams Plan to Evaluate[a]	Industry/Government Lab/Academic/Trucking Fleet Teams		
16	Aftertreatment	(Implied)	SCR and engine-out NO_x optimization	Minimize DPF regeneration; optimize PM-NO_x
17	Turbo technology	Turbocharger with its own CVT transmission[i]	Turbo compound	Turbo compound, dual turbos
18	EGR loop	(Implied)	(Implied)	Hybrid EGR (low/high pressure)
19	Variable valve technology	(Implied)	(Implied)	Compression ratio control, cylinder deactivation
20	Engine downsizing		Evaluation	

NOTE: A summary of the three SuperTruck Team plans is shown. These projects are in response to the DOE's FOA (DOE, 2010b). Entries described as "(Implied)" indicate that the companies had listed the topic, but without any elaboration to describe the approach they intend to take. It is not clear whether the "implied" technologies will be included in the final project, or what form the technologies might take.

Acronyms are listed in appendix I. TBD, to be determined.

[a] Includes technologies suggested by the DOE, as well as suggestions from the contracting companies.

[b] Donald Stanton, "Cummins-Peterbilt SuperTruck Program," presentation to the committee, September 9, 2010, Washington, D.C.

[c] Derek Rotz, "Daimler: DTNA/DDC R&D with DOE: PCC, NZ-50, Super Truck," presentation to the committee, September 9, 2010, Washington, D.C.

[d] Anthony Cook, "Navistar's SuperTruck Program," presentation to the committee, September 9, 2010, Washington, D.C.

[e] CRADA: "Integrated Thermal and External Aerodynamics—Cummins," Argonne National Laboratory, Jules Routbort, Merit Review, June 2010.

[f] Elmar Bockenhoff, "Daimler: Heavy Duty Diesels: The Road Ahead," September 27, 2010, DEER Conference, Detroit, Mich.

[g] Donald Stanton, "High Efficiency Clean Combustion for SuperTruck," September 29, 2010, DEER Conference, Detroit, Mich.

[h] Kevin Sisken, Detroit Diesel Corporation, "Increased Engine Efficiency via Advancements in Engine Combustion Systems," September 29, 2010, DEER Conference, Detroit, Mich.

[i] Cummins Partners with VanDyne SuperTurbo on Super Truck Program, August 17, 2010. Available at TruckingInfo.com.

and "ride height lowered at highway speed" as potential new contributions.

Navistar and Daimler will investigate the optimization of an electric hybrid system, noting that even a modest (circa 6 to 9 percent) contribution may justify the complexity and expense of such systems, given the high fuel consumption of highway-duty-cycle tractor-trailer combinations. It appears that the two teams exploring electric hybrids plan to use substantially different systems.

All three companies have specific plans for managing hotel loads during extended idle periods. Idle reduction regulations have been imposed by 46 states and jurisdictions on the heavy-truck industry (see Chapter 6 for details). The states' permitted idle time has an average limit of 5 minutes and a range of 0 to 15 minutes (usually per 6 to 8 hours, sometimes per hour) (ATRI, 2011). Cummins and Daimler plan to evaluate fuel cells to support hotel loads and reduce idle. As an alternative, Daimler is also pursuing a hybrid-electric solution that could manage the hotel loads. Thus, Daimler will evidently have two optional solutions to the idle-fuel-consumption problem. Navistar indicated that its hybrid-electric-system battery was expected to be able to handle the hotel load and idle reduction requirements.

Waste heat recovery utilizing a Rankine cycle was described by all three companies. Two different methods of energy utilization are being explored. Energy from the WHR system can be fed directly back to the crankshaft as mechanical energy, as shown in Figure 8-1. Another approach is to use the WHR energy to generate electricity for use by a hybrid system or to power electrical accessories. The WHR system as shown in Figure 8-1 is relatively complex and bulky, making packaging, reliability, and cost all significant issues to be overcome before this technology can be implemented in production.

Project Schedules

A project phase-timing chart is a typical management tool for assessing and tracking the resources needed and expended for even moderately complex projects. It is not known if detailed charts were submitted to the DOE during the competitive bidding on the FOA, but such charts should have been part of the application deliverables. Two of the companies included extremely brief planning charts in their public presentations at the committee's September 8-9, 2010, meeting. The chart from the Cummins presentation is nominally useful to perceive the project complexity and time frame but not adequately detailed to assess needed resources. For example, the Cummins schedule calls for a test cell demonstration of a 50 percent BTE engine by the end of 2011, a demonstration of the vehicle-level target at the end of 2012, a demonstration of more than 50 percent vehicle freight efficiency improvement over a 24-hour duty cycle by October 2013, and a demonstration of the 55

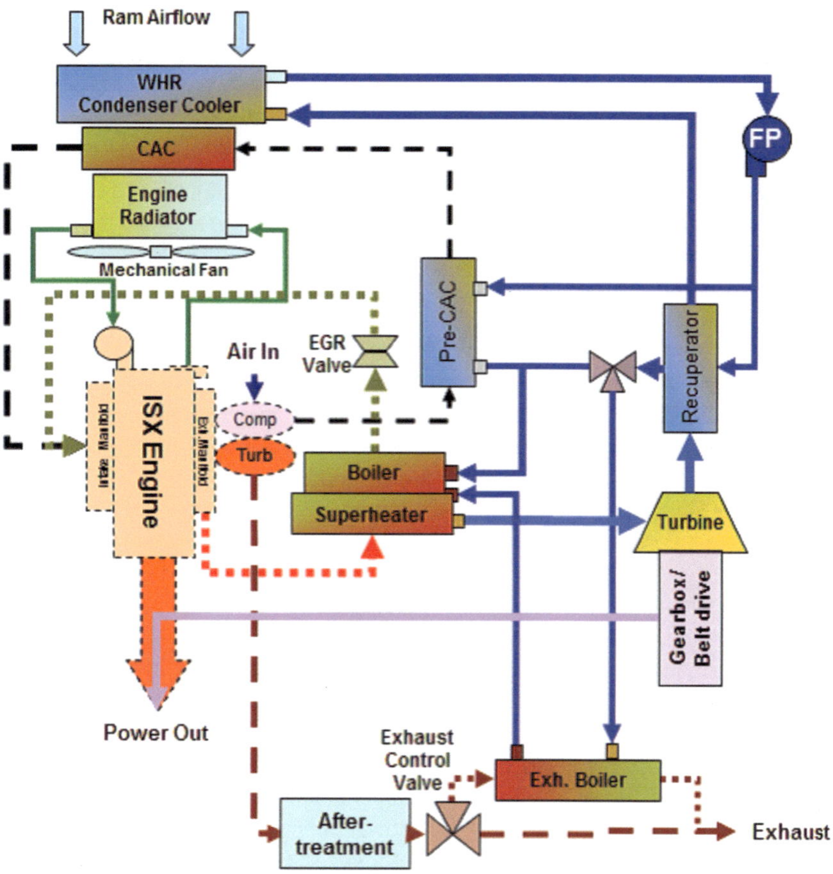

FIGURE 8-1 Cummins organic Rankine cycle waste heat recovery with energy returned mechanically to crankshaft. Acronyms are defined in Appendix I. SOURCE: Donald Stanton, Cummins, "Cummins-Peterbilt SuperTruck Program," presentation to the committee, September 9, 2010, Washington, D.C.

percent BTE engine objective by June 2014. The Cummins schedule did not provide any additional detail regarding the company's project plans. The chart from the Daimler presentation to the committee indicates only the overall project time frame but otherwise provides no additional details. Navistar did not include a project schedule in its presentation.

COMMITTEE EVALUATION OF SUPERTRUCK PROJECT PLANS

Evaluation of SuperTruck Program Goals

Although the SuperTruck program has well-defined goals, it leaves many very important details to the individual teams. The overall freight efficiency improvement needs to be measured on a specific duty cycle or combination of duty cycles. The selection of these duty cycles is left up to the contractors. Two of the contractors showed the committee plans to use a range of duty cycles rather than a single duty cycle in order to better evaluate the technologies. The SuperTruck FOA does specify that at least 75 percent of the duty cycle's distance traveled must be representative of highway operation and that the vehicle shall operate at a weight of 65,000 lb. It is the committee's understanding that the baseline truck will be tested at a gross combined weight (GCW) of 65,000 lb and that the same freight load will be carried by the SuperTruck prototype. This means that the test weight of the SuperTruck prototypes may be more or less than 65,000 lb, depending on changes in the weight of the empty vehicle. This means that the vehicle will be representative of a "cubed-out" operation, where the trailer is filled with low-density freight before the vehicle reaches the maximum legal weight, which is typically 80,000 lb. For long-haul trucks, about 60 percent cube-out, but there is a significant population (tankers, bulk haulers, haulers of high-density freight such as steel, etc.) that routinely "gross out." Given the fact that the majority of trucks cube out, the committee is satisfied with the DOE's selection of 65,000 lb total vehicle weight for the SuperTruck project.

The selection of duty cycle is very important, for several reasons. First, the selection of duty cycle can have a significant impact on the performance of specific fuel-saving technologies. For example, a constant 65 mph cruise cycle, with no grade, would highly favor aerodynamic improvements, but energy storage systems such as hybrid powertrains would have little or no benefit. A duty cycle that includes grades and lower-speed operation will give less emphasis to the benefits of aerodynamic improvements, but hybrid systems will offer a greater benefit. A duty cycle with no grades would also not capture the benefit of a feature such as predictive cruise control, which adjusts speed in anticipation of grade changes. One result of the requirement that the vehicle be tested at 65,000 lb is that there will be little fuel-consumption advantage in weight reduction, as well as little penalty for a weight increase. The committee's understanding is that since a cubed-out situation is being represented, the freight load will be held constant, regardless of any change in empty vehicle mass. As a result, a 1,000 lb change in vehicle weight can be expected to cause only a 0.4 to 1.0 percent change in fuel consumption (NRC, 2010).

A second possible effect of duty-cycle selection is to make it easier or more difficult to reach the 50 percent vehicle fuel economy (33 percent fuel consumption) improvement target. This issue follows from the issues regarding the performance of specific technologies on a given duty cycle.

The third, and perhaps most important, issue surrounding the choice of duty cycle is the question of how well the selected duty cycle will represent "real-world" operating performance. This issue is not easy to resolve, because no one duty cycle can possibly represent all real-world truck operations, even within a fairly narrow segment such as heavy-duty long haul. Some operators often run with light loads, and others often operate at (or above) the maximum legal load. Some operators work in areas with little traffic where sustained constant-speed operation is possible. Some operators spend most of their time in or near large cities where congestion often restricts operating speed and where speed fluctuation is substantial. Some operators stay in areas with little grade, and others spend a lot of time in the mountains. All of these variations will have a substantial effect on fuel consumption and on the relative benefits of various fuel-saving technologies. There is no "one size fits all" solution.

The NRC report on medium- and heavy-duty vehicles showed that a fuel-consumption reduction of 50 percent is possible for a Class 8 tractor-trailer truck (NRC, 2010). This figure makes the 33 percent fuel-consumption target of the SuperTruck program appear relatively modest. However, several factors need to be taken into account. First, the NRC report used an "average" 2008 model truck as its baseline, whereas the SuperTruck program uses a "best-in-class" 2009 model as the baseline. The "average 2008" tractor would have a higher drag coefficient and higher rolling resistance tires than the best-in-class 2009 tractor, and the average 2008 trailer is without aerodynamic devices. In the NRC (2010) report, the "average 2008" vehicle was defined as having a drag coefficient C_d = 0.635, and a rolling resistance C_{rr} = 0.0068. The SuperTruck contractors declined to reveal the C_d or C_{rr} values for their 2009 best-in-class vehicles, but these should be significantly better. Second, the 50 percent fuel savings value in the NRC (2010) report included benefits from fleet management technology and driver training and coaching. These factors are not included explicitly in the SuperTruck program. Once these differences in baseline and scope are taken into account, the SuperTruck program targets match well with the NRC (2010) report's projection of fuel savings available in the 2016 to 2020 time frame.

A number of 21CTP projects over the past several years have had as a goal the achievement of either 50 or 55 percent BTE from the engine. The engine section titled "Brake Thermal Efficiency Improvements" of Chapter 3 describes these projects and their results in detail. The 50 percent BTE target has proven to be a substantial challenge, which requires the use of expensive and complex technologies such as bottoming cycles. In 21CTP projects to date, technologies have been demonstrated individually which, if combined on a single engine, should provide a BTE of 50 percent (DOE, 2010a). However, there has not yet been a demonstration of a 50 percent BTE engine in a vehicle. In addition, the requirement for 50 percent BTE at cruise load poses an additional challenge, because the best point for BTE is typically at a higher load. This issue will be exacerbated by the fact that SuperTruck vehicle improvements will significantly reduce power demand at cruise, which will push the engine to a less efficient, lighter-load operating point at cruise. Given these considerations, the 50 percent BTE target appears to be a relatively risky, but not impossible, goal. The consensus of the committee is that the technical paths to achieving 55 percent BTE that the contractors will provide are indeed likely to include some "stretch" goals and some technologies that may prove impractical or extremely expensive. As technology progresses over time, the 55 percent target may become more feasible, but there are fundamental thermodynamics issues that will be difficult to overcome. The DOE Office of Vehicle Technologies Multi-Year Program Plan (DOE, 2010c, p. 2.3-2) states that "this activity will also conduct R&D on advanced thermodynamic strategies that may enable engines to approach 60 percent thermal efficiency." Any consideration of BTE targets beyond 55 percent should be carefully examined in light of the laws of thermodynamics.

Evaluation of SuperTruck Team Plans

In addition to public presentations made to the committee by the Cummins/Peterbilt and Navistar teams, the committee visited the Cummins/Peterbilt team on November 8, 2010, and the Navistar team on January 13, 2011. These visits were made on a confidential basis, and so details of the plans that were discussed during these visits cannot be included in this report. The Daimler team presented its preliminary plans to the committee on a public basis on September 8-9, 2010.

The Daimler presentation used predictive cruise control as an example of how it intends to create predictive controls for the engine, transmission, and vehicle.[7] No details were offered on what these controls might do. The presentation lists the technologies that are planned for the SuperTruck program and offers a preliminary risk/benefit evaluation for many of these technologies. Many of the features that are planned fit the list suggested by the DOE. Some additional items added by the Daimler team include predictive engine controls, predictive vehicle controls, route-management and driver-feedback features, and the use of solar panels.

The Cummins/Peterbilt team presented a comprehensive plan that addresses a number of technology areas, including engine, transmission, bottoming cycle, tractor and trailer aerodynamics, weight reduction (or at least compensation for the weight of new components and systems), and rolling resistance.[8] Additions to the DOE list of suggested technologies include a solid oxide fuel cell APU for idle elimination and a "smart axle," which evidently shifts to 6X2 operation when high traction is not required. Adaptive cruise control and unspecified driver-feedback features are also part of the Cummins/Peterbilt plan. The team will not use a hybrid system, because its simulation work indicates that waste heat recovery has more potential in long-haul applications.

The Navistar team's plan generally follows the list of technologies provided by the DOE. Notable additions include speed-adjusted ride height and an active 5th wheel for additional aerodynamic improvement.[9] Navistar also plans to use a series/parallel hybrid system, by which the vehicle operates in series mode (diesel-electric) at low vehicle speeds and operates in parallel mode (with direct drive from the engine to the axle) at high speeds. The dual-mode hybrid system also allows for electric-only operation for short distances. Navistar will evaluate a dual fuel system (gasoline and diesel fuels) as part of its program to demonstrate an engine capable of 55 percent BTE. The Navistar plan also includes a very extensive range of weight-reduction features.

All three contractors have plans for evaluating the cost-effectiveness and potential payback of the technologies in their plans. A business case analysis for commercialization is one of the deliverables for the SuperTruck programs (DOE, 2010b). The committee believes that this cost-effectiveness evaluation is a critical part of the project and hopes that comparable methods and approaches will be used so that the results from the three contractors can be compared.

In general, the committee believes that all three SuperTruck projects have plans that offer the potential for meeting program goals. The plans also cover a wide range of technologies and allow for the evaluation of technologies over a range of operating conditions. One concern is that current plans do not call for all three SuperTrucks to be evaluated on a common duty cycle, since each team will develop its own duty cycle or cycles. This lack of a common duty cycle will make it more difficult to compare the performance of the three vehicles and the benefits of individual features and technologies at the end of the program.

The SuperTruck projects include a high level of effort and a high level of technical risk. Areas of risk include the following:

- Many of the fuel-saving technical features, such as hybrid systems, aerodynamic features, and bottoming cycles, add significant amounts of weight. A substantial amount of weight-reduction effort (and cost) will be required just to maintain the baseline vehicle weight.
- Some technologies, particularly bottoming cycles, are likely to pose significant reliability issues.
- Some technologies may not prove cost-effective, such as extensive weight reduction.
- Because each contractor will develop its own duty cycles and test protocols, it will not be possible to compare the results of the three programs directly.

To deal with the last area of risk, the committee believes that the contractors should calculate the fuel consumption for the baseline vehicle and engine and for the fully developed SuperTruck vehicle and engine, using the EPA and NHTSA fuel-consumption regulations. This will allow comparison of the improvements achieved by the three contractors. In addition, the committee believes that the three contractors and the DOE should agree on a single, "real-world" vehicle fuel-consumption test protocol (duty cycle) that will be used by all three contractors, in addition to the tests developed independently by each contractor. This common test would provide another data point that could be used to compare the accomplishments of the three SuperTruck projects.

FINDINGS AND RECOMMENDATIONS

Finding 8-1. The three SuperTruck projects will be the flagship projects under the 21CTP for FY 2011 through FY 2014; the goals are in concert with recommendations made in the 2008 NRC Phase 1 report. A large portion of the DOE 21CTP budget will be devoted to these three projects. Each SuperTruck project integrates a wide range of technologies into a single demonstration vehicle (engine, waste heat recovery, driveline, rolling resistance, tractor and trailer aerodynamics, idle reduction, weight reduction technologies, etc.), and the contractors are pursuing sufficiently different technical paths to avoid excessive duplication of effort. The results will help determine which fuel-saving technologies are ready and cost-effective for OEM-level product development programs.

---

[7] Derek Rotz, Kevin Sisken, and David Kayes, Daimler and Detroit Diesel, "DTNA/DDC R&D with DOE; PCC, NZ-50, Super Truck," presentation to the committee, September 9, 2010, Washington, D.C.

[8] Donald Stanton, Cummins, "Cummins-Peterbilt SuperTruck Program," presentation to the committee, November 15, 2010, Washington, D.C.

[9] Anthony Cook, Navistar, "Navistar's SuperTruck Program," presentation to the committee, September 9, 2010, Washington, D.C.

Finding 8-2. Rather than have a number of targets for each subsystem, the SuperTruck projects have only two types of goals: one for the engine and one for overall vehicle fuel efficiency. This approach reflects the EPA and NHTSA approach to heavy-duty fuel efficiency regulations. Each project team is allowed to select a set of technologies that meet the project goals. The engine goal of 50 percent BTE for the demonstration vehicle appears to be feasible, although there is risk in being able to achieve it at a cruise condition. The engine goal of 55 percent BTE demonstrated in a test cell is very high risk but might be achievable. The overall vehicle goal of a 33 percent reduction in load-specific fuel consumption appears to be feasible.

Finding 8-3. Unfortunately, the SuperTruck program expresses vehicle efficiency targets in terms of fuel economy rather than fuel consumption. The vehicle target is stated as a 50 percent improvement in fuel economy rather than as a 33 percent reduction in fuel consumption. This can lead to confusion regarding the actual benefits of the program.

Recommendation 8-1. The DOE should state the SuperTruck program vehicle efficiency goals in terms of load-specific fuel consumption and track progress on this basis—that is gallons per 1,000 ton-miles, which is the metric used in the EPA/NHTSA fuel consumption regulations.

Finding 8-4. The committee believes that the SuperTruck project teams have developed plans that address the needs of the SuperTruck program and that have a reasonable chance for success. The keys to success include proper implementation of the plans along with the flexibility to adapt to new information and intermediate results during the course of the project.

Finding 8-5. The SuperTruck projects allow each team to design its own test duty cycle(s) within certain constraints. One negative consequence of this approach is that the three trucks may never be tested using a common cycle for comparison.

Recommendation 8-2. The DOE and the SuperTruck contractors should agree on at least one common vehicle duty cycle that will be used to compare the performance of all three SuperTruck vehicles. In addition, fuel consumption improvements should be calculated on the basis of the EPA and NHTSA fuel consumption regulations.

Finding 8-6. The SuperTruck projects go beyond the scope of previous 21CTP projects. Instead of relying entirely on simulations and laboratory testing, each of these projects will result in a drivable truck. The committee believes that it is important to take technologies that have been developed to date and implement them in a real vehicle. Often, the implementation of new technologies in real-world applications yields unexpected results, and these results must be explored before any new technology can be considered ready for production implementation.

REFERENCES

ATRI (American Transportation Research Institute). 2011. Idling Regulations Compendium. Available at http://www.atri-online.org/index.php?option=com_content&view=article&id=164&Itemid=70. Accessed January 18, 2011.

DOE (U.S. Department of Energy). 2010a. Secretary Chu Announces $187 Million to Improve Vehicle Efficiency for Heavy-Duty Trucks and Passenger Vehicles. Washington, D.C. Available at http://www.energy.gov/8506.htm.

DOE. 2010b. Funding Opportunity Number: DE-FOA-0000079. U.S. Dept of Energy, Recovery Act—Systems Level Technology Development, Integration, and Demonstration for Efficient Class 8 Trucks (SuperTruck) and Advanced Technology Powertrains for Light Duty Vehicles (ATP-LD), DE-FOA-0000079. August 24. Washington, D.C.: Department of Energy. Available at http://www.energy.gov/recovery/documents/zeDE-FOA-0000079.pdf.

DOE. 2010c. Multi-Year Program Plan 2011-2015. December. Washington, D.C.: Office of Vehicle Technologies.

NRC (National Research Council). 2008. Review of the 21st Century Truck Partnership. Washington, D.C.: The National Academies Press.

NRC. 2010. Technologies and Approaches to Reducing the Fuel Consumption of Medium- and Heavy-Duty Vehicles. Washington, D.C.: The National Academies Press.

9

Efficient Operations

INTRODUCTION AND BACKGROUND

There are many available technologies that can be applied to reduce the fuel consumption of trucks, but there are other ways of saving fuel that do not require any changes to vehicle or engine technologies. The overall fuel consumption of a truck fleet can be influenced significantly by the ways in which the vehicles are operated and maintained. Factors such as how close to full payload the trucks operate, whether they run on the most efficient routes, and even driver training can play a role in determining overall fleet fuel consumption. In addition, regulations can directly affect fuel consumption by constraining or promoting technology implementation and efficient operations. The infrastructure in which trucks operate also affects fuel consumption, through factors such as speed fluctuation and congestion. Electronic features can be added to the truck to modify the performance of the engine or vehicle in ways that can save fuel. All of these possibilities fall under the category of "efficient operations."

The U.S. Department of Energy (DOE) and U.S. Department of Transportation (DOT) have recently proposed "efficient operations" as a new area for work under the 21st Century Truck Partnership (21CTP). Although no programs or work as yet have been initiated in the 21CTP in this new area of efficient operations, the committee's statement of task charges the committee with examining and commenting on issues related to the 21CTP strategy. The 21CTP's proposal for this new area is laid out in a February 2011 draft white paper, "Reducing Fuel Consumption in U.S. Trucking—A DOE-DOT Joint Study and Whitepaper" (DOE-DOT, 2011). In this draft, the two agencies explore opportunities to improve the efficiency of trucking operations. The paper focuses on two areas:

- Opportunities for joint research and development (R&D) effort between the DOE and the DOT, and
- Opportunities for modifying regulations (primarily DOT regulations) in ways that could permit more efficient operations.

This chapter provides the committee's review of the identified opportunities for joint R&D. Because efficient operations is a new area for the 21CTP, the committee first offers its own review of the opportunities for reducing trucking fuel consumption by increasing the efficiency of operations, based on available literature and its own expertise. This information is provided as input to the DOE and DOT for their consideration as they revise the white paper and decide how to move forward in this area.

The committee notes that addressing freight fuel consumption requires a systems approach. Changes made to improve vehicle efficiency can have a negative impact on other aspects of trucking operations. For example, aerodynamic features may impede access to some loading docks. Also, changes made to vehicle specifications may have an effect on road wear or safety. The complete set of outcomes from a given proposed change must be evaluated in order to make good decisions about whether the change should be implemented.

Although it is beyond the purview of this committee to suggest changes to regulations, the committee offers its perspectives on R&D that would be needed in order to determine whether some of the commonly discussed changes to regulations would permit more efficient operations.

EFFICIENT-OPERATION OPPORTUNITIES

The objective of the draft white paper is as follows: "A key objective of this white paper is to identify specific opportunities and challenges with respect to advancing the state-of-the-art of truck fuel consumption and to highlight particular research needs that are seen as critical to maximizing the overall efficiency of freight movement and of trucks in general. It is proposed to develop a set of specific topics on which DOE, DOT and EPA (U.S. Environmental Protection Agency) can—and should—work together to further the goal of improved truck efficiency" (DOE-DOT, 2011, p.4).

Below, the committee has assembled a list of opportunities for more efficient truck operations. This list is by no means exhaustive, but it is intended to provide a survey of the opportunities available. As noted above, this information may be useful to the DOE and DOT as they revise the white paper and identify the highest-priority areas for joint R&D. It is indicated whether each item is covered in the current draft of the white paper.

1. *Vehicle maintenance.* Tire pressure has been shown to have a measurable effect on fuel consumption (NRC, 2010). This factor can be addressed by frequent, scheduled maintenance checks or by automated tire-pressure-maintenance systems that are mounted on the vehicle. In addition, factors such as axle alignment have an effect on rolling resistance and thus on fuel consumption (NRC, 2010). Axle alignment can be checked and adjusted as needed as a part of routine maintenance. As the diesel particulate filter (DPF) fills with ash, regenerations become more frequent, costing additional fuel consumption. Proper maintenance of DPFs can limit fuel consumption. Several other maintenance factors can affect fuel consumption. Maintenance practices vary widely, so some operators may have little to gain from improved maintenance practices, whereas others might see a significant benefit. The DOT-DOE white paper addresses the topic of trailer maintenance, noting that research into parasitic losses could help determine whether some type of maintenance regulations aimed at reducing fuel consumption would be useful (DOE-DOT, 2011, p. 7).
2. *Packaging optimization.* Some goods are shipped in bulk or with minimal packaging, but many products have extensive packaging. In the case of many consumer products, the size of the package can be several times the size of the actual product. Because trucks carrying these types of loads are typically filled by volume before reaching their maximum weight, a change in packaging can allow a given truck to carry substantially more product. This means that fewer loads are required to deliver the same amount of product, directly reducing both the number of trucks in operation and fuel consumption. Some companies, such as Walmart and a range of consumer-product manufacturers, have put considerable effort into packaging optimization, but there is still scope for improvement. This topic is addressed briefly in the DOT-DOE white paper (DOE-DOT, 2011, p. 1).
3. *Load management optimization.* This term describes efforts by trucking companies to ensure that trucks run as close to full payload as possible over the shortest distance needed to make deliveries. "Deadheading" (running empty on the way to pick up a load) needs to be minimized. Sophisticated software is used by many fleets to optimize pickup and delivery routes, both in terms of distance and in terms of maximizing vehicle capacity utilization. Because trucking is a low-margin business, competition places intense pressure on trucking companies to improve their load management. The difference between profitable operation and bankruptcy can be a few percent difference in average load factor. Over time, more companies are implementing ever-more sophisticated load-management systems. However, it must be recognized that there are instances in which a truck needs to run empty or with a very small load, or else a given customer cannot be served at all. This topic is addressed in the draft white paper (DOE-DOT, 2011, p. 14).
4. *Routing optimization.* Routing determination is normally done using the same system used for load management. The goal here is to select the most time- and fuel-efficient route, which may not always be the shortest distance. Factors that are considered include congestion (which varies by time of day), speed limits, the number of traffic signals or other situations requiring stops, and hills. Planned routing can also be adapted in real time during operation to take into account special conditions such as weather or accidents. As with load management, this is an area in which the industry is investing a significant amount of effort, so there is little if any productive role for agencies to play. One exception may be in the area of providing real-time information on road conditions in order to allow continuous optimization of routes, perhaps by the use of vehicle-to-infrastructure (V2I) communications. Such communications for fuel economy purposes are mentioned in the white paper (DOE-DOT, 2011, pp. 7-8). Another area of opportunity for the agencies is to develop and maintain a database that includes information on road restrictions, road construction, hazardous materials routes, preferred truck routes, and so on, so that the route-planning software used by trucking companies has up-to-date data to work with. This idea is not mentioned in the DOE-DOT white paper.
5. *Supply-chain optimization.* It is not unusual for a product to contain many individual components that are shipped from all over the world to a final assembly point before the final product is shipped to the customer. For example, raw materials or specialized components may be shipped from the United States to China, combined with other parts from many countries, and then shipped back to the United States for sale. Manufacturers consider the cost of shipping when they set up a supply chain, but shipping is only one of many costs that are considered. If the cost of shipping increases significantly (for example, during a spike in fuel prices), then companies may reconsider their supply chains, especially if the increase in shipping cost appears to be permanent. When companies choose supply chains that involve extensive shipping,

this results in higher fuel consumption, and more just-in-time shipping can lead to dramatically higher fuel consumption (for example, by favoring shipping by air or truck instead of by ship or train). The white paper addresses this topic (DOE-DOT, 2011, pp. 1-2).

6. *Infrastructure improvements.* Changes to increase road capacity and reduce congestion have a direct effect on truck fuel consumption. Congestion leads to more frequent speed changes and additional idling, both of which cost fuel.
7. *Intelligent transportation systems (ITSs).* A wide range of features and technologies falls under the term "ITS." These include driver information systems that can provide warning of accidents or congestion and suggest alternative routes. They can involve demand-management features that restrict access to highways prone to congestion. ITSs also include optimization of traffic-signal operation that can increase capacity, reduce congestion, and reduce speed fluctuation, all of which reduce fuel consumption. "On demand" traffic-signal switching using vehicle presence detection for control is increasingly used to reduce wait times (and fuel consumption due to idling) at stops. Changes such as traffic-signal optimization are particularly attractive, because no road construction or change to the vehicle fleet is needed for all vehicles on the route to benefit. Other ITS applications include ramp metering to reduce congestion, electronic on-road toll collection, automated electronic screening such NORPASS and PrePass for weight and safety inspections, credential checking and border clearance, among others (NRC, 2010, pp. 168-171). These applications can save fuel by reducing congestion and eliminating the need for starting and stopping. ITSs can include vehicle-to-vehicle (V2V) communication as well as V2I communication. ITS features that require V2V or V2I communication will take time to implement, because not all vehicles will have the required systems onboard. However, in many cases the overall traffic flow will benefit even if only a portion of the vehicles involved are actively using ITS. The white paper addresses ITS issues (DOE-DOT, 2011, pp. 7-9).
8. *Driver training.* In order to obtain a commercial driver's license (CDL), drivers of commercial vehicles go through more rigorous training than is required for drivers of light-duty vehicles. The primary focus of this training is safety, which misses an opportunity to train drivers in how to operate their trucks in the most fuel-efficient manner. Some truck fleets do have their own driver-training programs. These programs usually go beyond CDL requirements, and often fuel efficiency is part of this training. In some cases, engine and truck manufacturers work directly with fleet operators to help them provide the best possible information to drivers. However, at this time, there is no standardized curriculum for this kind of training, and many drivers do not receive any fuel efficiency training at all. Also, it is unclear how far such training could be standardized, in view of the many manufacturer-specific features and characteristics that drivers need to understand. Driver training is mentioned in the introduction of the white paper, but is not discussed in any detail. However, the white paper does suggest that research into aspects of driver behavior that might affect the fuel-saving performance of specific driver management features (see item 9 below) would be useful (DOE-DOT, 2011, pp. 12-13).
9. *Driver-management features.* Engine and vehicle manufacturers have developed a wide range of control features aimed at encouraging (or forcing) drivers to operate in a way that reduces fuel consumption. Driver-management features are not discussed in the white paper in any detail, but they offer significant potential for fuel savings. Examples of these features include the following:
   a. *Progressive shift.* This feature reduces the maximum engine speed available in the lower gears, which forces the driver to shift to a higher gear earlier than he or she might otherwise choose. Cummins offers a more sophisticated version of progressive shift called load-based speed control. The control algorithm estimates the vehicle mass and the grade and adjusts the engine-speed governor to be appropriate for the vehicle load.
   b. *Gear-down protection.* At cruising speed, it is typically possible to operate in more than one gear. For example, it may be possible to run at 65 mph in both 9th and 10th gears with a 10-speed transmission. The gear-down protection algorithm reduces the maximum vehicle speed that can be attained in 9th gear, forcing the driver to shift to the more economical 10th gear if he or she wants to run at 65 mph. With gear-down protection, the lower gears are reserved for conditions that require their use, such as climbing hills or coming up to cruising speed.
   c. *Road-speed governors.* Virtually every truck sold today includes a road-speed governor, but the use of the governor and the governor setting are left up to the owner. The road-speed governor limits cruising speed. Because fuel consumption increases with cruising speed, the use of a governor saves fuel. Note that there are tradeoffs: to the extent that speed is reduced, trip times will increase. This will lead to a need for more trucks to deliver the same quantity of freight per day. Speed governors also have no effect on fuel consumption when other constraints such as congestion or road conditions limit vehicle speed so that it is at or below the speed-governor setting (NRC, 2010). Many owners of large fleets set their road speed governors in the range of 62 to

67 mph. Most owners of smaller fleets and most owner-operators do not use road-speed governors. The Owner-Operator Independent Drivers Association has registered strong objection to the regulation of road-speed governor settings, citing safety concerns caused by differential car-truck speeds.

d. *Smart cruise control*. Cruise control helps the vehicle maintain a constant speed when road conditions permit, and reduces driver effort. Cruise controls can be developed with features that help save fuel. For example, cruise control can allow small speed fluctuations that permit the engine to operate at its most efficient load point a higher percentage of the time than would be the case if precise speed control were the goal. There are also cruise-control systems that automatically adjust truck speed to maintain a safe distance from the vehicle ahead. A more recent example is the introduction by Daimler Trucks of predictive cruise control. This feature works with a Global Positioning System (GPS) unit to adjust cruise speed based on speed limits and terrain. The truck may speed up before climbing a hill, for example. Tuning of these algorithms has to balance driver acceptance and fuel-savings potential. For example, drivers might object if large speed fluctuations around the set point selected by the driver are used by the cruise-control algorithm.

e. *Driver reward systems*. Features are available to track aspects of driver performance that can affect safety and fuel consumption. Some of these systems can be programmed to provide the driver with a direct financial interest in reducing fuel consumption by awarding bonuses for fuel-sensitive driving styles. Some of the rewards are determined in real time by algorithms in the engine-control module, such as access to higher engine power and torque (and possibly even a change in the speed-governor setting) when the vehicle is being operated in a fuel-efficient manner. The goal of these systems is to provide drivers with an incentive to operate in a fuel-efficient way, by sharing some of the fuel cost savings with the driver.

10. *Barriers to the application of fuel-saving technology.* As the white paper notes, several technologies that could save fuel run into regulatory constraints. Examples in this category include the following:

a. DOT mirror regulations specify the size, number, and locations of mirrors. Mirrors are a significant source of vehicle aerodynamic drag, and the replacement of mirrors with a system of cameras and in-cab displays has been proposed. These vision systems could not only save fuel but also could improve safety by reducing or eliminating blind spots. However, regulations would need to change to allow use of camera-based systems. This issue is addressed by the white paper (DOE-DOT, 2011, p. 6).

b. The aerodynamic skirts used on trucks today do not extend to cover the drive axles of tractors or the trailer axles. The reason for this is that if skirts are built around existing axles, the truck may violate regulations governing maximum width. Other aerodynamic improvement features, such as boat-tails, can run into issues with length regulations at the local level, even when operators are granted a federal exemption. This issue is addressed by the white paper (DOE-DOT, 2011, p. 5).

c. Weight limits discourage the use of heavy fuel-saving features such as auxiliary power units (APUs) or aerodynamic devices, especially on vehicles that frequently run at or near the maximum weight limit. In these cases, the fuel-saving feature directly reduces the load-carrying capacity of the truck, which defeats the purpose of the fuel-saving feature. This issue could be addressed by providing a weight allowance (i.e., an increase in legal weight) for vehicles with specific fuel-saving features. This issue is not mentioned in the white paper.

d. Federal vehicle-size and -weight regulations have not changed since 1983. These regulations prevent the use of high-productivity vehicles, including long combination vehicles (often called, respectively, HPVs and LCVs). HPVs involve greater freight volume or weight than is allowed under existing regulations. Current regulations forbid the use of HPVs in many areas, regardless of whether safety and road damage considerations are adequately addressed in the vehicle design. Some very large fuel savings are possible if regulatory barriers to the appropriate use of HPVs are removed (see the section "Fuel-Saving Opportunities from Efficient Operations" in this chapter, as well as Table 9-2). For example, a bill has been introduced in Congress that would allow weights of up to 97,000 lb, compared to the current limit of 80,000 lb, for trucks using a 6th axle.[1] In addition to saving fuel, the use of HPVs could reduce truck vehicle miles traveled (VMT), congestion, accident rates, shipping costs, driver shortages, and road damage by combining longer and heavier vehicles with appropriate operational restrictions and performance requirements. Certain trailer and dolly configurations are inherently more stable than others, providing opportunities to upgrade braking and stability-control requirements. The white paper has a discussion of some options for heavier vehicles (DOE-DOT, 2011, pp. 13-15),

[1] The Safe and Efficient Transportation Act., S. 747, was introduced on April 6, 2011. The proposed legislation is available at http://www.gpo.gov/fdsys/pkg/BILLS-109s747is/pdf/BILLS-109s747is.pdf.

but the range of options considered is limited, and options to increase the allowable volumetric capacity of trucks are discussed in a very limited way (DOE-DOT, 2011, p. 15).

e. As noted in previous chapters, the EPA and the National Highway Traffic Safety Administration (NHTSA) have implemented fuel-consumption and greenhouse gas (GHG) regulations for trucks. So far, these regulations are aimed only at engines and at vehicles (tractors in the case of tractor-trailer vehicles). Trailers were left out of the regulations because some trailer manufacturers are small businesses. However, the trailer can have a significant effect on both the aerodynamic drag and the rolling resistance of a tractor-trailer vehicle. The idea of expanding the EPA/NHTSA regulations to include trailers is not discussed in the white paper.

11. *Vehicle specification.* Purchasers of new trucks have the opportunity to select a vehicle specification that is optimized to achieve a good combination of productivity and low fuel consumption. There are many features that buyers can select from, including engine type, power rating, transmission type, wheelbase, axle ratio and number of drive axles, tire type, aerodynamic improvement features, lightweight features such as aluminum wheels, and so on. Because profit margins in trucking are normally very small, there is intense interest from larger trucking companies in choosing the best possible vehicle specification for their application. Vehicle specification tools are a key technology for controlling operating costs, including fuel consumption.

a. Truck-specification tools have been developed by both vehicle and engine manufacturers. These tools can be quite sophisticated, allowing comparison of various vehicle specifications on a huge variety of routes, with any load specified by the customer. A wealth of detail is available, such as data to answer a question like this: "How many times would the driver need to shift gears on a run from Chicago to Los Angeles?" The primary use of the specification tools, however, is fuel-consumption optimization. These tools are widely used by larger trucking companies but are often not used or not fully taken advantage of by smaller fleets and by owner-operators. In many cases, this may be a result of a lack of awareness of how the tools can help, but in other cases the buyer lacks the information on loads and routes that is required to feed the tool. There are some limitations to these tools: many trucks are used to carry widely varying loads over their lifetime, so it may not be practical to specify a truck with a narrow focus on the first use. Trucking companies generally specify trucks with resale value as one of the parameters that they consider. There may be an opportunity for the agencies to encourage the more widespread use of existing vehicle-specification tools. These tools are not discussed in the white paper.

As the list above makes clear, a wide variety of technologies, training practices, and regulatory changes falls under the umbrella term "efficient operations." All of the items listed above have some potential for saving fuel, and some of them have potential to save a substantial amount of fuel—as much as vehicle technologies that may cost tens of thousands of dollars per truck. The cost of efficient operations technologies can range from several thousand dollars per vehicle to actual cost reductions. For example, HPVs are more expensive on a per truck basis but less expensive on a unit freight capacity basis.

FUEL-SAVING OPPORTUNITIES FROM EFFICIENT OPERATIONS

As noted in the draft white paper as well as in an earlier NRC (2010) report, the data that would be needed to quantify the possible fuel savings in some of the areas above are sparse or do not exist. This is especially an issue when the potential savings depend strongly on the details of drive cycles. The white paper proposes R&D and testing to develop some of the needed data. In this section, the committee uses data available in the literature to quantify some of the fuel-saving opportunities available for features discussed above. This assessment should be useful to the DOE and DOT as they set R&D priorities for efficient operations.

Much of the data listed below references the NRC (2010) report titled *Technologies and Approaches to Reducing the Fuel Consumption of Medium- and Heavy-Duty Vehicles*, which in turn references a wide variety of sources. In this section, the term "fuel savings" means fuel consumption in units of gallons per 1,000 ton-miles or per cube-mile. Note that the fuel savings quoted below are typically determined for a specific operating condition and are not likely to be representative of the savings potential possible for trucks operating under a wide range of conditions. It is common practice to evaluate a fuel-saving technology under the operating condition for which it provides the largest benefit.

Tire Underinflation and Axle Misalignment

The NRC (2010) report referred to above estimates that the effect of running with all tires underinflated by 20 percent is a 2 to 3 percent increase in fuel consumption. Typical inflation pressures for tractor-trailer tires are 100 to 120 psi. Thus a tire that is 20 percent underinflated would be 20 to 25 psi below the recommended inflation pressure. Inflation pressures for medium-duty trucks vary but typically fall between those for light-duty and heavy-duty vehicles.

Axle misalignment caused by suspension maladjustment produces a slip angle, which increases rolling resistance. This effect increases with the square of the slip angle. NRC (2010) estimates the effect as an approximately 0.1 percent increase in fuel consumption for an average slip angle of 0.1 degree, and 0.4 percent increase for a slip angle of 0.2 degrees. Other studies have assessed the extent of tire underinflation in the field, such as a Federal Motor Carrier Safety Administration (FMCSA, 2005) report, but the committee is not aware of studies on the extent of axle misalignment to be found in the field. The FMCSA (2005) report says that only 44 percent of all tires checked had inflation pressures on all tires within 5 psi of the target value, and about 7 percent of tires were off by 20 psi or more. There is also a lack of data about other areas, such as the extent to which fuel consumption needed for DPF regeneration increases as miles accumulate.

Packaging

As noted in item 2 above in the committee's list in the section "Efficient-Operation Opportunities," packaging can significantly affect the cost of shipping a product, as well as the fuel consumed in shipping. Some products have packages that are several times larger than the product itself, which results in trucks running with the trailer full (capping the number of products that can be shipped in a single load), but well below the legal weight limit. Walmart has been working for more than 10 years now to use sustainable packaging materials by reducing the size and the energy and natural resources needed to produce packaging. Its target, as stated in its 2010 Sustainability Report, is to reduce total packaging by 5 percent from 2008 to 2013 (Walmart, 2010). The report does not specify whether this reduction is in the amount of packaging material or in the volume of packages, or whether some other metric is being used. Although the fuel savings from modified packaging of certain products can be substantial, the committee is not aware of any published research that quantifies the potential overall fuel-consumption opportunity in this area.

Load Management, Routing Optimization, and Supply-Chain Optimization

The benefits of load management and routing optimization can be substantial. However, this is an area of intense competition among trucking companies as well as among suppliers of logistics software and systems. As far as the committee is aware, there is no published research that quantifies the overall fuel-saving opportunity. The same comments apply to supply-chain optimization, except that in this case, manufacturers and distributors are also part of the overall picture.

Infrastructure Improvements

Infrastructure improvements (road expansion and construction) are the subject of intense debate. For a given volume of traffic, congestion causes increased fuel consumption, and the increase can be dramatic. A truck may consume twice the fuel in urban traffic that it consumes traveling the same distance on an uncongested road. However, congestion also affects demand. As congestion becomes intolerable, road users find other ways of moving freight (alternative routes, night operations, etc.), or they may simply give up (which has the effect of limiting economic growth). Therefore, there is debate over whether new or expanded roads save fuel, or whether they lead to increased traffic and thus higher fuel consumption.

Intelligent Transportation Systems

As the draft white paper (DOE-DOT, 2011) notes, ITS programs have historically been driven largely by safety concerns. There has been limited research to quantify the fuel-saving opportunities available through ITSs, or even to explore all the ways in which ITSs could be used to save fuel. The white paper discusses the potential fuel savings from ITSs primarily in relation to the avoidance and mitigation of congestion, because concepts such as route management are dealt with elsewhere. The NRC (2010) report estimates the fuel-saving potential of ITSs in the 2015-2020 time frame as 8 to 15 percent. This figure includes some technologies that are described elsewhere in this chapter, such as predictive and adaptive cruise control, as well as other ITS technologies such as predictive control of hybrid systems, use of electronic tow bars, and real-time route optimization.

The white paper asserts that research on V2V and V2I communications is the most critical aspect of the ITS program, and that this work is crucial for obtaining significant improvements in fuel consumption, not just safety. The paper states, "With the magnitude of potential energy savings, it is important that DOE and DOT work very closely in this domain so that the magnitude of energy savings is not overlooked and research and applications development is sufficiently funded for assuring the maximum benefits in fuel savings, in addition to safety" (DOE-DOT, 2011, p. 8). Given the 8 to 15 percent potential energy savings estimated for ITSs (NRC, 2010), the committee concurs with the high importance of determining fuel savings as part of any project that involves ITS-related technologies. Further, the potential for ITS technologies to reduce fuel consumption would seem to argue for the development of one or more goals by the 21CTP.

Driver Training and Management

Driver training and driver management features have significant potential for saving fuel. According to the NRC (2010, p. 175) report, several case studies report fuel savings from 1.9

to 17 percent for driver-training programs. Generally, there is more opportunity for savings in urban and congested environments than there is for savings on the open road. Large fleets often develop their own driver-training programs, many of which include fuel-efficient driving techniques. Smaller fleets and owner-operators typically do not have driver-training programs. Driver-management features also vary in their effects based on operating conditions. Progressive shift algorithms are useful in urban driving but make little difference in long-haul operation. Conversely, cruise control is not used in urban driving but can contribute to fuel savings on the open road. Studies reviewed in the NRC (2010, p. 126) report show benefits of 1 to 5 percent for predictive cruise control, depending on the vehicle and duty cycle.

Road-speed governors are discussed in the NRC (2010, pp. 166 ff.) report. Vehicles that currently run at 70 mph would see fuel savings of 7 to 10 percent by running at 60 mph, and vehicles slowing from 65 to 60 mph would save 3.5 to 5 percent. However, there are disadvantages to lower speeds that must be carefully considered before any mandatory road-speed governor setting is implemented. Road-speed governors could include the use of GPS devices or wireless systems that would change the governor setting based on local speed limits and road conditions.

Aerodynamic Features That Exceed Width or Length Limits

A report by the Truck Manufacturers Association (NETL, 2007) shows that replacing mirrors with cameras has the potential to reduce fuel consumption at high-speed cruise by up to 3 percent, depending on the type of mirrors being replaced. The committee is not aware of any studies estimating the potential benefit from tractor and trailer skirts that extend over the drive axles and trailer bogies, but the value is likely to be a few percent.

Size and Weight Limits

The NRC (2010, p. 163) report discusses the effects of size and weight limits. Fuel savings of up to 13 percent and truck VMT reductions of up to 23 percent are projected, depending on the nature of regulatory changes. The American Transportation Research Institute (ATRI) estimated potential fuel savings of up to 25 percent for weight-limited trucks (grossed-out) and 28 percent for volume-limited trucks (cubed-out) (ATRI, 2008). In Figure 9-1, reproduced from the ATRI report, the metric used is ton-miles per gallon, which is a fuel economy metric. The changes in fuel consumption are smaller, but still significant.

ATRI Estimated Efficiency Improvement

Percentage Change in TM/gal – Weight-Limited Scenario

Configuration	Gross Vehicle Weight			
	97,000*	100,000	120,000	140,000
Compared to 5-axle @ 80,000 lbs GVW				
6-axle	17%	--	--	
RMD	--	1%	25%	--
TPD	--	-10%	15%	33%
Compared to Double @ 80,000 lbs GVW				
TRPL	--	3%	31%	--

* Estimate based on straight-line extrapolation of fuel consumption data

Percentage Change in TM/gal – Cube-Limited Scenario

Over 60% of long haul trucks

Configuration	Gross Vehicle Weight		
	83,175*	84,040*	100,370*
Compared to 5-axle @ 60,000 lbs GVW			
RMD	20%	--	--
TPD	--	--	39%
Compared to Double @ 60,000 lbs GVW			
TRPL	--	29%	--

* Estimate based on straight-line extrapolation of fuel consumption data

FIGURE 9-1 High-productivity vehicle descriptions and operational efficiency potentials compared to 5-axle tractor-semitrailer or double baselines. Values given are for fuel economy (FE) in ton-miles per gallon. These values can be converted to fuel consumption (FC; gallons per ton-mile) using the following equation: Percentage change in FC = 100 × [percentage change in FE/(100 + percentage change in FE)]. For example, a 33 percent increase in FE equals a 25 percent decrease in FC. TM, ton-miles; GVW, gross vehicle weight. SOURCE: Courtesy of ATRI (2008).

The topic of longer and heavier vehicles (called high-productivity vehicles or long combination vehicles by different authors) raises public concern about safety, so it is essential that discussion and research are based on high-quality data rather than on emotional arguments. The authors of a study of Canadian experience with LCVs (Regehr et al., 2009) concluded: "An analysis of the safety performance of LCVs relative to other types of articulated trucks operating on Alberta's rural LCV network revealed that, from a collision rate perspective, LCVs as a group had better safety performance than other articulated trucks…." The conclusion is intensified considering that, for a fixed quantity and density of freight transported by articulated trucks, each LCV replaces 1.5 to 2 standard 5-axle semitrailers, which, over the same period, had higher collision rates than those of LCVs. In this sense, LCVs operating in Alberta in this period provided increased freight productivity and reduced the number of collisions that would have occurred if standard configurations had been used to haul the same freight. The longest and heaviest LCV in the CSC (2003) study, the turnpike double, had the best safety record of any articulated vehicle.

A recent report from the Organization for Economic Cooperation and Development (OECD), *Moving Freight with Better Trucks* (OECD, 2010), provides an insight into the good safety performance of LCVs. According to the report, many LCV configurations are more dynamically stable than typical truck configurations used on the road today. This makes it easier for a driver to control an LCV in difficult high-speed maneuvers and in strong crosswinds. In addition, regulations limiting LCV operation to specific roads and conditions, along with rules requiring additional driver qualifications, appear to improve LCV safety. The OECD report is the most comprehensive review of the opportunities and challenges of LCV implementation of which the committee is aware. Results from Europe, Australia, Canada, and the United States are all considered in this report.

Longer, heavier trucks require an increase in power in order to maintain existing levels of vehicle performance (acceleration rates and speed on grades). This increase in power can be accommodated to some extent with existing truck technology, but large changes in weight, such as with turnpike doubles, would result in lower vehicle performance. The impact of lower speeds on other traffic could be minimized by restricting or preventing the operation of these vehicles on routes where traffic flow becomes an issue. There may also be value in setting minimum power-to-weight ratio requirements for longer and heavier trucks, to limit the potential for impeding traffic flow.

Road wear and damage are known to be controlled by contact (Hertz) stress. Relative fatigue damage of pavement is approximately a fourth-power function of axle load, which is related to contact stress (FHWA, 2000). As a result, vehicle weight and the distribution of that weight over the tire contact patches are critical factors in determining the durability of road surfaces. LCVs typically have lower weight per axle than do current trucks, which allows these vehicles to provide reduced road surface wear, compared to conventional trucks.

The fuel-saving potential of high-productivity vehicles makes the removal of barriers to their use one of the most powerful fuel-saving tools available. In addition, the economic benefits obtained by higher-productivity vehicles could be used to support the addition of safety technologies that would improve the acceptability of these larger vehicles. Currently, the major barriers to HPVs are (1) regulatory limits on vehicle size and weight, and (2) lack of adequate data and experience on which to base analyses of benefits and drawbacks. Before 1983, states had almost complete freedom to regulate truck length and weight. In 1983, Congress passed legislation that standardized the trailer length limit at 53 ft and the maximum weight at 80,000 lb. The 1983 legislation "grandfathered" in state regulations that had previously allowed vehicles longer or heavier than provided for in the 1983 national standard, but no additional states are allowed to set standards above the federal limits. Also, states with grandfathered length or weight limits (see Table 9-1) cannot increase these limits. This prevents neighboring states from standardizing their regulations on anything beyond the federal limits or the limits that applied in the states in question before 1983. Twenty U.S. states have pre-1983 regulations that allow a wide range of vehicle lengths and weights, while the remaining 30 states are denied the option of longer or heavier vehicles by the 1983 law. Table 9-1 shows the regulations for the states. The diversity of these regulations means that trucks complying with different state standards can only be used for intrastate commerce, which greatly reduces the fuel-saving potential of HPVs in the United States.

Canada, Australia, and Scandinavian countries have extensive experience in the operation of trucks that are longer and heavier than traditional American trucks, and their experience, in addition to that of some U.S. states, can be drawn upon by the DOT and DOE in analyzing the effects of using these vehicles in the United States (see Appendix H in this report). The European Union (EU) is now considering an increase in maximum truck weight and length from 40 to 60 metric tons (88,000 lb to 132,000 lb) and from 16.5 to 25 meters (54 ft to 82 ft) as a measure to reduce both greenhouse gas emissions and fuel consumption (Die Zeit, 2010). If this change is implemented, it would effectively extend the current Scandinavian regulations across Europe. There may be opportunities for joint research with European regulators, and perhaps even standardization of some regulatory requirements. This would allow manufacturers that sell trucks in both the United States and the EU to use a single design for both markets. Current designs of trucks for use in the United States and those for use in the EU diverge widely, driven primarily by different regulatory requirements.

Given President Obama's emphasis on energy security, the DOE and DOT, working with the Congress, should consider the recommendations of the NRC's Transportation Research

TABLE 9-1 Maximum Truck Size and Weight Limits for 13 of 20 States Subject to the ISTEA Freeze

	Truck Tractor and Two Trailing Units		Truck Tractor and Three Trailing Units		Other[a]
State	Length (ft)	Weight (lb)	Length (ft)	Weight (lb)	Length
Colorado	111	110,000	115.5	110,000	78
Idaho	95	105,500	95	105,500	78-98
Kansas	109	120,000	109	120,000	No
Montana	93	137,800	100	131,060	88-103
Nebraska	95	95,000	95	See note[b]	68
Nevada	95	129,000	95	129,000	98
North Dakota	103	105,500	100	105,500	103
Oklahoma	110	90,000	95	90,000	No
Oregon	68	105,500	96	105,500	70 ft 5 in
South Dakota	100	129,000	100	129,000	73-78
Utah	95	129,000	95	129,000	88-105
Washington	68	105,500	No	--	68
Wyoming	81	117,000	No	--	78-85

[a] A commercial motor vehicle combination with two or more cargo-carrying units not included in descriptions "truck tractor and two trailer units" or "truck tractor and three trailer units."

[b] No maximum weight is established because this vehicle combination is not considered an "LCV" per the ISTEA definition because it is only allowed up to 80,000 lb.

SOURCE: FHWA (2010).

Board (TRB) in TRB Special Report 267 (TRB, 2002). The TRB proposed a way forward for high-productivity vehicles in the United States that would deal responsibly with the issues raised by different stakeholders. Fundamental to the TRB's proposed process is the creation of a Commercial Traffic Effects Institute (CTEI) by Congress that would manage all issues related to high-productivity vehicles. Extensive pilot studies would be used to evaluate the consequences of increased truck size and weight, following accepted practices for test design and for analysis. In addition, all states would be authorized to begin permitting for heavier, 6-axle combination trucks and various other HPVs. Because having a range of permitted vehicle sizes and weights that varies from state to state effectively restricts the use of HPVs across state lines, the DOE and DOT should consider assigning high priority to further research and analysis on the potential for fuel savings from uniform, nation-wide size and weight regulations.

Trailers

The benefit of trailer aerodynamics is estimated NRC (2010, p. 100) as 9 percent today, growing to 12 percent by 2020. These savings were based on a baseline tractor-trailer coefficient of drag (C_d) of 0.625, which is lower than the baseline of 0.69 used in the EPA and NHTSA GHG and fuel-consumption regulations. As the earlier NRC (2010) report notes, there are also rolling resistance benefits that could be obtained if regulations were extended to cover trailer tires. For tractor-trailer combination vehicles, a significant potential fuel-savings has been neglected by leaving trailers out of the EPA and NHTSA regulations.

Summary of Fuel-Saving Opportunities

The values of the fuel saving opportunities discussed above are summarized in Table 9-2. Note that these fuel savings are often estimated for a specific operating condition of a specific vehicle type, not for the average operation of all trucks. Thus the fuel savings opportunities in the table are not valid for the entire truck population. There are many other fuel-saving opportunities not shown in Table 9-2, such as improved engine and driveline efficiency, reduced aerodynamic drag of the tractor, and others. This list only includes items related to efficient operations or to removal of regulatory barriers. The trucking industry normally operates on very narrow profit margins, so efficient operation is an area of intense interest and competition among trucking firms. Larger fleets invest significantly in finding ways to improve their operational efficiency. As a result, the industry can supply a wealth of ideas and data to help improve its own efficiency. The industry is also in a position to point out potential drawbacks and unintended consequences of regulatory changes that are intended to save fuel by improving the efficiency of trucking. The DOE, DOT, and other government

TABLE 9-2 Summary of Fuel Saving Opportunities

Source of Savings	Fuel Savings Opportunity (%)	Notes
Prevent tire underinflation	2-3	For a truck with 20% underinflation of all tires.
Prevention of axle misalignment	0.1 to 0.4	For 0.1 to 0.2 degree misalignment.
Packaging reduction	Unknown	Research needed.
Load management, routing optimization, and supply chain management	Unknown	Areas of competition among trucking companies.
Infrastructure improvements	Varies with severity of congestion	
Intelligent transportation systems	8 to 15	Primarily in urban conditions.
Driver training	1.9 to 17	Higher potential in urban driving.
Predictive cruise control	1 to 5	Useful only in rural driving.
Road-speed governors	3.5 to 5	For trucks slowed from 65 to 60 mph. Only effective where other factors do not limit speed.
Replacement mirrors with cameras	1.5 to 3	Only a factor at higher speeds.
Remove regulatory barriers to increased size and weight	0 to 28	Depends on what new configurations are allowed and where they can operate.
Trailer aerodynamic improvements	Up to 12	Only effective at highway speeds.

agencies interested in reducing fuel consumption by improving the efficiency of trucking will also need to work closely with members of the industry to gain the advantage of the ideas, data, and practical experience that are available. There may also be advantages in working with large customers of the trucking industry (shippers), to see if there is potential for shippers to make changes in their operations that would improve the overall freight system efficiency. The existing white paper (DOE-DOT, 2011) does not discuss industry involvement, and the committee would like to see this oversight remedied.

OPPORTUNITIES FOR COLLABORATION BY DEPARTMENTS OF ENERGY AND OF TRANSPORTATION

The DOE and DOT have opportunities to collaborate in facilitating the development and use of efficient operations for trucking. There are also some areas in which other agencies, such as the EPA and NHTSA (a part of DOT), need to be involved. This section considers some of the opportunities that are available.

Research and Development

One large set of opportunities for collaboration involves research, development, and information dissemination. In many cases, the potential benefits of fuel-saving technologies are not well understood across a wide range of operations. As noted at the beginning of the section above on "Fuel-Saving Opportunities from Efficient Operations," for most of the technologies and approaches discussed above, the range of potential fuel savings possible in actual field use is very wide. It is very difficult to estimate even an approximate fuel savings for the implementation of a given technology across a wide range of truck applications, even in a relatively narrow segment such as long haul. There is considerable scope for further research to evaluate options under real-world operating conditions, to provide both regulators and the industry with data of better quality. In particular, research and development efforts are needed to identify, develop, and quantify the performance of items such as ITS features that can save fuel, and to assess the effects of removing regulatory barriers that prevent the application of some fuel-saving technology. Before new regulations are implemented, however, research is needed to help define potential benefits and avoid possible negative side effects, including effects on trucking company operations and on other concerns such as safety and road damage.

Based on the discussion above in the sections "Efficient-Operation Opportunities" and "Fuel-Saving Opportunities from Efficient Operations," the committee believes that the DOE and DOT could usefully work together on research in the following areas:

1. Assessing the effects of vehicle maintenance practices on fuel consumption and the effects of potential maintenance standards aimed at reducing fuel consumption.
2. Evaluating the effects of infrastructure improvements (increased road capacity) on fuel consumption, traffic growth, and economic growth.
3. Analyzing the potential for ITSs to reduce fuel consumption and for the development of new ITS features specifically intended to save fuel.
4. Identifying what constitutes "best practice" driving techniques from the point of view of fuel consumption, as limited by other issues such as the possibility of delaying traffic. Developing driver-training curricula, allowing for flexibility to adapt to various technologies and the differing characteristics of vehicles in the field.
5. Determining the magnitude of fuel savings available from modifications to regulations, such as mirror requirements, size limitations that inhibit the use of aerodynamic features, and weight limits that prevent or constrain the use of features such as APUs or aerodynamic improvements that add weight to the vehicle.

Investigating potential concerns that could be raised by changes in regulations and ways of attaining the benefits of the modified regulations while minimizing negative effects.

6. Evaluating alternatives for modifying size and weight restrictions in ways that would allow for the use of high-productivity vehicles while improving safety and avoiding road-damage issues. For example, certain trailer and dolly configurations are inherently more stable than others, providing opportunities to upgrade braking and stability-control requirements. In addition, training requirements for drivers of high-productivity vehicles need to be addressed. This research could be done by the DOE and DOT or by the Commercial Traffic Effects Institute proposed by the TRB. The committee suggests that this research consider the Canadian experience with management of LCVs, where certain equipment requirements and restrictions in use result in major improvements in safety compared with more traditional tractor-trailer operations (NRC, 2010).[2]

Regulatory Matters

Another area for possible cooperation among the DOT and DOE is regulatory matters, as they are informed by the R&D described above. The committee believes that if the DOT and DOE decide to work on such regulatory matters, they should consider cooperating in the following areas:

- Vehicle maintenance practices that affect fuel consumption;
- Road design specifications that affect fuel consumption;
- Implementation of ITS features designed specifically to reduce fuel consumption (some features are likely to require changes in regulation);
- Driver-training requirements related to fuel consumption;
- Modification and standardization of regulations that constrain the use of fuel-saving features, such as regulations related to mirrors and limitations on maximum vehicle width and weight that inhibit the use of APUs and certain aerodynamic improvement features;
- Implementation of aerodynamic and rolling resistance requirements for trailers (requires the involvement of the NHTSA and EPA); and
- Modification of vehicle size and weight restrictions on a national basis, to allow for the use of high-productivity vehicle configurations on a national scale.

[2] Government of British Columbia requirements for LCV operation are listed at http://www.th.gov.bc.ca/cvse/LCV/faqs.htm; Ontario requirements for operating 53-ft doubles are listed at http://www.mto.gov.on.ca/english/trucks/lcv/program-conditions/index.shtml.

The draft white paper emphasizes the importance of implementing the systematic validation of technologies stating: "Developing drive cycles that are relevant for the many different types of trucks in use will allow much more accurate estimates of the fuel savings that are possible with specific technologies, and understanding the variations in drive cycles will allow an accurate assessment of how robust a technology will be in actual use" (DOE-DOT, 2011, p. 12). The paper also notes: "Component testing of heavy duty truck technologies is currently being performed in Canada under the EnergoTest campaign (McCormick, 2009), and experience from this testing could be used as guidance for similar testing in the U.S." (DOE-DOT, 2011, p. 12). The committee notes that it agrees with these principles and believes that it would be helpful for all involved agencies to collaborate among themselves as well as with industry. Should the work in the draft (and later, modified) white paper go forward, an appropriate set of improvement targets (goals) should be set.

The committee believes that regulatory intervention in the following two areas could prove counterproductive: (1) Packaging—It is difficult to imagine how packaging regulations might work, or how they could properly cover the vast range of products that are shipped every day; and (2) Load management, routing optimization, and supply-chain management. These are areas of fierce competition among manufacturers, trucking companies, and their logistics solutions suppliers. Although there is a useful role for the DOT and DOE in doing research and development on driver-management features, the committee believes that the proposed regulation for road-speed governing will save fuel. There is substantial competition among engine and truck manufacturers on other driver-management features such as progressive shift, gear-down protection, and bonuses or power increases for fuel-sensitive driving. The situation today is that truck owners are free to select the features that work best for their operation and to adjust the parameters that these features use. Large fleets tend to be fairly sophisticated in terms of figuring out what features work in their operations and how to set the parameters. For any given driver-management feature there will be situations in which its use is not appropriate or in which a given set of parameters will lead to inefficient, or even unsafe, operation. There is an opportunity to provide training to small fleets and owner-operators so that they can see how using these features can make their operations more fuel-efficient and thus more profitable.

FINDINGS AND RECOMMENDATIONS

The following findings and recommendations are the result of the committee's review of the draft DOE-DOT (2011) white paper on efficient operations. They describe what the committee believes should be added to or changed in the white paper to help the 21CTP promote and enable more efficient trucking operations.

Finding 9-1. The DOE-DOT draft white paper proposes "efficient operations" as a new direction for the 21CTP. The committee agrees that this is an important area for R&D under the umbrella of the 21CTP. It also agrees that cooperation among the DOE and DOT and other agencies would be beneficial, particularly for assessing the possible effects of removing regulatory barriers to the use of fuel-saving measures.

Recommendation 9-1. As suggested in the draft white paper on efficient operations, the DOE and DOT, in cooperation with the EPA and other agencies, should conduct joint research on efficient operations and should cooperate as appropriate on any regulations that affect fuel use and safety.

Finding 9-2. Extensive information is available regarding the importance of trailer aerodynamics and rolling resistance. The data show that trailer aerodynamic-improvement features and rolling resistance contribute significantly to overall vehicle fuel consumption.

Recommendation 9-2. The available data show that trailer aerodynamic-improvement features and rolling resistance contribute significantly to overall vehicle fuel consumption. Therefore, the DOE and DOT should look in detail at options for trailer improvement.

Finding 9-3. The application of intelligent transportation systems has the potential to reduce fuel consumption substantially, particularly in urban areas. Certain elements of ITSs, such as adaptive traffic signals, do not require new vehicle technology, so they can be rolled out much faster than other elements.

Recommendation 9-3. Traditionally, ITSs have been viewed as a way of improving safety. As suggested in the draft white paper on efficient operations, the DOT and DOE should conduct additional research and development devoted to exploiting the potential for reduced fuel consumption.

Finding 9-4. Driver-management features must be carefully researched and developed in cooperation with vehicle manufacturers and operators. There are important concerns with driver-management features that need to be addressed, regarding unintended consequences stemming from allowing the vehicle (or its controller) to ignore or modify driver input. Consideration must also be given to identifying the types of intervention that drivers would accept.

Finding 9-5. Trucking companies already have very strong economic incentives to improve operational efficiency and average load factors. As a result, they are making significant investment in logistics technology. In addition, shippers have an economic incentive to reduce the size and weight of packaging materials. The trucking industry is a valuable source of ideas, data, and experience regarding efficiency, and the industry can help agencies avoid unintended negative consequences of efforts to improve efficiency.

Recommendation 9-4. The DOE and DOT should work with the trucking industry to take advantage of the ideas, data, and experience that the industry can provide to accelerate efficiency improvements and to avoid unintended negative outcomes of efforts to improve trucking efficiency.

Finding 9-6. High-productivity vehicles, known as HPVs or LCVs, as currently configured and using current technology, can reduce fuel consumption by up to 28 percent. In addition, HPVs can reduce greenhouse gas emissions, truck vehicle miles traveled, congestion, shipper costs, truck highway accidents, road damage, and truck driver shortages.

Finding 9-7. High-productivity vehicles have proven to be a highly controversial and emotional topic. Some U.S. states, as well as countries including Canada, Australia, and the Scandinavian countries, have extensive experience with HPV operations and safety performance. Operational limitations and equipment policy used for decades in Canada have significantly increased safety for HPVs compared with that of more conventional tractor-trailers. In 2002, the NRC's Transportation Research Board proposed a process, to be led by a congressionally chartered Commercial Traffic Effects Institute, to make decisions regarding a number of critical and historically controversial issues that effectively have prevented the growth of HPV use for nearly three decades. As far as the committee can determine, no action on the CTEI recommendation has been considered by Congress.

Finding 9-8. The draft white paper on efficient operations brings up the topic of high-productivity vehicles and the possibility of raising weight and size limits to accommodate them. However, the white paper focuses narrowly on 6-axle tractor-trailer combinations with weights up to 100,000 lb (45.5 metric tons) and does not address other options that increase volumetric freight capacity or that allow weights beyond 100,000 lb .

Finding 9-9. The committee finds the case for fuel savings of HPVs compelling, and the case for improved safety of HPVs compared to that of standard 5-axle semitractor trucks is also strong.

Recommendation 9-5. The DOT and DOE should look at the full range of high productivity vehicles in use in some U.S. states and around the world and review the literature available on the safety and fuel-saving performance of these vehicles. The assessment should take into consideration that the higher productivity of these vehicles can also be used to justify the implementation of additional safety technologies.

Recommendation 9-6. The DOT and DOE, in discussion with the Congress, should consider the recommendations of the Transportation Research Board regarding the establishment of a Commercial Traffic Effects Institute or a similar approach.

Finding 9-10. The DOE-DOT draft white paper on efficient operations in its current form does not include any goals that could be used to prioritize and drive R&D efforts on efficient operations.

Recommendation 9-7. Specific goals for efficient operations should be developed, with strong consideration given to exploiting the potential for intelligent transportation systems to reduce fuel consumption. In addition, priorities should be set for the R&D, testing, and data collection needed to analyze the benefits, drawbacks, and potential unintended consequences of removing barriers, including regulatory barriers, to the application of fuel-saving features. The draft white paper on efficient operations should be rewritten to take the findings and recommendations of the committee into account. The 21CTP partners, trucking fleets, and major suppliers should be involved in setting goals and research priorities.

Finding 9-11. There is a need for a more detailed evaluation of the large potential for fuel savings from efficient operations than is provided in the existing DOE-DOT draft white paper of February 25, 2011. This more detailed study can be used to set goals, targets, and timetables for fuel savings from efficient operations.

Recommendation 9-8. The DOE and DOT should study the potential fuel savings from efficient operations in more detail, including a review of cost-effectiveness and ease of implementation. Once this information is available, goals, targets, and timetables for fuel savings from efficient operations should be established. Programs should then be developed and implemented to realize the available fuel savings.

REFERENCES

ATRI (American Transportation Research Institute). 2008. Energy and Emissions Impact of Operating Higher Productivity Vehicles. Revised Edition. American Transportation Research Institute.

CSC (Canada Safety Council). 2003. Literature Review of the Safety of Long Combination Vehicles and Their Operation in Canada. Available at http://archive.safety-council.org/info/traffic/LCVs.pdf. Accessed May 18, 2011.

Die Zeit. 2010. Die Giga-Brummis. Die Zeit Newspaper (Germany), No. 43, October 21, 2010, pp. 39-40.

DOE-DOT (Department of Energy and Department of Transportation). 2011. Reducing Fuel Consumption in U.S. Trucking—A DOE-DOT Joint Study and White Paper. 21st Century Truck Partnership Efficient Operations White Paper. Draft. February 15. Washington, D.C.

FHWA (Federal Highway Administration). 2010. Federal Size Regulations for Commercial Motor Vehicles. FHWA-MC-96-03.

FMCSA (Federal Motor Carrier Safety Administration). 2005. Motor Carrier Safety Progress Report. Available at http://www.fmcsa.dot.gov/facts-research/research-technology/tech/tired-pressure-sensors.htm. Accessed September 23, 2011.

McCormick, C. 2009. Proof Positive. Motortruck. September-October.

NETL (National Energy Technology Laboratory). 2007. Test, Evaluation, and Demonstration of Practical Devices/Systems to Reduce Aerodynamic Drag of Tractor/Semitrailer Combination Unit Trucks. Prepared for NETL by the Truck Manufacturers Association, Contract DE-FC26-04NT42117. April. Washington, D.C.

NRC (National Research Council). 2010. Technologies and Approaches to Reducing the Fuel Consumption of Medium- and Heavy-Duty Vehicles. Washington, D.C.: The National Academies Press.

OECD (Organization for Economic Cooperation and Development). 2011. Moving Freight with Better Trucks: Improving Safety, Productivity and Sustainability (Summary). Paris: International Transport Forum. Available at http://www.internationaltransportforum.org/jtrc/infrastructure/heavyveh/TrucksSum.pdf.

Regehr, J., J. Montufar, and G. Rempel. 2009. Safety Performance of Longer Combination Vehicles Relative to Other Articulated Trucks. Canadian Journal of Civil Engineering 36:40-49. Available at http://www.nrcresearchpress.com/doi/abs/10.1139/L08-109.

TRB (Transportation Research Board). 2002. Regulation of Weights, Lengths and Widths of Commercial Motor Vehicles. Special Report 267. Washington, D.C.

Walmart. 2010. Walmart 2010 Sustainability Report. Available at http://walmartstores.com/sites/sustainabilityreport/2010/environment_waste.aspx. Accessed April 10, 2011.

Appendix A

Biographical Sketches of Committee Members

John H. Johnson, *Chair,* is a presidential professor emeritus in the Department of Mechanical Engineering-Engineering Mechanics at Michigan Technological University (MTU) and a fellow of the SAE and the American Society of Mechanical Engineers (ASME). His experience spans a wide range of analysis and experimental work on advanced engine concepts, diesel and other internal engine emissions studies, fuel systems, and engine simulation. He was previously project engineer at the U.S. Army Tank Automotive Center, and chief engineer in applied engine research at the International Harvester Company before joining the MTU mechanical engineering faculty. He served as chairman of the MTU mechanical engineering and engineering mechanics department from 1986 to 1993. He has served on many committees related to engine technology, engine emissions, and health effects—for example, committees of the SAE, the NRC, the Combustion Institute, the Health Effects Institute, and the Environmental Protection Agency—and consults to a number of government and private sector institutions. In particular, he served on many NRC committees, including the Committee on Fuel Economy of Automobiles and Light Trucks, the Committee on Advanced Automotive Technologies Plan, the Committee on the Impact and Effectiveness of Corporate Average Fuel Economy (CAFE) Standards, and the Committee to Assess Fuel Economy for Medium and Heavy-Duty Vehicles. He chaired the NRC Committee on Review of DOE's Office of Heavy Vehicle Technologies and the NRC Committee on Review of the 21st Century Truck partnership, Phase 1. Dr. Johnson received from SAE the Horning Memorial Award, Colwell Merit Award (two), McFarland Award, Myers Award for Outstanding Student Paper, the Franz Pischinger Powertrain Innovation Award, and the ASME Honda Medal. He received his Ph.D. in mechanical engineering from the University of Wisconsin.

Joseph M. Colucci (NAE) is president, Automotive Fuels Consulting, Inc., and retired executive director, Materials Research, General Motors Research and Development Center. His previous positions include serving as head, assistant head, research engineer, and senior research engineer, Fuel and Lubricants Department, General Motors Research and Development Laboratories. His research interest focuses on vehicle emissions and fuel economy and on the interactions among the engine, fuel system, fuel, and emissions-control system. Conventional engines (spark-ignition and diesel) and fuels (gasoline and diesel fuel), alternative fuels, and new vehicle propulsion systems (hybrids and fuel cells) are also among his current interests. These research topics have societal benefits for improved air quality and reduced vehicular energy consumption. Mr. Colucci has served on numerous technical advisory committees. He has a B.S.M.E. from Michigan State University and an M.S.M.E. from the California Institute of Technology.

David E. Foster is the Phil and Jean Myers professor of mechanical engineering, University of Wisconsin, Madison, and director of the Engine Research Center, which has won two center of excellence competitions for engine research and has extensive facilities for research on internal combustion engines. A member of the faculty at the University of Wisconsin since he completed his Ph.D., Dr. Foster teaches and conducts research in thermodynamics, fluid mechanics, internal combustion engines, and emission formation processes. His work has focused specifically on combustion kinetics, emission formation processes, and the incorporation of simplified or phenomenological models of emission formation processes into engineering simulations. He has published extensively in this field throughout the world and for leading societies in this country. He is a recipient of the Ralph R. Teetor Award, the Forest R. McFarland Award, and the Lloyd L. Withrow Distinguished Speaker Award of the Society of Automotive Engineers (SAE) and he is an SAE Fellow. He has served on a number of NRC committees, including the Committee on Review of the Research Program of the Partnership for a New Generation of Vehicles. He is a registered professional engineer in the

State of Wisconsin and has won departmental, engineering society, and university awards for his classroom teaching. He received a B.S. and M.S. in mechanical engineering from the University of Wisconsin and a Ph.D. in mechanical engineering from the Massachusetts Institute of Technology.

Larry J. Howell is a consultant to industry and government, specializing in the management of research for business innovation, automotive technology, and vehicle structures and materials. He retired from General Motors (GM) in 2001 as Executive Director, Science, of the General Motors R&D Center's six science labs (Thermal and Energy Systems' Electrical and Controls Integration; Manufacturing Systems; Materials and Processes; Chemical and Environmental Sciences; and Vehicle Analysis and Dynamics). Dr. Howell had global responsibility for joint research with universities, government agencies, and GM's alliance partners. He also served as secretary to GM's Science Advisory Committee, which reports to GM's Board of Directors on technology matters.

Prior to his promotion to executive director, he served as department head of the Engineering Mechanics Department at GM R&D. In this position, he had responsibility for research in vehicle structures and materials, vehicle noise and vibration, vehicle aerodynamics, and vehicle safety including vehicle crashworthiness and occupant protection. For many years, he was a member of GM's safety subcommittee. Later, as executive director, he had responsibility for all of GM's safety research including stability and control technology (e.g. GM's StabiliTrak system for reducing the potential for spin-out and roll-over) and accident avoidance systems such as adaptive cruise control. He was also a member of GM's manufacturing manager council. He received GM's john M. Campbell Award in 2000 for outstanding contributions to: "Advancements in the Engineering Capability of General Motors and Leadership Excellence in all Phases of GM R&D Activities." Prior to joining GM, Dr. Howell worked for General Dynamics Corporation as senior dynamics engineer and was a principal investigator on NASA contracts focused on the structural dynamics of the Space Shuttle. He has served on the College on Engineering advisory board of the University of Illinois and Western Michigan University. He represented GM as a member of the Industrial Research Institute (IRI), has served on the Board of Directors, and is an emeritus member of the IRI. Dr. Howell has served on several National Research Council panels, including: Use of Lightweight Materials in 21st Century Army Trucks; Benefits of DOE's Light-Duty Hybrid Vehicle R&D Program; and Review of the 21st Century Truck Partnership. He has also served as a reviewer of several NRC reports. Dr. Howell received a B.S., M.S., and Ph.D. in aeronautical and astronautical engineering from the University of Illinois, Urbana, Illinois. He also completed the Executive program at Dartmouth's Amos Tuck School of Business Administration. He has published 27 journal articles and 25 internal company reports at General Motors and General Dynamics. His more recent publications are "Globalization Within the Auto Industry" and "Adapting GM Research to a New Corporate Strategy," both published in IRI's *Research Technology Management.*

John G. Kassakian (NAE) is professor of electrical engineering and former director of the Massachusetts Institute of Technology's (MIT's) Laboratory for Electromagnetic and Electronic Systems. His expertise is in the use of electronics for the control and conversion of electrical energy, industrial and utility applications of power electronics, electronic manufacturing technologies, and automotive electrical and electronic systems. Before joining the MIT faculty, he served in the U.S. Navy. Dr. Kassakian is on the boards of directors of a number of companies and has held numerous positions with the IEEE, including founding president of the IEEE Power Electronics Society. He is a member of the NAE, a fellow of the IEEE, and a recipient of the IEEE's William E. Newell Award for Outstanding Achievements in Power Electronics (1987), the IEEE Centennial Medal (1984), and the IEEE Power Electronics Society's Distinguished Service Award (1998). He is a co-author of the textbook *Principles of Power Electronics* and has served on a number of NRC committees, including the Committee on Review of the Research Program of the Partnership for a New Generation of Vehicles and the Review of the FreedomCAR and Fuel Research Program. He has an Sc.D. in electrical engineering from MIT.

David F. Merrion is chairman of David F. Merrion LLC; and chairman of Truck Emission Control Technologies, Inc. He is the retired executive vice president of engineering for Detroit Diesel Corporation (DDC). His positions at DDC included staff engineer, Emissions and Combustion; staff engineer, Research and Development; chief engineer, Applications; director, diesel engineering; general director, Engineering (Engines and Transmissions); and senior vice president, Engineering. Mr. Merrion has extensive expertise in the research, development, and manufacturing of advanced diesel engines, including alternative-fueled engines. He is a Society of Automotive Engineers fellow and a member of American Society of Mechanical Engineers. He served as president of the Engine Manufacturers Association, a member of the Environmental Protection Agency's (EPA's) Mobile Sources Technical Advisory Committee, a member of the Coordinating Research Council, and a member of the U.S. Alternate Fuels Council. He has served on a number of National Research Council committees, including the Standing Committee to Review the Research Program of the Partnership for a New Generation of Vehicles; the Committee on Review of the 21st Century Truck Partnership, Phase 1; and the Committee to Assess Fuel Economy Technologies for Medium- and Heavy-Duty Vehicles. He has a B.S. in mechanical engineering from General Motors Institute (Kettering University) and an M.S. degree in mechanical engineering from the Massachusetts Institute of Technology.

Thomas E. Reinhart is program manager, Engine Design & Development, Engine, Emissions and Vehicle Research Division, Southwest Research Institute. His previous positions include: Cummins, Inc., Columbus, Indiana, 1980-2000 (NVH Engineer, 1980-1984, Senior Engineer, Midrange Engine NVH, 1984-1987, Manager, Noise & Vibration Technology, 1987-1994, Director, Noise & Vibration Technology, 1994-2000); Roush Industries, Inc., Livonia, Michigan (Program Manager–Powertrain NVH, 2001-2004), Visteon Corporation, Van Buren Township, Michigan (Senior Manager–Chassis Systems NVH, 2004-2005). He is leading projects in engine design, performance and emissions development, as well as in gasoline and diesel engine NVH improvement. He has led a number of programs, including several emissions reduction projects, as well as the clean sheet design and development of a new off-highway diesel engine. Mr. Reinhart has more than 25 years of experience in diesel engine and powertrain design, analysis, and development, with particular expertise in noise and vibration testing and analysis. He has published 14 technical papers on a range of diesel NVH topics. He has a wide range of experience in the NVH issues of applications ranging from trucks through agricultural equipment, construction, forestry, marine, rail, and military vehicles. Mr. Reinhart has worked with customers on a range of issues, including NVH, drivability, fuel consumption, and adaptation of engines to a wide range of applications. Mr. Reinhart holds four patents for ideas related to diesel engine NVH control. For several years, he was a member of Cummins' patent review committee. His work experience also covers a wide range of development projects on gasoline and diesel engines, as well as on transmissions and on fuel cell vehicle powertrains. He is a member of the Institute of Noise Control Engineering (INCE), International Institute of Acoustics and Vibration (IIAV), SAE, and has been a member of the Board of Directors of INCE since April 2008. He has also been a member of the organizing committee for the SAE Noise & Vibration Conference since 2002, and chairman of the Diesel Noise session at this conference since 2003. He has an M.S. in mechanical engineering, Purdue University, and a B.S. in mechanical engineering, Purdue University.

Bernard Robertson (NAE) is the president of BIR1, LLC, an engineering consultancy specializing in transportation and energy matters that he founded in January 2004, upon his retirement from DaimlerChrysler Corporation. During the latter part of his 38-year career in the automotive industry, Mr. Robertson was elected an officer of Chrysler Corporation in February 1992. He was appointed senior vice president coincident with the merger of Chrysler Corporation and Daimler-Benz AG in November 1998, and was named senior vice president of engineering technologies and regulatory affairs in January 2001. In his last position, he led the Liberty and Technical Affairs Research Group, Advanced Technology Management and FreedomCAR activities, and hybrid electric, battery electric, fuel cell, and military vehicle development. In addition, he was responsible for regulatory analysis and compliance for safety and emissions. Mr. Robertson holds an M.B.A. degree from Michigan State University, a master's degree in automotive engineering from the Chrysler Institute, and a master's degree in mechanical sciences from Cambridge University, England. He is a member of the National Academy of Engineering, a fellow of the Institute of Mechanical Engineers (U.K.), a chartered engineer (U.K.), and a fellow of the Society of Automotive Engineers.

Charles K. Salter is retired after working 39 years with Mack Trucks, Inc./Volvo PowerTrain NA (3.5 years). His experience covers a wide range of heavy-duty diesel engine engineering and development. His most recent position was as executive director of engine development, where he was responsible for all engine/system functions (design and analysis; emissions control/fuel economy optimization; electronics system development; performance durability testing; manufacturing, supplier, sales and service liaison). This responsibility included design and production introduction of the world's first fully electronically controlled diesel unit pumps for 12-liter, six-cylinder engines in 1990. He jointly initiated (with Detroit Diesel) and developed, with the Environmental Protection Agency (EPA) and various industry participants, a urea infrastructure for targeted 2007 calendar year engine production (then delayed to 2010). He participated in industry collaborative research through the U.S. Department of Energy Diesel Crosscut Committee, which was part of the 21st Century Truck Partnership. He was a consultant to Volvo PowerTrain NA from 2005 to 2007 on an advanced large truck diesel exhaust gas recirculation cooler vibration study/amelioration and on heavy-duty truck hybrid powertrain duty cycle test procedure development for comparative fuel consumption (EPA/industry/Hybrid Truck Users Forum). He has been a member of the Society of Automotive Engineers for 43 years; an organizer for World Congress technical sessions on heavy-duty diesel fuel injection systems for several years; and company representative to the Engine Manufacturers Association for 25 years, including 13 years on its board of directors, where he has been treasurer, vice president, and president. He holds a B.S. in mechanical engineering from Pennsylvania State University and an M.S. in engineering, solid mechanics, from the University of Maryland.

Kathleen C. Taylor (NAE) is retired director of the Materials and Processes Laboratory at General Motors Research and Development and Planning Center in Warren, Michigan. Dr. Taylor was simultaneously chief scientist for General Motors of Canada, Ltd. in Oshawa, Ontario. Earlier Dr. Taylor was department head for physics and physical chemistry and department head for environmental sciences. Currently, Dr. Taylor serves on the DOE Hydrogen Technology

Advisory Committee, the Transportation Research Board Committee for a Study of Potential Energy Savings and Greenhouse Gas Reduction from Transportation, the board of the National Inventors Hall of Fame, and the Advisory Committee for Columbia University Center for Electron Transport in Molecular Nanostructures. Dr. Taylor was awarded the Garvan Medal from the American Chemical Society. She is a member of the National Academy of Engineering, the American Academy of Arts and Sciences, and the Indian National Academy of Engineering and a fellow of SAE International and the American Association for the Advancement of Science. She was the president of the Materials Research Society and chair of the board of directors of the Gordon Research Conferences. She has expertise in R&D management, fuel cells, batteries, catalysis, exhaust emission control, and automotive materials. She received an A.B. in chemistry from Douglass College and a Ph.D. in physical chemistry from Northwestern University.

Wallace R. Wade (NAE) was chief engineer and technical fellow, Powertrain Systems Technology and Processes, Ford Motor Company, Dearborn, Michigan, where he served for 32 years prior to his retirement. He was responsible for the development, application, and certification of emission and powertrain control system technologies for all Ford Motor Company's North American vehicles. His technical responsibilities have included low emission technologies for internal combustion engines; analytical and laboratory based powertrain calibration with objective measures of driveability, the first domestic production OBD II (On-Board Diagnostic) system; technology for diesel particulate filters (DPF) with active regeneration; electronic control systems for gasoline and diesel engines; low heat rejection and low friction, direct injection diesel engines; and an ultra low emission, gas turbine combustion system. Today he is a consultant to industry and government. Mr. Wade was elected to the National Academy of Engineering in 2011 for implementation of low-emission technologies in the automotive industry. He is a fellow member of the Society of Automotive Engineers and the American Society of Mechanical Engineers. He received the Henry Ford Technology Award and has been recognized as a Distinguished Corporate Inventor by the National Inventors Hall of Fame. He has received five SAE Arch T. Colwell Awards and the SAE Vincent Bendix Automotive Electronics Engineering Award. He has received 26 patents related to improvements in powertains and has written 25 published technical papers on powertrain research and development. He has an M.S.M.E. degree from the University of Michigan, and a B.M.E. degree from Rensselaer Polytechnic Institute, both in mechanical engineering.

Appendix B

Presentations and Committee Meetings

SEPTEMBER 8-9, 2010, WASHINGTON, D.C.

Overview of 21st Century Truck Partnership
Patrick Davis, DOE Office of Vehicle Technologies

Overview of DOE's Vehicle Technologies Program R&D
Patrick Davis, DOE Office of Vehicle Technologies

Overview of U.S. Army (National Automotive Center [NAC]/Tank Automotive Research and Development Engineering Center [TARDEC]) Activities
Paul Skalny, U.S. Army

Overview of EPA SmartWay Transport and Clean Automotive Technology Activities
Cheryl Bynum and John Kargul, U.S. EPA

Overview of DOT Truck Safety and Productivity Activities
Rolf Schmitt, Federal Highway Administration (FHWA)
Luke Loy, Federal Motor Carrier Safety Administration (FMCSA)
Bob Kreeb, National Highway Traffic Safety Administration (NHTSA)

Review of Previous NAS Review Recommendations, with Responses
Ken Howden, DOE Office of Vehicle Technologies

Heavy Duty Vehicle Industry Perspective
Susan Alt, VP, Customer and Industry Relations, Volvo Trucks N. America

SuperTruck Project Team Presentations
Donald Stanton, Cummins
Anthony (Tony) J. Cook, Navistar, Inc.
Derek Rotz, Kevin Sisken, and David Kayes, Daimler

NOVEMBER 15-16, 2010, WASHINGTON, D.C.

Engines Area
Gurpreet Singh and James Eberhardt, DOE

Fuels Area
Kevin Stork, DOE

Safety
Bob Kreeb, NHTSA; Luke Loy, FMCSA; and John Nicholas, FHWA

Idle Reduction
Glenn Keller, Argonne National Laboratory (ANL)

Department of Defense Related Activities: Overview of U.S. Army Medium/Heavy Vehicle Efforts
Bill Haris, U.S. Army

Power Demands
Mark Smith, PNNL; Carol Schutte, DOE; Dean Paxton, PNNL; David Warren, ORNL; Jerry Gibbs, DOE; and Kambiz Salari, LLNL

Heavy Duty Hybrids Area
Kevin Walkowicz, NREL, and Bill Batten, Eaton Corporation

Previous NAS Findings and Recommendations (2008) and 21CTP Responses Discussion
Ken Howden, DOE

Summary of Committee Site Visit to EPA Ann Arbor Laboratory
Wally Wade, Committee Member

Greenhouse Gas Emissions/Fuel Economy Update
Byron Bunker, Environmental Protection Agency (EPA), and Jim Tamm, National Highway Traffic Safety Administration (NHTSA)

JANUARY 31, 2011, WASHINGTON, D.C.

DOE EERE Biomass Program—Outlook and Costs for Biofuels
Zia Haq, DOE Biomass Program

Status and Prospects for Energy Storage Technologies for Medium- and Heavy-Duty Hybrids
Jim Miller, Argonne National Laboratory

DOE Aftertreatment Activities, and Combustion Memorandum of Understanding
Ron Graves, ORNL; Dennis Siebers, SNL; George Muntean (by phone), PNNL

Summary of Visit to TARDEC
Dave Merrion and Wally Wade, Committee Members

Solid Oxide Fuel Cell Technology Status
Dan Hennessy, Delphi Corporation

Cutting Edge Developments for Heavy-Duty Aftertreatment Technologies and Research Needs
Tim Johnson, Corning

Current Status of Electric Medium-Duty Truck Technology
Jan Friesner, Navista, and Scott Carson, Smith Electric

Questions and Answers on Program Activities with 21CTP
Ken Howden, DOE

ADDITIONAL INFORMATION GATHERING MEETINGS

Subgroups of the committee also made site visits to the Environmental Protection Agency, Ann Arbor, Michigan, October 26, 2010; the U.S. Army Tank Automotive Research, Development and Engineering Center, Warren, Michigan, January 10, 2011; and the Oak Ridge National Laboratory, Oak Ridge, Tennessee, February 17, 2011.

Proprietary-information-gathering site visits were also made to Cummins Technical Center, Columbus, Indiana, November 8, 2010; Navistar, Inc., Truck Development and Technology Center, Fort Wayne, Indiana, January 13, 2011; and Eaton Corporation's Eaton Innovation Center, January 14, 2011.

Appendix C

Responses from the 21st Century Truck Partnership to the Findings and Recommendations from the National Research Council Phase 1 Review

Update: November 10, 2010

☑ = Finding/recommendation was highlighted in presentation of findings and is considered of greater importance by the committee.

Finding Number and Subject Matter	Finding and Recommendation	21CTP Response (Updated November 10, 2010)
☑ 1-1 OVERALL REPORT	*FINDING:* The key benefit of the 21CTP is the coordination of research programs directed toward the goal of reducing fuel usage and emissions while increasing heavy vehicle safety. Federal involvement is bringing stakeholders to the table and accelerating the pace of development. Very few U.S. manufacturers of trucks and buses or heavy-duty vehicle components have the R&D resources to develop new technologies individually. Thus, the 21CTP is giving some of those companies access to extraordinary expertise and equipment in federal laboratories, in addition to seed funding that draws financial commitment from the companies to push forward in new technology areas. The Partnership provides the United States with a forum in which the various agencies, in combination with industry and academia, can better coordinate their programs. Research funding of the 21CTP has been declining steadily in recent years, and this decline is threatening the attainment of program goals. The current level is not in proportion to the importance of the goal of reducing fuel consumption of heavy-duty vehicles. *RECOMMENDATION*: The 21st Century Truck Partnership should be continued, but the future program should be revised and better balanced based on the recommendations of this report. In addition, more manufacturers should be recruited as participants, such as the major truck manufacturers and suppliers that are not in the Partnership. Research funding should be commensurate with well-formulated goals that are strategic to reducing fuel consumption of heavy-duty vehicles while improving safety. The 21CTP should also conduct an assessment of heavy-truck research activities overseas and determine if any changes in the future program would be appropriate based on foreign programs.	The Partnership concurs with the recommendation that it should continue, and the members of the Partnership are committed to conducting the joint research efforts necessary to achieve the efficiency, emissions, and safety goals set forth for such research. Independently of the NRC review, the Partnership has begun to re-examine the Partnership's structure and processes, and will take the NRC's recommendations into consideration as part of this assessment. Furthermore, the Partnership has expanded its membership, and welcomes the addition of ArvinMeritor, Inc., and the increased participation by both industry and agency partners.
☑ 1-2 OVERALL REPORT	*FINDING*: Many of the program goals were not met. Some of the goals were not plausible, from either an engineering or funding perspective. Other goals were not met because some of the technologies proposed for meeting the goals were not applied. Notable failures of that kind are discussed in Chapter 3, under the headings of "Goal of Thermal Efficiency of 55 Percent" and "Goals Involving Fuels." *RECOMMENDATION*: A clearer goal setting strategy should be developed, and the goals should be clearly stated in measurable engineering terms and reviewed periodically so as to be based on the available funds.	The Partnership is pleased that the NRC panel pointed out a number of successes within the work of the Partnership, including the work to develop hybrid medium-duty and heavy-duty components and systems and the work done to implement idle reduction technologies in the market. The Partnership would like to highlight the accomplishments by the major engine manufacturers in meeting stringent 2007 emissions regulations with no degradation in fuel economy. The efforts of the Partnership played an essential role in this accomplishment. The Partnership has completed several projects relevant to the program's goals and recently launched, through the SuperTruck opportunity, a major effort largely in line with those goals.
☑ 2-1 MANAGEMENT	*FINDING*: The 21CTP is operated as a virtual network of agencies and government laboratories, with an unwieldy structure and budgetary process. Agency personnel meet frequently and industry partners meet periodically for limited sharing and communication. This has been the extent of the coordination. Both government agencies and industry partners, per their remarks to the committee, have found the arrangement less than effective. The program was most productive when a full-	The Partnership continues to examine its organization and management structure as part of its ongoing self assessment efforts. The NRC panel's recommendations are a key guide in this assessment.

	time person from industry was assigned to coordinate the cross-agency efforts. Oversight of the 21CTP is provided through an Executive Committee with representation from DOE, DOT (the U.S. Department of Transportation), EPA (the U.S. Environmental Protection Agency), DOD, and the industry partners. Although that committee lacks authority to make cross-agency decisions and implement firm actions, it has been most effective when chaired by a full-time executive. This seemed to be an effective measure to ensure cooperation among agencies and address program challenges. *RECOMMENDATION*: A full-time, technically capable leader with consensus building skills should be appointed to coordinate the 21CTP program among industry partners and government agencies. This person could chair the Executive Committee and would be authorized to make recommendations to the committee on behalf of the entire program on stopping or redirecting existing research, on setting research priorities, and on future funding levels.	
☑ 2-2 MANAGEMENT	*FINDING*: As confirmed in meetings with the DOE and other agencies there is no single source of funds for the 21CTP, as perhaps intended by its creators. Instead, each of the four agencies has its own stream of funds. DOE, DOT, DOD, and EPA budget and optimize funding based on their own priorities. In addition, they maintain funding to companies with multiyear cooperative agreements. Thus, managing the 21CTP program and projects across multiple agencies has been challenging. The result has created difficulties in setting program priorities, especially in aligning budgets to programmatic requirements. A result has been difficulty in balancing between near- and long-term projects and setting appropriate metrics and measures. In addition, variation in funding levels year to year has diminished the impact of project achievements and results and reduced the probability of success and commercialization. The result of this complexity and lack of transparency is that some federal funds were spent by industry partners and by other federal agencies in ways that cannot be accounted for in the funding structure by fiscal year." *RECOMMENDATION*: A portfolio management process that sets priorities and aligns budgets among the agencies and industrial partners is recommended. A proposed Table of Project Priorities (Figure 2-5) would provide an objective way of ranking research and development projects according to their expected outcomes. This could evolve into a budgeting process that ensures support for programs of merit beyond a single year. Pre-competitive, collaborative technology and concept development could receive proper focus for successful programs.	The recommendations of the NRC panel on management processes will be considered as part of the ongoing Partnership assessment. Indeed, as the NRC points out, the four federal agencies have very different funding structures in and of themselves. Their participation in the Partnership allows for otherwise unlikely coordination of some research goals and achievements. The ability to directly align budgetary decisions across the agencies, however desirable, may be outside the scope of this voluntarily collaborative organization.
☑ 3-1 ENGINES	*FINDING*: Although DOE has concluded that the 50 percent thermal efficiency goal has been achieved, the experimental test results show that none of the industry partners achieved the goal of 50 percent thermal efficiency at 2010 emissions standards with a complete engine system. Each partner either failed to test a complete engine system on an engine dynamometer and used analysis to project results or failed to achieve 50 percent thermal efficiency at 2010 emissions standards with a complete system. Details of the analytical projections were proprietary and were not provided to the committee. Moreover, the work that was accomplished was at the intrinsically more efficient peak torque condition rather than at an engine speed and load representative of 65 mile per hour (mph) road load. *RECOMMENDATION*: Objective and consistent criteria should be used to assess the success or failure of achieving a key goal of the 21CTP such as the attainment of 50 percent thermal efficiency. Detailed periodic technical reviews of progress against the program plan should be conducted so that deficiencies can be identified early and corrective actions implemented to ensure success in accomplishing program goals. DOE should continue to work toward	DOE's public- private partnerships are to facilitate technologies which would demonstrate feasibility of stretch goals. Achieving 50 % engine efficiency for heavy duty engines is an overarching national goal. The industry must also meet prevailing emissions with the efficiency goals. The EPA 2010 near-zero emissions requirements were mandated in the middle of the industry contracts. A goal of 50% engine efficiency was a high-risk goal, especially when concurrent with the near-zero emissions achievement. Looking forward, DOE's SuperTruck program, launched in 2010, includes the demonstration of 50% thermal efficiency. In effect, it represents the culmination of the modeling, design, analysis, subsystem development and hardware testing completed by both publicly and privately funded efforts within the overall 21CTP umbrella over the past decade. DOE funding

Update: November 10, 2010

	demonstrating 50 percent thermal efficiency at the peak efficiency condition as well as at a representative 65-mph road load engine speed and torque condition. DOE should also consider reducing the number of industry contracts on specific engine projects that are funded so that only the engine systems most likely to meet the goal, based on system modeling and analytical projections, will be developed and tested experimentally.	levels have limitations and this therefore represents a significant success. Industry observed that meeting 2002/2004 emissions were at a cost of approximately 3-4 % more fuel use per truck. The partnership reiterates that because of learning from these programs, the industry was able to introduce 2007 emissions compliant heavy-duty products without a fuel economy penalty from the previous year's products–a major achievement. DOE EERE is placing more emphasis now on demonstration and commercialization activities, and the panel's recommendations are in line with this new philosophy. The SuperTruck projects involve demonstration strategies for 50 percent thermal efficiency goals and pursuit of real-world demonstration of 50 percent thermal efficiency engines in Class 8 trucks, building on previously-developed technologies and developing new technologies with partners. . Lessons learned from this effort will guide the requirements for a future solicitation involving 55 percent thermal efficiency engines in truck demonstrations.
3-2 ENGINES	*FINDING*: The goal of achieving 50 percent thermal efficiency at 2010 emissions was not clearly specified by the 21CTP. Each of the three industry partners used a different test procedure for measuring thermal efficiency (see Table 3-4). Likewise, none of the industry partners demonstrated 2010 emissions using the required EPA test procedures with aged engine and aftertreatment systems. A goal of this importance should be specified by standard test procedures so that the results are verifiable and compatible with industry standards. *RECOMMENDATION*: Future work to achieve the goal of 50 percent thermal efficiency at 2010 emissions should be specified by industry standard test procedures. SAE J1349 Engine Power Test Code is the industry standard for net power ratings and should be specified for the thermal efficiency portion of this goal (SAE, 2004). Test results should clearly provide all of the engineering details required to interpret the results.	Adherence to the full conditions of an SAE standard requires the level of technology maturity beyond that expected from pre-competitive R&D projects. The goal of these pre-competitive research projects was to develop stretch capabilities of engine and emissions systems. It was not intended to reach the preproduction technology level (aged engine and aftertreatment systems).
3-3 ENGINES	*FINDING*: Some of the technical features used to approach the goal of 50 percent thermal efficiency, as shown in Table 3-4, differed among the three industry partners and no explanation or technical analysis was provided to justify the different approaches. Furthermore, the effectiveness of the individual features used on the demonstration engines could not be determined due to the lack of analysis or system modeling. A validated system model should have been used to compare test data with analytical projections to determine if each feature was performing as expected. *RECOMMENDATION*: Prior to beginning future test phases of this program to achieve 50 percent thermal efficiency, system modeling should be used so that the preferred technical approaches could be selected and test data could be compared with analytical projections to determine if the expected results have been obtained.	The Partnership concurs with this recommendation. Each of the industry participants used their proprietary, internal analytical rigor, codes and modeling approaches to choose the technical path that showed the most promise of achieving the target of 50% thermal efficiency. In the SuperTruck solicitation, proposers were "encouraged to make extensive use of modeling and simulation to make technology choices and determine potential benefits."
3-4 ENGINES	*FINDING*: Although DOE stated that the 2010 emissions standard was achieved in the demonstrator engines attempting to achieve 50 percent thermal efficiency, only steady state emissions at one test condition were reported rather than test results from the EPA specified test procedures for the 2010 emissions standard. In some cases, the emissions were estimated from engine-out emissions and assumed aftertreatment efficiency. *RECOMMENDATION*: Achieving compliance with 2010 emissions with a "one-off" prototype	Industry must ensure feasibility of meeting prevailing emission levels in any thermal efficiency improvement project. Emission compliance is a hard requirement, while thermal efficiency goal is an R&D target.

	engine designed to demonstrate 50 percent thermal efficiency may be too stringent a goal for the 21CTP. The emission objective levels should be revised to be the demonstration of emissions at a single point, where the emission level selected to be demonstrated should have the potential for meeting the 2010 emissions as specified by EPA test procedures.	
3-5 ENGINES	*FINDING*: Although industrial partners reported on their progress, the presentations were high level summaries with critical engineering information omitted, thereby making the assessment of accomplishments relative to goals difficult. *RECOMMENDATION*: DOE should work to develop a review process that will allow future review committees to evaluate "sensitive" information so quantitative assessments of progress can be made.	The Partnership concurs with this recommendation. A process developed by the Partnership to address Committee's recommendation is to include site visits by the Committee members to the Partnership member companies and agencies. This process is being implemented as part of the 2010 NAS review of the Partnership.
3-6 ENGINES	*FINDING*: Achieving the 21stCTP's goal of 50 percent peak thermal efficiency is not expected to result in the Partnership's goal of 50 percent thermal efficiency for a typical Class 8 tractor-trailer combination on a level road at a constant speed of 65 mph and a GVW of 80,000 lbs. Even if 50-percent thermal efficiency were to be achieved at, or near, the peak torque condition, up to a 7 percent improvement (3.4 percentage point improvement) task would still remain to achieve 50 percent thermal efficiency at the 65 mph road-load condition. *RECOMMENDATION*: The 21CTP should clearly define, in addition to the peak thermal efficiency condition, the specific 65 mph road-load condition for demonstrating 50 percent thermal efficiency. The committee suggests using one of the 13-mode steady state emission test points for approximating the 65-mph road load condition. For typical engines, drivetrains, and vehicles, emission test point A50 (60 percent engine speed, 50 percent load) would be appropriate, although the most appropriate point (or multiple points, if necessary) should be determined for the specific engine, powertrain, and vehicle configuration under consideration, although this should be confirmed for each engine under consideration. The 21CTP should request each of the three current industry partners to test their experimental demonstration engines according to this recommendation. A recent CRC study (CRC ACES-1, July 2007) has proposed new cycles under development that may correlate better with actual in-use emissions and, possibly fuel usage, for heavy-duty diesel trucks). This study found that their in-use operation could be partitioned into the following four modes (with associated maximum speeds noted): creep (9 mph), transient (48 mph), cruise (59 mph) and high-speed cruise (65 mph). Each mode was highly transient and only the high-speed cruise mode reached 65 mph. The 21CTP should monitor this work and consider the possible future application of these cycles for assessing thermal efficiency improvements for HHDDEs.	The Partnership concurs with this recommendation. The SuperTruck solicitation included "the development of a heavy-duty diesel engine capable of achieving 50% Brake Thermal Efficiency (BTE) on a dynamometer under a load representative of a level road at 65 mph." Partnership closely monitors CRC ACES efforts. Results of this work, together with other inputs, will contribute to the selection of test cycles for demonstrating vehicle level goals in the SuperTruck projects.
3-7 ENGINES	*FINDING*: DOE and the industry partners did not appear to address the potential commercial viability of the technologies or the potential costs required to achieve cost effective solutions, as illustrated in Table 3-10. *RECOMMENDATION*: DOE should request the industry partners to make an assessment of cost objectives required to achieve commercial viability.	The Partnership concurs with this recommendation. The SuperTruck solicitation included the following requirement: "The vehicle freight efficiency improvement must be achieved while meeting prevailing emission standards and Class 8 tractor-trailers vehicle safety and regulatory requirements. The systems developed shall be validated as cost effective via a business case analysis and will be reviewed for commercialization potential in later project phases as part of the phase gate review process."
☑ 3-8 ENGINES	*FINDING*: DOE is shifting prematurely to component research to support the 2013 stretch goal of 55 percent thermal efficiency before completely demonstrating the earlier 2010 goal of 50 percent.	Completion of the demonstration of the 50 percent thermal efficiency would entail a full scale pre-production development of technologies

	Importantly, after analyzing the results of the lengthy and extensive efforts carried out in the area of Low Temperature Combustion LTC), it is considered unlikely that this technology will be a successful enabler of the 55 percent stretch goal at any time in the near term, as it cannot be adequately controlled over the full range of operating conditions of heavy-duty engines and has not demonstrated inherent fuel-consumption advantages. Based on the open literature, the chances for success of LTC as a practical technology appear limited. *RECOMMENDATION*: DOE should complete the demonstration of the 50 percent thermal efficiency goal before embarking on the 55 percent goal. With respect to ongoing work on Low Temperature Combustion, DOE should objectively analyze the potential viability of this combustion concept for heavy-duty engine applications, recognizing the many issues that would need to be resolved to achieve commercial viability.	expected to perform across the wide range of engine speeds and loads including 65 mph representative points. A project of such magnitude is deemed financially prohibitive, outside DOE's mandate of public–private partnerships, and well beyond the pre-competitive R&D supported by DOE. Heavy Truck Engine program budgets have been well below industry recommended levels for many years. The feasibility of reaching a 2013 stretch goal of 55% thermal efficiency demonstration goal may need to be reconsidered, and the partnership has commenced the process to revise the program goals. The original 21CTP timeline will also be re-evaluated. A feasible, while stretch thermal efficiency demonstration, goal for the next set of projects may well be somewhere between 50% and 55%. Such stretch goal will encourage continuing introduction of high risk, high reward technologies into the program, while allowing each company to leverage its individual technical strengths and teaming capabilities. With respect to Low Temperature Combustion (LTC), DOE agrees that LTC may not be a sole enabler for the efficiency targets. Rather, DOE has stated that advanced combustion approaches will be a critical element, along with other engine and aftertreatment advances, of the path toward simultaneously higher efficiency and emission compliant heavy-duty engines (See 21CTP Roadmap). Engine manufacturers (world-wide) continue to strongly support the need for LTC research and pursue implementation. The advances that have already resulted from the combustion R&D have been an essential part of achieving 2004 and 2007 emission goals with minimal degradation of engine efficiency, and were an important part of reaching 2010 emission goals.
3-9 ENGINES	*FINDING*: Information on the effects of fuel formulations on LTC operation was not presented to the committee by the 21CTP. However, the Committee's opinion is that any single diesel fuel formulation is unlikely to optimize LTC over all modes of operation. The optimum fuel for light-load operation will likely have different properties than the optimum fuel for heavy-load operation. *RECOMMENDATION*: DOE should try to specifically confirm whether or not a single non-specialty diesel fuel formulation will optimize LTC over all modes of operation and, modify its priorities accordingly based on the data.	The Partnership concurs with this recommendation. Currently, such work cannot be prioritized since DOE's petroleum based fuel budget has been zeroed out. One example of the relevant work in this area is modeling of fuel kinetics at the Lawrence Livermore National Laboratory.
3-10 ENGINES	*FINDING*: Even if LTC is successful at light loads, traditional diesel operation will likely be necessary at cold start and higher loads. Due to the different emission issues at light loads and heavy loads, it is very implausible that heavy-duty diesel engines will require no aftertreatment. *RECOMMENDATION*: DOE should undertake an analysis of a mixed-mode scenario to determine whether unburned HC and CO control in the LTC regime and DPF and NOx control in the traditional diesel combustion regime is not more complex and costly than aftertreatment for traditional diesel alone.	The Partnership concurs with this recommendation. As part of the SuperTruck projects, DOE will compare potential of mixed-mode operation to that of the traditional aftertreatment-supported combustion.
3-11	*FINDING*: At the reduced budget levels for FY 2008 and beyond, the inclusion of five engine	Recovery Act has afforded DOE an opportunity to fund three new cost-

ENGINES	manufacturers as cost-sharing participants reduces the ability of funding projects of "critical mass," which is not in keeping with the national interest. *RECOMMENDATION*: DOE should fund only one or, possibly, two manufacturers during the next phase of the program so that only the most promising projects of a significant scope can be accomplished.	shared vehicle and engine efficiency R&D projects – two teams under Recovery Act funding, and one team under regular program budget.
3-12 ENGINES	*FINDING*: The thermoelectric conversion systems are at a very basic stage and seem to have been "lumped" into the 21CTP as a matter of budgetary convenience for more basic work going on primarily at the National Laboratories. *RECOMMENDATION*: The thermoelectric conversion research should be removed from the 21CTP program until a more advanced level of technical maturity is attained. At the very least, a technical analysis of the candidate waste energy recovery systems is needed to determine if future efforts on thermoelectric conversion systems within the framework of the 21CTP are justified.	While the majority of DOE's thermoelectric work is done on light duty vehicles, we have seen continuing interest from the heavy-duty community in this technical area. Therefore, DOE will consider funding heavy-duty thermoelectric work on a case-by-case basis. Currently, the only heavy duty project is a work with Michigan State University and Cummins.
☑ 3-13 FUELS	*FINDING*: It is unlikely that the goal of identifying and validating nonpetroleum fuel formulations, optimized for use in advanced combustion engines, will be achieved by 2010. DOE's nonpetroleum fuels effort is focused on resolving biodiesel operational issues and commercialization barriers, but DOE did not provide a timetable for successful resolution of these efforts. DOE is also investigating oil sands and shale oil as other sources of petroleum fuel replacement. DOE did not present a plan for 5 percent replacement of petroleum fuels. The Renewable Fuels Standard of the Energy Policy Act of 2005 is likely to have a role in accelerating the availability of non-petroleum fuels. *RECOMMENDATION*: DOE should continue to work with biodiesel developers and users to assure compatibility when biodiesel is blended with conventional diesel fuel and problem-free use of biodiesel fuels in diesel engines. Successful deployment will require resolving operational issues and updating the biofuel specifications. Development of refining technology to make acceptable diesel from shale oil or tar sands is not high-risk research suitable for federal funding and should be left to the private sector. DOE should develop specific plans, including key actions and timetables, for 5 percent replacement of petroleum fuels.	Much emphasis is being placed on resolving issues and remaining questions with utilization of biodiesel. With respect to non-conventional petroleum sources (e.g., oil sands and shale oil), the DOE Vehicle Technologies Program (wherein 21CTP resides) has no mission to sponsor research in fuel production processes and has no plans to do so. The national labs under DOE funding are studying the effects of blended fuels utilizing oil sands to determine whether there could be impacts of non-conventional petroleum fuels on combustion and emissions from advanced LTC strategies for high-efficiency engine combustion systems.
3-14 FUELS	*FINDING*: DOE is exploring fuel properties of petroleum-based fuels that could have beneficial effects on engine efficiency and emissions, including aftertreatment. The committee is concerned about the viability of low temperature combustion regimes used in this effort, and that the applicability of the results of this project may be of limited value. The committee is also concerned that DOE's work may define optimum fuel properties for an engine with a new combustion regime that are not consistent with the properties of conventional diesel fuel defined in the ASTM specification for No. 2 diesel fuel. A potential implication of such a result is that a future engine with a new combustion regime may require a separate fuel, which would entail significant problems in the refining, distribution, storage, availability and cost of a special diesel fuel for these engines. *RECOMMENDATION*: The committee recommends against assuming that specialized fuels will be commercially available for future engines with new combustion regimes. Due to the issues concerning the viability of low temperature combustion regimes and commercially available specialized fuels, DOE should consider redirecting these efforts towards work with greater probability of contributing to the overall goals of the 21CTP.	The DOE efforts to understand the effects of fuel properties on combustion support two principal objectives; first to assess the robustness of combustion strategies to fuel property variations found within modern commercial fuels, and second to develop better understanding of fuel effects to support longer term co-development of fuels and engines by industry. There is a third potential benefit in defining improved combustion rating criteria. The committee report referred to FACE for example. FACE is administered through CRC, chaired by an energy company representative, and controlled by a mission statement that prohibits optimization objectives. The engine and energy industries have been participants in FACE and related projects for several years. The projects have been generally well-received in previously conducted peer reviews.
3-15 FUELS	*FINDING*: DOE provided little insight into the scope and magnitude of the effort to address the goal of developing non-petroleum fuel formulations beyond biodiesel that could provide	Most of the DOE-sponsored efforts in this area have been focused on biodiesel as the committee noted. Otherwise, the emphasis has been on

Update: November 10, 2010

	additional fuel economy improvements and near-zero emissions. DOE did not report any specific work plans, results, or timetables addressing this objective. *RECOMMENDATION*: DOE should reaffirm that this goal should continue to be pursued. If the goal is considered to strongly contribute to the overall 21CTP goals, DOE should develop specific work plans and timetables for addressing this goal.	developing a fundamental understanding of a wide range of fuel property effects on combustion and emission controls, encompassing the expected properties of non-petroleum fuels. Note the goal says "identify and validate" and not "develop." One intent was to look for non-petrol fuel property effects that would incentivize their use (eg, the apparent DPF regeneration advantage with biodiesel). Going forward, DOE and the partners will re-evaluate this objective and will need to bring plans in line with the EISA.
3-16 AFTER-TREATMENT	*FINDING*: No specific goals have been outlined for 21CTP diesel engine aftertreatment systems but some goals have been set for eliminating aftertreatment. However, as discussed in this chapter, the goal of eliminating aftertreatment does not appear to be achievable in the foreseeable future. *RECOMMENDATION*: Specific goals should be set for aftertreatment systems (improved efficiency, lower fuel consumption, lower cost of substrates, lower cost catalyst, etc.).	Engine manufacturers' selection of aftertreatment system components depends on trade-offs of many technical and commercial factors, including combustion strategy, available fuel injection equipment, supplier preferences, etc. This leads to different approaches selected by different manufacturers, as happened with each of the recent emission regulations in 2004, 2007 and 2010. The Partnership prefers not to constrain manufacturers' technical choices. Therefore, we have technical goals for the overall engine/aftertreatment system, rather than for its individual sub-systems.
3-17 AFTER-TREATMENT	*FINDING*: The CLEERS, DCT, and CRADAs have contributed to many successful projects and programs. *RECOMMENDATION*: The 21 CTP should continue with the CLEERS, DCT and CRADA activities for aftertreatment systems.	The Partnership concurs with this recommendation. We appreciate the recognition and agree that these collaborative projects and working groups have been effective and productive.
3-18 HTML	*FINDING*: The High Temperature Materials Laboratory is a valuable resource, providing specialized instrumentation and professional expertise in support of materials research. 21CTP projects have utilized the laboratory extensively; it has provided support to 35 different 21CTP projects since 2001. Whereas few advanced materials were actually utilized in the 21CTP project to demonstrate the major 50 percent thermal efficiency goal, it is expected to contribute to the 21CTP in valuable ways in the future. *RECOMMENDATION*: The DOE should continue to provide 21CTP projects access to the HTML. Although HTML's budget is not explicitly linked to the 21CTP, DOE should make every effort to maintain a stable budget for the HTML, in order to keep it at the "state of the art" level, and able to respond to the needs of the broader research community.	The Partnership concurs with this recommendation. In 2007 the HTML discussed a five-year plan with the director of the Vehicle Technologies Program that addresses the needs of the HTML and the HTML User Programs to effectively support the missions of the Vehicle Technologies Program. Specifically the plan included recommendations for investments in instrumentation and human resources that are both related and relevant to the goals of the Vehicle Technology Program.
3-19 HEALTH EFFECTS	*FINDING*: ACES is a cooperative, multi-party effort to characterize the emissions and assess the safety and potential health effects of new, advanced engine systems, aftertreatment, fuels and lubricants. It is an animal study using rats and not focusing on the direct effects on humans. DOE is providing the major funding for this program. *RECOMMENDATION*: The committee endorses the DOE funding of this study and recommends that this continue for the remainder of the study until results become available in the 2012-2013 time period.	The Partnership concurs with this recommendation. We appreciate the endorsement of the ACES project and agree with the NRC panel on its value.
☑ 4-1 HYBRIDS	*FINDING*: Challenges with lithium-ion anode/cathode materials and chemical stability under high power conditions will likely preclude achieving the 15-year durability targets by 2012. *RECOMMENDATION*: Much closer interaction between military and commercial suppliers is recommended to identify the highest-priority areas for further research in an attempt to expedite the development of commercially viable battery or battery/ultracapacitor systems that can	The Partnership agrees that more R&D in this area - especially in the area of Power Density and Battery Life, is needed. Indeed, subsequent to the NRC's first report and as part of the ARRA 2009, battery research, development and production capacity have greatly increased through DOE. The Partnership would be happy to provide additional

	accomplish the unique high-power needs of heavy-duty vehicles.	documentation with regard to these technologies.
4-2 HYBRIDS	*FINDING*: There are significant differences associated with the use of battery energy storage systems in heavy-duty vs. light-duty applications. *RECOMMENDATION*: Due to these differences and the much lower production volumes for heavy-duty applications, it is appropriate to continue funding and conduct sufficient research and development to demonstrate prototypical success in heavy duty applications, or identify areas for continued research.	The Partnership concurs with this finding. Current research focused on Light-duty applications may not necessarily apply to Heavy-duty applications. A crosscut light-duty / heavy-duty meeting on energy storage is planned for January 2011.
4-3 HYBRIDS	*FINDING*: The information exchange between DOD, DOT, DOE appears to be rather casual due to completely separate funding mechanisms, priorities, and testing methods. *RECOMMENDATION*: Jointly-funded programs that prioritize research, build upon the success of each agency's programs and thereby necessitate technology transfer between the partners would significantly improve the technology transfer and reduce the chance for "reinventing the wheel," or duplicating other's mistakes.	The Partnership works within the constraints of current funding mechanisms, which differ among the member agencies. These limitations make it difficult to devise a jointly-funded R&D program. We believe that our role in fostering communication among agencies goes a long way toward improving technology transfer and learning from each other's mistakes and successes.
4-4 HYBRIDS	*FINDING*: The metrics used for comparing battery technologies differ from manufacturer-to-manufacturer, agency-to-agency, and even for different evaluations within a given agency. Terminologies also vary in definition. Many existing standards for measuring battery parameters are technology specific, making accurate comparison of different technologies difficult or impossible. *RECOMMENDATION*: Metrics should be standardized or modified to enable more accurate comparisons across different battery technologies for transportation use. Universal terminologies should be defined, published, and recommended for adoption by the various battery manufacturers.	The Partnership concurs with this recommendation. It is our intent to increase activities in Codes and Standards for hybrid components including battery storage.
4-5 HYBRIDS	*FINDING*: Very little data are published about batteries when used in conjunction with ultracapacitors for heavy-duty HEV applications in this program. Recent developments show great promise with this technology, especially for heavy-duty applications requiring high power output for acceleration and fast charging for braking energy recovery. *RECOMMENDATION*: Expanded research effort and associated funding focus should be focused on ultracapacitors or supercapacitors as "hybrid" storage systems, in combination with batteries.	The Partnership concurs with this recommendation. Additional power density needs associated with heavy-duty applications could benefit from this technology.
☑ 4-6 HYBRIDS	*FINDING*: R&D on heavy-duty hybrid trucks and buses has demonstrated significant progress, achieving 35 to 47 percent fuel economy improvements in hybrid-electric delivery vans and urban buses, with specialized applications and the hydraulic hybrid delivery van in the 50 to 70 percent range (60 percent is the present 21CTP target). Commercial success has already been achieved with hybrid urban buses, albeit with major governmental subsidies. Despite the promising progress, significant hurdles still remain to achieving the fuel economy improvement targets for a broader range of heavy-duty hybrid vehicle (HHV) applications, reducing the cost, and improving HHV reliability sufficiently to achieve broader commercial success. In addition, there are opportunities for achieving significant system-level improvements that would make HHVs more attractive to original equipment manufacturers (OEMs) and users, such as the merging of hybrid propulsion and idle reduction features, including start-stop operation and creeping under all-electric power. *RECOMMENDATION*: Development and demonstration of heavy-duty hybrid truck technology should be continued as part of the 21CTP program in order to reduce barriers to	The Partnership agrees with this finding. The SuperTruck demonstration projects may be an excellent venue to develop and commercialize hybrid technologies. Hydraulic hybrids could also use additional research funding to achieve their maximum system level performance. Furthermore, the Partnership is in the process of bringing together the hybrid electric vehicle energy storage experts from light vehicle and medium & heavy vehicle industries for a crosscut discussion to identify possible research areas of mutual interest. A meeting is planned for October 2010.

	NRC Finding / Recommendation	Response
	commercialization. These development projects should include efforts to capitalize on opportunities for system-level improvements made possible by HHV technology in order to extract the maximum possible value from any new hybridized propulsion equipment that is installed in future trucks and buses.	
☑ 4-7 HYBRIDS	*FINDING*: Progress in the development of HHV technology under the 21CTP program has been hindered by the decision to focus on component-level technology rather than on systems. Successful development and commercialization of HHV technology requires coordinated, customized development of the combustion engine, electrical/hydraulic drive equipment, mechanical powertrain, and controls, as components of an integrated system, in order to realize its full potential. In addition, the coordination of HHV project activities among the 21CTP's federal partners (DOD, EPA, and DOE) has not matched the level achieved in other 21CTP programs such as nighttime idle reduction, making it more difficult to achieve ambitious HHV technology targets. *RECOMMENDATION*: Coordination of all 21CTP heavy-duty hybrid truck development and demonstration activities should be strengthened across components, programs, and agencies to maximize the system benefits of this technology and to accelerate its successful deployment in commercial trucks and buses. In addition to improved cross-agency coordination, HHV stakeholder-based organizations including the Validation Working Group and the Hybrid Truck Users Forum should be engaged more aggressively to assist in identifying and overcoming key hurdles to the successful commercialization of HHV technology.	The Partnership concurs. The Partnership is exploring strategic alliances with HTUF as part of the ongoing assessment of its operations. DOE VTP director Pat Davis will be presenting at the HTUF event in late September. This will be an opportunity for engaging HHV component suppliers and hybrid truck users. DOE has met for a series of meetings with CALSTART to discuss and consider a number of areas where there is opportunity for cooperation. Likewise, DOE has met with representatives from the U.S. Army to consider cooperative research opportunities. A memorandum of understanding between the Department of Defense and the DOE was established. As was presented September 8, the U.S. EPA has a hydraulic hybrid vehicle demonstration effort that complements 21CTP hybrid efforts. Finally, DOT has been funding the purchase of heavy duty hybrid electric vehicles.
☑ 4-8 HYBRIDS	*FINDING*: Emissions of heavy-duty trucks are currently measured and certified by EPA for each engine type rather than for any truck as a complete unit. Since current procedures do not allow either the fuel economy or emissions of complete hybrid propulsion systems to be certified, neither the fuel economy or emissions gains of hybrid trucks are recognized by this procedure. This approach serves as a deterrent to commercialization of HHV technology since there is at present no practical way for truck purchasers to derive any direct tax credits for buying hybrid trucks as called for in the U.S. Energy Policy Act of 2005, which expires in 2009. Developing the necessary test procedures to address this shortcoming is expected to be a complex and lengthy process, and EPA has not been able to devote sufficient resources to resolve this problem in a timely manner. *RECOMMENDATION*: Since tax credits for hybrid trucks established in the Energy Policy Act of 2005 expire at the end of 2009, DOE should work with EPA and stakeholders to accelerate the development of fuel economy and emissions certification procedures for heavy-duty hybrid vehicles so that the benefits of hybridization can be rewarded to encourage commercial adoption	The Partnership agrees with this finding. In fact, we believe that 21CTP, EMA, TMA and SAE organizations should be utilized in an advisory capacity during the development of these new regulations. The Partnership is exploring the possibility of developing strategic alliances with several of these organizations. It is appropriate for EPA to maintain responsibility for developing these types of procedures based on the Agency's past experience with fuel economy and emission measurement protocols. EPA agrees that DOE involvement in this process would be beneficial and could help accelerate development of procedures more suitable than the interim procedures currently specified by the IRS for obtaining tax credit certification. Under the SmartWay program, the EPA is working on drive cycle development for different vocations. DOE participated in this effort through the workshops held by EPA.
4-9 HYBRIDS	*FINDING*: Recent statements by representatives of some heavy-duty truck OEMs have reported that there are opportunities for fuel economy improvements between 5 and 7 percent in hybridized versions of Class 8 long-haul trucks, yielding annual fuel cost savings exceeding $9,000 per year. This result runs counter to generally-held opinions about the low potential of hybrid versions of	The Partnership concurs with this recommendation. Two of the three SuperTruck projects include hybridization as part of their work. This effort will include a combination of analytical simulation and

	Class 8 long-haul trucks for substantial fuel savings, and no documented study results have been made available to the committee to firmly substantiate the recent claims. *RECOMMENDATION*: The committee recommends that the potential benefits of hybrid Class 8 long-haul trucks be evaluated as part of the 21CTP program by conducting a documented study using a combination of analytical simulation and experimental data. If the results of the study confirm the recent claims of substantial fuel economy opportunities in hybrid long-haul trucks, the 21CTP program management is encouraged to find ways to contribute directly to the accelerated development of the necessary hybrid technology and its successful demonstration in prototype vehicles.	experimental prototype work. In appreciation of the growing importance of hybrid work to the Partnership, we have accepted a request of ArvinMeritor (one of the leading players in hybrid technology for heavy-duty applications) to join the Partnership as a member.
☑ 5-1 PARASITICS	*FINDING*: The More Electric Truck program demonstrated an integrated system to reduce idling emissions and fuel consumption. The test program showed significant progress toward achieving the objectives of both Goal 2 in Chapter 5 ("Develop and demonstrate technologies that reduce essential auxiliary loads by 50 percent, from the current 20 hp to 10 hp, for class 8 tractor-trailers) and Goal 6 in Chapter 6 ("Produce by 2012 a truck with a fully-integrated idling-reduction system to reduce component duplication, weight, and cost."). It did so by demonstrating 1 to 2 percent estimated reduction in fuel use including significant truck idling reductions. According to DOE, this translates into an overall fuel savings for the U.S. fleet of 710 to 824 million gallons of diesel fuel (about $2 billion per year at 2.75 per gallon). *RECOMMENDATION*: Given the potential of this program to save fuel, the committee recommends that the 21CTP continue the R&D of the identified system components that will provide additional improvements in idle reduction and parasitic losses related to engine components that are more efficient and provide better control of energy use. The program should focus also on the cost-effectiveness of the technologies.	DOE is exploring the cost and benefits of continuing R&D in this area compared to the potential for efficiency improvements by other heavy truck technologies. DOE funded an effort through Oak Ridge National Laboratory to characterize Class 8 duty cycles by recording over 100 parameters from six tractors and ten trailers for a 15-month period. DOE and DOT jointly funded a similar Class 7 duty cycle assessment. DOT led a cooperative effort to establish an idle reduction equipment weight exemption. DOE also funded a project to develop a factory integrated and installed idle reduction auxiliary power unit to efficiently prevent unnecessary truck idling.
5-2 PARASITICS	*FINDING*: The 21CTP lightweight materials research was terminated as a result of the 2007 budget reduction. *RECOMMENDATION*: The committee agrees with the decision to terminate lightweight materials research in order to provide as much budget resource as possible to continue research in engine efficiency and emissions reduction technologies, as improvements in engine efficiency offer greater potential for overall gains in vehicle fuel efficiency.	Thanks to the availability of Recovery Act funding to support two SuperTruck teams, DOE program budget is now available to support continuation of lightweighting efforts in heavy-duty area.
5-3 PARASITICS	*FINDING*: Prior to termination of the lightweight materials program, several lightweight material projects demonstrated weight reduction potential for truck components. However, the program did not achieve the longer term objective (planned for 2012) of demonstrating a 5,000-pound weight reduction for a complete class 8 tractor trailer combination. *RECOMMENDATION*: Due to the termination of the project in 2007, it will be the responsibility of truck manufacturers to take the next steps of system integration, product validation, and ultimately production of a lightweight truck. Although an interim step of system integration at the pre-production stage would have been useful, it is not inappropriate that the OEMs now assume responsibility for continuation of the work, as the next steps will require development of a business case which comprehends material costs and the costs of modifying existing manufacturing systems to accommodate the introduction of advanced materials.	The Partnership concurs with this recommendation. Integration of new materials into commercial offerings is the responsibility of the industrial partners.
5-4 PARASITICS	*FINDING*: The committee noted that the above list of research areas was extensive and comprehensive. However, the list appeared to be significantly more ambitious than the budget for	The Partnership concurs with this recommendation.

Update: November 10, 2010

	the 21st CTP could fund. The committee assumed that this was the case since no projects or results from any of the above research areas were provided. *RECOMMENDATION*: In addition to identifying a list of research areas that could provide solutions to thermal management challenges, DOE should develop, fund and implement plans for pursuing the key areas that will lead to the successful accomplishment of the specific 21st CTP Goal 4A. DOE's first step should be to assess the candidate technology, or technologies, that have the highest potential for meeting the requirements of Goal 4A. This goal and its status were briefly discussed with the committee and the following information was provided: "Track and laboratory tests met or exceeded goals, validation test is underway."Unfortunately, a description of the track and laboratory tests that had been performed, the engineering details and the results from these tests, or a description and timetable for the validation test reported to be under way were not described for the committee.	This technical area has been renamed "Power Demands" in the current list of Partnership goals to reflect broader range of the relevant technologies and stronger emphasis on those technologies that are capable of reducing engine power requirements and may lead to engine downsizing.
5-5 PARASITICS	*FINDING*: Based on the above observations, the committee was not able to accurately assess the progress on this goal or the expectation of whether this goal can be successfully achieved. *RECOMMENDATION*: DOE should provide periodic status reports on the 21CTP goals that include the technical status vs. the program plan, funding vs. budget, and the expected future accomplishments vs. the program plan.	The Partnership concurs with this recommendation. We will examine methods for providing status reporting for projects as part of its overall assessment of the Partnership's processes and methods (as outlined in the Management responses above).
5-6 PARASITICS	*FINDING*: The achievement of present program targets would require the involvement of a wide range of new program participants and the sharing of responsibilities among new program partners, inherently incorporating higher technical and durability risks than the present approaches. Truck manufacturers are assemblers of components specified by the truck buyer, and cooperative design and development relationships may not exist between suppliers. *RECOMMENDATION*: DOE should determine if the above approach for achieving Goal 4A is feasible within the scope of the 21CTP and containable within the available budget. DOE should take a strong leadership role with appropriate funds to bring manufacturers and suppliers together for systems research and development for Goal 4A and Goal 3.	The Partnership concurs with this recommendation. The SuperTruck projects for the first time integrate vehicle, engine, and component goals into an integrated set of objectives. The SuperTruck teams bring together manufacturers and suppliers for coordinated systems R&D.
5-7 PARASITICS	*FINDING*: The committee noted that the DOE list of research topics in friction, wear and lubrication was extensive and comprehensive. However, the list appeared to be significantly more ambitious than the budget for the 21CTP could fund. The committee assumes that this was the case since no projects or results from any of the above research areas were provided. *RECOMMENDATION*: In addition to identifying a list of topics addressing friction, wear, and lubrication technologies, DOE should develop, fund and implement plans for pursuing key areas that will lead to the successful accomplishment of the specific 21CTP Goal 4B. DOE's first step should be to conduct detailed friction testing of a range of heavy-duty diesel engines, transmissions and final drives to determine those with best-in-class friction. With respect to engines, previous industry light- and heavy-duty engine friction reduction investigations that included lightweight-low friction piston and piston ring designs, low friction coatings and surface finishes, reduced engine bearing sizes and other design modifications should be reviewed to determine opportunities for reducing engine friction below best-in-class levels. From this assessment, other candidate technologies with the highest potential for meeting the requirements of the engine portion of Goal 4B should be identified. Likewise, the efficiencies of transmissions and final drives on heavy-duty trucks should be measured and compared with the efficiencies of best-in-class light-duty vehicles, normalized for load differences, thereby providing insight for friction reductions in heavy-duty truck transmissions and final drives. From this assessment, other	The Partnership concurs with this recommendation. The SuperTruck projects will continue the research in the areas of friction, wear and lubrication. This work will include efforts in reducing engine and transmission losses.

	candidate technologies with the highest potential for meeting the requirements of the driveline portion of Goal 4B should be identified.	
5-8 PARASITICS	*FINDING*: In contrast to the report by DOE to the committee, the analysis of the basis of this goal by the committee indicates that it is very unlikely that this goal can be achieved within the scope of the 21CTP. The achievement of the goal's projected fuel savings appears to be very unlikely with accompanying high risks relative to component life. *RECOMMENDATION*: DOE should reassess the basis of this goal and determine if 50 percent reductions in powertrain and drivetrain losses are technically feasible. Based on this assessment of technical feasibility, DOE should determine if this goal should be pursued based on its potential fuel savings vs. other competing programs within the 21CTP. If DOE determines that this goal should be pursued, they should then develop specific program plans, timing and funding.	The Partnership concurs with this recommendation. The SuperTruck teams will conduct three independent assessments of the Parasitics reduction goal. Based on the results of this work, the partnership goal will be reassessed.
5-9 PARASITICS	*FINDING*: There is a precedent for government to establish performance measures for tires as illustrated by the Uniform Tire Quality Grading System (UTQGS) adopted by NHTSA in 1980 [Part 575.104 of the Consumer Information Regulations]. The UTGS applies to passenger car tires and requires manufacturers to grade new tires for tread wear, wet traction and temperature resistance. Tread wear is graded on a numerical scale, while traction and temperature resistance are graded on an alphabetic scale. There is no current requirement for grading rolling resistance, or for grading truck tires. *RECOMMENDATION*: DOE, EPA, and DOT should arrange to gather and report information on the influence of individual truck tires on vehicle fuel consumption; to convey such tire information to both buyers and sellers; and to periodically reassess the effectiveness of this consumer information and the methods used for communicating it.	The Partnership concurs with this recommendation. EPA's SmartWay Transport Partnership tested and publicly reported the fuel savings impact of low rolling resistance truck tires. Tire impact on fuel consumption is also part of the Agency's criteria to designate the most fuel-efficient heavy-duty trucks and trailers as "SmartWay." EPA supports the continued assessment of methods to assess tire rolling resistance and its impact on heavy truck fuel efficiency. DOE recently conducted similar evaluations of the use of new generation wide base single tires and is sharing this information with the 21CT Partnership.
☑ 6-1 IDLE REDUCTION	*FINDING*: Idle reduction is one of the most effective ways to reduce pollutant emissions (especially locally) and improve fuel consumption. As a result of the Energy Policy Act of 2005, the authority for this effort now rests with EPA and DOT. Several important lines of research are carried on in the 21CTP. In addition, the EPA SmartWay Transport Partnership voluntary program is effective at promoting the use of electrified parking spaces. The 21CTP, in cooperation with several major shippers, has demonstrated a number of cost-effective technologies (such as fuel-fired cab heaters and coolers) that are being used by existing fleets. (One fleet is installing more than 6,000 heaters, and another is installing more than 7,000.) One trucking company reported that diesel-fired heaters provided 2.4 percent fuel savings and a payback in less than 2 years at $2.40 per gallon. *RECOMMENDATION*: The 21CTP should continue to support R&D for the technologies that reduce idle time and address the remaining technical challenges (including California emission requirements, completely integrated APU/HVAC (auxiliary power unit/heating, ventilation, and air-conditioning) systems, and creep devices).	The Partnership concurs with the NRC panel's recommendation on continued R&D efforts for idle reduction technologies. It should be noted that EPA, through its SmartWay Transport Partnership, does not promote, recommend, or endorse idle reduction technologies. Rather, the SmartWay program demonstrates and deploys idle reduction technologies through its grant authority under the Clean Air Act, Section 103, and the Energy Policy Act of 2005, Section 791. The SmartWay program has demonstrated and deployed both mobile and stationary idle reduction technologies. As discussed in the 5-1 response, DOE also funded a project to develop a factory integrated and installed idle reduction auxiliary power unit to efficiently prevent unnecessary truck idling.
6-2 IDLE REDUCTION	*FINDING*: An effective government-industry cooperative program has been established to examine idle-reduction technologies, which have been successfully employed for nighttime truck operation. *RECOMMENDATION*: The success of the nighttime anti-idling measure and deployment should be the basis for expanding to technologies that can be applied for daytime operation, which will then lead to greater fuel savings than nighttime operation.	The Partnership concurs with this recommendation. Both nighttime and daytime idling contribute to avoidable non-productive use of fuel. Truck drivers will idle for a variety of reasons, but the primary reason for long duration idling is to rest comfortably during the driver's federally mandated rest period.

		Long duration idling (more than 6 hours) occurs during the driver's rest period. Short duration idling occurs while truck drivers are waiting to load or off-load cargo. Cargo loading or unloading wait times can last up to 3 hours. Technologies that are appropriate for long duration idling might not be cost-effective for short duration idling. For example, a $7,000-$9,000 idle reduction device that supplies heat and air conditioning will provide a much faster payback for a truck operator who idles 8-10 hours per day, than for one who idles only 3 hours per day. Additionally, conventional idle reduction devices (e.g., auxiliary power units) designed to reduce long duration idling supply heat or air conditioning to the sleeper compartment. However, many truck drivers whose operations involve short duration idling do not have sleeper compartments that justify the need for such technologies. Reducing short-duration idling can be achieved through operational strategies, rather than necessarily with technologies. For example, improved logistics can reduce idling times by scheduling loading/unloading during non-peak hours, or contacting the truck driver when the loading dock is available. Instead of waiting in a line idling, the truck driver can find a more convenient location and wait without the need to idle.
6-3 IDLE REDUCTION	***FINDING***: DOE has built an effective outreach instrument in its monthly publication, "The National Idling Reduction News." This publication and education through conferences and other agencies such as the EPA provides stakeholders with significant information and guidelines for idle reduction. ***RECOMMENDATION***: DOE should continue its current successful education and outreach program as currently operated.	The Partnership concurs with this recommendation. The Partnership appreciates the acknowledgement of its success in education and outreach related to idling reduction, and agrees that this work has been useful to the community.
6-4 IDLE REDUCTION	***FINDING***: Progress on the incentive part of this goal has been excellent as evidenced by the SmartWay Transport Partnership between EPA and industry. The patchwork of anti-idling regulations nationally is an impediment to broader use of anti-idling measures. ***RECOMMENDATION***: EPA should renew its efforts to promulgate national anti-idling regulations, and DOE should review whether additional R&D is needed to implement those regulations.	EPA has no legal authority to promulgate anti-idling laws, or any time or behavioral limits on truck drivers. EPA's legal authority rests with promulgating emission standards for vehicles and engines. The idle reduction efforts of the EPA SmartWay Transport Partnership, which resulted in a model for a state or local idling law, are part of a larger educational campaign for states and local governments. The purpose of this campaign is to inform the public about the need for more consistent and practical anti-idling laws. EPA intent is to educate stakeholders, rather than to promulgate national legislation on idle-reduction.
6-6 IDLE REDUCTION	***FINDING***: The DOE-sponsored demonstrations with two major trucking fleets resulted in deployment of several idle-reduction devices. Greater success was achieved with cab heating than with cab cooling. It appears that only one device met the goal of less than 2-year payback. It is unclear whether the emissions requirement of the goal was met. ***RECOMMENDATION***: Given that funding and responsibility for idle-reduction technologies was redirected in the Energy Policy Act of 2005 to EPA and DOT, there is no requirement for DOE to pursue this area. However, given the progress to date and potential attractive returns on investment, it would be desirable for DOE, EPA, and DOT to continue to advance this aspect of	The Partnership concurs with this recommendation. EPA's SmartWay Transport Partnership program has a robust deployment program for idle reduction technologies. Under the Energy Policy Act of 2005, coupled with recent Congressional appropriations, EPA intends to fund many grant projects to deploy both mobile and stationary idle reduction technology projects nationally. The wide variety of idle reduction technologies offers many options. Since trucking operations

	fuel reduction and environmental mitigation.	vary widely, no single technology addresses all idle situations; many companies will employ a variety of technologies and strategies throughout their fleet. The federal government organizations involved in this area should coordinate efforts so as to avoid duplicative projects. Navistar has recently completed a DOE-funded APU development project.
6-7 IDLE REDUCTION	*FINDING*: The More Electric Truck program demonstrated an integrated system to reduce idling emissions and fuel consumption. The test program showed significant progress to achieve the objectives of goal 6 by demonstrating 1-2 percent estimated reduction in fuel use including significant truck idling reductions. According to DOE, this translates into an overall annual fuel savings for the U.S. fleet of 710 to 824 million gallons of diesel fuel (about $2 billion/year at $2.75/gallon). *RECOMMENDATION*: Given the potential of this program to save fuel, the committee recommends that the 21CTP continue the R&D of the identified system components that will provide more improvements in idle reduction and parasitic losses related to engine components that are more efficient and provide better control of energy use.	The Partnership concurs with this recommendation. Electrification of accessory loads is being pursued by the SuperTruck project teams. This task becomes even more relevant in light of the efforts by manufacturers to develop waste heat recovery (WHR) technologies. Availability of electricity consumers onboard will allow WHR devices to avoid the losses involved in converting electricity obtained from exhaust heat energy into mechanical energy.
6-8 IDLE REDUCTION	*FINDING*: The work on fuel cell APU is being carried out by the DOD and a number of contractors are being supported. There is no evidence that goal 7 has been met at this time. *RECOMMENDATION*: The DOE's 21CTP should continue to monitor and interact with the DOD program. As DOD reaches its goals, DOE should explore with major truck operators the possibility of bringing appropriate fuel cell APU technologies into commercial use.	The Partnership concurs with this recommendation. DOE will coordinate more closely with DOD to track progress in this area.
☑ 7-1 SAFETY	*FINDING*: The program manager of the 21st Century Truck Partnership has little or no direct authority for heavy-duty truck safety projects because there is no budget in the program itself to support safety projects. The program manager will need to continue to work with DOT, because DOT has several initiatives with the goal of making improvements in heavy-duty truck safety. They range from driver education to accident avoidance technology. However, the committee was unable to determine whether the goals would be met as a result of these initiatives. *RECOMMENDATION*: DOT should develop a complete and comprehensive list of current and planned heavy-duty truck safety projects and initiatives, and prioritize them in order of potential benefit in reducing heavy-duty truck-related fatalities. The list should provide quantitative projections of fatality reduction potential attributable to each project. The list should also be used to prioritize budget and resource allocation, in order to expedite heavy-duty truck safety progress.	The 21st Century Truck (21CT) Partnership agrees that a main factor to be used to determine research priorities in the safety area of the program is potential benefit in terms of fatality reduction. For DOT, improving safety is the Department's No. 1 goal. Specific to the 21CT program, potential safety benefit is a main factor used to determine safety focus areas. These areas include braking, rollover, vehicle position (safe following distance and in-lane tracking), visibility (driver vision enhancement), and tire safety. These areas were identified in 21CTP first safety white paper.
☑ 7-2 SAFETY	*FINDING*: Programs are underway to develop and implement technologies and vehicle systems to support safety goals. Indeed, private industry, through internal research and commercial product development has produced commercially available systems for enhanced braking, roll stability, and lane departure warning. They are beginning to be used in the field. It is now important to determine to what extent these accident avoidance technologies will reduce the number of accidents and therefore fatalities and injuries. *RECOMMENDATION*: DOT should continue programs in support of heavy-duty truck on-board safety systems, with emphasis on accident avoidance and with priority set by a comprehensive potential cost/benefit analysis (Recommendation 7-1). Particular emphasis should be placed on monitoring the accident experience of heavy-duty trucks, as these systems begin to be deployed in the field (for example, as electronic stability control systems begin to penetrate the fleet). It is the role of the manufacturers to develop safety systems for commercial application. DOT can play	The Partnership agrees with the recommendation. Within the priority areas identified for 21CT in the safety area, these steps are largely being implemented. However, similar to the above recommendation, as currently written, this recommendation is too broad and seems to be for DOT research as a whole. Again, it is not within the scope of the 21CT program to monitor and coordinate all DOT programs related to onboard safety systems.

Update: November 10, 2010

	important roles in (1) providing support for field tests (known to DOT as Field Operational Tests, in (2) monitoring field data to help substantiate benefit analyses used to prioritize resources, and (3S) implementing regulations that would require the adoption of safety systems that were proved to be effective. With adequate field data, DOT should refine and more rigorously specify and prioritize goals for accident avoidance technologies.	
7-3 SAFETY	***FINDING***: In spite of extensive improvements in light vehicle crashworthiness made during the past decade, the number of fatalities caused by heavy-duty truck accidents has remained nearly constant, at approximately 5,000 per year, although the fatality rate has decreased showing that progress is being made. In most cases, the occupant(s) of the light vehicle is the one fatally injured. It appears that to make significant safety progress, it will be necessary to reduce the number of accidents substantially by implementing accident avoidance technologies as well as methods for improving driver behavior. In light of this need, DOT future plans have been directed largely at accident avoidance technologies. ***RECOMMENDATION***: The committee agrees with the apparent decision by DOT to put more emphasis on accident avoidance technologies than on additional crashworthiness research. In additional, DOT should continue to focus on driver education and law enforcement. Furthermore, DOE and DOT should work collaboratively, as there often are trade-offs between vehicle safety and fuel economy, for example, as new fuel efficient systems emerge. There are obvious trade-offs between safety and fuel economy in many areas of research such as tire mechanics, braking (especially with respect to hybrid vehicles). Of course, any additional work in aerodynamics or weight reduction might alter the vehicle configuration and therefore its crashworthiness. Moreover, as new fuel efficient systems emerge, such as hybrid electric systems, and vehicles using alternate fuels including, for example, hydrogen, it will be imperative that DOE and DOT work closely to ensure continued progress toward more fuel efficient vehicles but without compromising highway safety.	The Partnership concurs with this recommendation. Opening up lines of communication between DOE and DOT has been one of the very positive outcomes of the program. It provides a great opportunity to exchange information such that the various areas within the program can be discussed and coordination among DOT and DOE. DOE and DOT will continue to share information and coordinate on various topics such as hydrogen, hybrid electrics, and additional alternative fuels, as they relate to tradeoffs between safety and fuel economy.

Appendix D

Highlights of Selected Propulsion Material Programs

This appendix presents a brief summary of activities in some selected programs on propulsion materials. The presentations cited are from the U.S. Department of Energy's Office of Vehicle Technologies Annual Merit Review held June 7-11, 2010, in Washington, D.C. The last project noted, "High Strength Light Weight Engines for Heavy Duty Diesel Trucks," was in a formative stage at the time of this report.

- *Catalyst Characterization for Exhaust Aftertreatment* (http://www1.eere.energy.gov/vehiclesandfuels/pdfs/merit_review_2010/propulsion_materials/pm028_watkins_2010_p.pdf. Accessed April 5, 2011.)
  - —*Objective*: To develop a quantitative understanding of process/product interdependence leading to catalyst systems with improved final quality, meeting prevailing emissions requirements.
  - —*Status*: Evaluated the feasibility of the advanced tools at Oak Ridge National Laboratory (ORNL) for quantitative analysis of materials changes underlying the selective catalytic reduction catalyst performance degradation with age. Initiated evaluation of Ammonia Oxidation Catalyst (selected by Cummins).
  - —*CRADA*: ORNL with Cummins and Johnson Matthey; completion date: September 2012.
  - —*DOE funding*: $196,000 in 2009; $147,000 in 2010.

- *Materials Issues Associated with EGR Systems* (http://www1.eere.energy.gov/vehiclesandfuels/pdfs/merit_review_2010/propulsion_materials/pm009_lance_2010_o.pdf. Accessed April 5, 2011.)
  - —*Objective*: To enable improved models and potential design improvements to reduce fouling and its impact on performance:
    1. Characterize thermophysical properties of the deposit under different operating conditions on model exhaust gas recirculation (EGR) cooler tubes.
    2. Determine long-term changes in deposit properties due to thermal cycling and water/hydrocarbon (HC) condensation.
  - —*Status*: Assembled industry team; collected coolers; established experimental setup.
  - —*CRADA*: ORNL with Caterpillar, Cummins, Detroit Diesel, GM, John Deere, Modine, Navistar, PACCAR, and Volvo/Mack; completion date: September 2011.
  - —*DOE funding*: $300,000 in 2009; $270,000 in 2010.

- *Materials for High Pressure Fuel Injection Systems* (http://www1.eere.energy.gov/vehiclesandfuels/pdfs/merit_review_2010/propulsion_materials/pm021_blau_2010_p.pdf. Accessed April 5, 2011.)
  - —*Objective*: To evaluate spray hole microstructures, nozzle residual stress states, and fatigue properties of current and future materials for high-pressure fuel injector nozzles for diesel engines.
  - —*Status*: A fatigue test plan has been used to study the effects of holes on fatigue crack initiation and propagation in current and future nozzle materials (new materials were not identified).
  - —*CRADA*: ORNL and Caterpillar; completion date: September 2011.
  - —*DOE funding*: $225,000 from 2008 through 2011.

- *Durability of Diesel Engine Particulate Filters* (http://www1.eere.energy.gov/vehiclesandfuels/pdfs/merit_review_2010/propulsion_materials/pm010_watkins_2010_o.pdf. Accessed April 5, 2011.)
  - —*Objective*: To develop test methods for characterizing the properties of ceramic diesel particulate filters (DPFs) and to develop analysis and inspection methods for assessing their reliability and durability.
  - —*Status*: A procedure has been developed for rank ordering the thermal shock resistance of DPF substrates. Proposed future work would characterize

field returned DPFs and compare their properties to virgin filters, using this information to refine lifetime prediction of filters.
— *CRADA*: ORNL with Cummins and Corning; first phase completed September, 2010; 3-year renewal in progress.
— *DOE funding*: \$318,000 in 2009; \$238,000 in 2010.

- *Low Cost Titanium—Propulsion Applications* (http://www1.eere.energy.gov/vehiclesandfuels/pdfs/merit_review_2010/propulsion_materials/pm006_lavender_2010_o.pdf. Accessed April 5, 2011.)
  — *Objective*: To reduce the cost to manufacture titanium components for reciprocating and rotating applications.
  — *Status*: A lower cost titanium bar made of sintered titanium powder (TiH_2) appears to meet performance requirements at lower cost (perhaps as much as 50 percent lower than ingot processed forgings). Cummins has identified an engine application (not specified) for final evaluation.
  — *CRADA*: Pacific Northwest National Laboratory (PNNL) with Cummins; completion date: October 2012.
  — *DOE funding*: \$300,000 in 2010.

- *Fatigue Enhancements by Shock Peening* (http://www1.eere.energy.gov/vehiclesandfuels/pdfs/merit_review_2010/propulsion_materials/pm002_lavender_2010_o.pdf. Accessed April 5, 2011.)
  — *Objective*: To evaluate the capability for surface modification to improve fatigue performance of steel, aluminum, and cast iron engine components to enable improved efficiencies by increasing injection pressures and the overall durability of reciprocating parts.
  — *Status*: Fatigue life of laser shock peened 52100 steel showed approximately 50 percent increase in rolling contact fatigue life; Cummins is moving to deployment.
  — *CRADA*: PNNL with Cummins; project completed in September 2010.
  — *DOE funding*: \$350,000 in 2008; \$340,000 in 2009; \$223,000 in 2010.

- *Proactive Strategies for Designing Thermoelectric Materials for Power Generation* (http://www1.eere.energy.gov/vehiclesandfuels/pdfs/merit_review_2010/propulsion_materials/pm014_hendricks_2010_o.pdf. Accessed April 5, 2011.)
  — *Objective*: Develop new high-performance n-type and p-type thermoelectric material (TE) compositions to enable: 17 percent on-highway efficiency of directly converting engine waste heat to electricity to help enable improved heavy-truck efficiencies to 50 percent by 2015.
  — *Status*: Determined that n-type Skutterudite materials show excellent thermoelectric properties; p-type Skutterudite materials are more challenging. Future work includes characterizing TE properties and validating with third-party testing (ORNL), and determining structural properties such as Young's modulus, Poisson's ratio, and mechanical strength.
  — PNNL with the Oregon Nanoscience and Microtechnologies Institute, Oregon State University; project is ongoing.

- *Low-Friction Hard Coatings* (http://www1.eere.energy.gov/vehiclesandfuels/pdfs/merit_review_2010/propulsion_materials/pm030_erdemir_2010_p.pdf. Accessed April 5, 2011.)
  — *Objective*: To design, develop, and implement low-friction and superhard coatings to increase the durability and fuel economy of engine systems.
  — *Status*: Argonne National Laboratory (ANL), in cooperation with Istanbul Technical University, has developed a superhard nanocomposite coating that provides friction coefficients of between 0.02 to 0.05 (compared with steel on steel at 0.10 to 0.15), and a production-scale deposition system. Future work will attempt to validate performance under fired engine conditions.
  — *CRADA*: ANL with Galleon International; Hauzer Techno Coating, and several engine original equipment manufacturers; completion date: September 2012.
  — *DOE funding*: \$125,000 in 2009; \$200,000 in 2010.

- *High Strength Light Weight Engines for Heavy Duty Diesel Trucks*
  — *Objective*: To develop durable lightweight engine components (for example, the use of aluminum in the block and head) for heavy-duty diesel engines. The project would include the development of a prototype engine.
  — *Status*: CRADA under development between ORNL and Cummins.
  — *Proposed budget for 2011*: \$500,000.

Appendix E

Available Models of Medium- to Heavy-Duty Hybrid and Electric Trucks

TABLE E-1 Available Models of Medium- and Heavy-Duty Hybid Vehicles and Electric Trucks

Truck OEM/ Chassis	Model	Powertrain	Body Type/ Application	Engine	Fuel	GVW	Class
Bright Automotive	IDEA	Plug-in Hybrid Electric: Bright	Cargo Van		Gasoline	3,200	1
Ford	Transit Connect	Force Drive Electric: Azure Dynamics	Cargo Van	Electric	Electric	5,005	1
Ford	E450	Hybrid Electric: Azure Dynamics Balance	Step Van, Shuttle Bus	5.4L EFI FFV V8	Gasoline	14,000	3
Freightliner	Business Class M2e Hybrid	Hybrid Electric: Eaton	City Delivery, Utility, Delivery Tractor		Diesel	33,000-55,000	7, 8
Freightliner Custom Chassis Corporation (FCCC)	MT-45, MT-55	Hybrid Electric: Eaton	Walk-in Van		Diesel		
GMC	TC5500	Plug-in Hybrid Electric: Dueco Odyne	Utility	Vortec 8.1L MD	Gasoline	19,500	5
Kenworth	T270	Hybrid Electric: Eaton	Delivery, Utility	PACCAR PX-6	Diesel	25,000	6
Kenworth	T370	Hybrid Electric: Eaton	Delivery, Utility	PACCAR PX-6	Diesel	33,000	7
Mack/Volvo	TerraPro Hybrid		Refuse		Diesel		8
Modec	Chassis Cab, Dropside & Box Van	Electric: Modec	Chassis Cab, Dropside, Box Van, Refrigerated Box Van, Tail Lift, Tipper	Electric	Electric	12,000	3
Navistar, Inc.	DuraStar Hybrid (Truck)	Hybrid Electric: Eaton	Beverage, Box Van, Refrigeration, Landscape Dump, Utility, Crane, Tree Trimmer, Recovery Towing, Armored Vehicle, Stake Flat, Grapple, Road Patch Truck, Refined Fuels, Propane Tank	MaxxForce DT	Diesel	23,500-39,000	6, 7
Navistar, Inc.	4300	Plug-in Hybrid Electric: Dueco Odyne	Utility, Digger Derrick, Air Compressor	MaxxForce DT	Diesel	25,000-37,000	6, 7, 8
Navistar, Inc.	DuraStar Hybrid (4x2) Tractor	Hybrid Electric: Eaton	Beverage Diminishing Load	MaxxForce DT	Diesel	Up to 55,000	7
Navistar, Inc.	WorkStar Hybrid (Truck)	Hybrid Electric: Eaton	4x4 Utility, Landscape Dump, Snowplow, Digger Derrick, Utility, Crane, Stake Flat, Box Van, Recovery Towing, Refined Fuels, Propane Tank	MaxxForce DT	Diesel	23,500-39,000	6, 7
Navistar, Inc.	eStar	All Electric: Modec	Delivery Van	Electric	Electric	12,100	3
Peterbilt	320 Hybrid	Hydraulic Launch Assist (HLA): Eaton	Refuse		Diesel		8
Peterbilt	330 Hybrid	Hybrid Electric: Eaton	Delivery Van	PACCAR PX-6	Diesel	26,000	6
Peterbilt	337 Hybrid	Hybrid Electric: Eaton	City Delivery, Fire & Rescue, Beverage, Municipal, Refuse, Utility	PACCAR PX-6	Diesel		6, 7

continues

TABLE E-1 Continued

Truck OEM/ Chassis	Model	Powertrain	Body Type/ Application	Engine	Fuel	GVW	Class
Peterbilt	348 Hybrid	Hybrid Electric: Eaton	Municipal, Service, Utility	PACCAR PX-6	Diesel	33,000+	7, 8
Peterbilt	386 Hybrid	Hybrid Electric: Eaton	Long Haul		Diesel		8
Smith Electric Vehicles	Newton	Electric: Smith	Food Distribution, Parcel Delivery, Chilled Food Distribution, Short Haul, Utility, Airport Operations, Public Sector	120 kW Induction Motor	Electric	16,535-26,455	5, 6, 7

SOURCE: Available at http://business.edf.org/projects/fleet-vehicles/hybrid-trucks-financial-incentives-guide/available-models-medium-heavy-duty-. Accessed April 8, 2011.

F

Proposed Mechanism for Obtaining Hybrid Vehicle Credits

The U.S. Environmental Protection Agency (EPA) and the U.S. Department of Transportation National Highway Traffic Safety Administration (DOT/NHTSA) issued proposed greenhouse gas emissions standards and fuel efficiency standards for medium- and heavy-duty engines in a notice of proposed rulemaking (NPRM) on October 26, 2010 (EPA/NHTSA, 2010a,b). On September 15, 2011, the EPA and NHTSA issued final greenhouse gas emissions standards and fuel efficiency standards for medium- and heavy-duty engines and vehicles that are tailored to each of three regulatory categories of heavy-duty vehicles: (1) Heavy-Duty Pickup Trucks and Vans, (2) Vocational Vehicles, and (3) Combination Tractors (EPA/NHTSA, 2011a). The agencies are providing credits for the use of hybrid powertrain technology as an incentive (EPA/NHTSA 2011a,b). The approach to account for the use of a hybrid powertrain when evaluating compliance with the standards is described below.

HEAVY-DUTY PICKUP AND VAN HYBRIDS

Test Procedure

For the heavy-duty pickup truck and van hybrid class of vehicles with gross vehicle weight ratings [GVWRs] between 8,500 and 14,000 lb (that are not already covered under the Model Year 2012-2016 light-duty truck and medium-duty passenger vehicle greenhouse gas [GHG] standards), the agencies have proposed that testing would be done using adjustments to the test procedures developed for light-duty hybrids, using the light-duty Federal Test Procedure (FTP) and the Highway Fuel Economy Test (HWFET), but extending the requirement for chassis certification for CO_2 emissions to diesel-powered vehicles. Currently, chassis certification is a gasoline requirement and a diesel option. Manufacturers would be allowed to continue to certify engines for criteria pollutant (non-GHG) requirements as they do today.

Fuel-Consumption Credits

The EPA and NHTSA fuel-consumption standards are expressed on a gal/100 mile basis, and that would apply to a manufacturer's fleet of heavy-duty pickup trucks and vans with a GVWR from 8,500 to 14,000 lb. The credits for the hybrid vehicle would be calculated according to the Averaging, Banking, and Trading (ABT) program described by an equation for fuel consumption credits given later in this section.

A manufacturer's credit or debit balance will be determined by calculating its fleet average performance using the data from the FTP and HWFET tests and comparing this data to the manufacturer's fuel-consumption standards, as determined by its fleet mix, for a given model year. A target standard is determined for each vehicle with a unique payload, towing capacity, and drive configuration (two-wheel versus four-wheel drive). These unique targets, weighted by their associated production volumes, are summed at the end of the model year to derive the production volume-weighted manufacturer annual fleet average standard. A manufacturer would generate credits if its fleet average fuel-consumption level were lower than its standard and would generate debits if its fleet average fuel-consumption level were above that standard.

In addition to production weighting, the credit calculations include a factor for the useful life, in miles, in order to allow the expression of credits in gallons. The following equation is used to calculate credits (debits) and account for the amount that the family limit is below (above) the standard, the payload tons, the sales volume, and the useful life.

Fuel Consumption Credits (gallons) = (FC Std − FC Act) × Volume × UL × 100,

where:

FC Std = Fleet average fuel-consumption standard (gal/100 mile)

FC Act = Fleet average actual fuel-consumption value (gal/100 mile)

Volume = Total production of vehicles in the regulatory category

UL = Useful life for the regulatory category (miles)

VOCATIONAL VEHICLE AND TRACTOR HYBRIDS

Test Procedure

For vocational vehicles and combination tractors incorporating hybrid powertrains, the agencies specify two methods for establishing credits. The first method uses chassis dynamometer evaluation of the vehicle, and the second method uses engine dynamometer evaluation with the powerpack in either a (1) pre-transmission format or a (2) post-transmission format. Each method requires a comparison test of the actual physical product, because the agencies are not aware of analytical models that can assess the technology.

Chassis Dynamometer Evaluation

Similar to heavy-duty pickup and van hybrids, to generate credits for hybrid vocational vehicles, full vehicle chassis dynamometer testing is a straightforward basis for comparing fuel consumption performance of hybrid vehicles to conventional vehicles. The agencies specify two sets of duty cycles for vocational trucks to evaluate the benefit depending on the vehicle application. The key difference between the two sets is that one does not include operation of a power take-off (PTO) unit while the other does. For example, delivery trucks do not operate a PTO while bucket and refuse trucks do.

The duty cycles that apply to hybrid powertrains without a PTO system are shown in Table F-1.

The transient cycle, derived from the California Air Resources Board (CARB) Heavy-Duty Truck 5 Mode Cycle, is 668 seconds long and travels 2.84 miles. The cycle contains 5 stops and contains 112 seconds of idling. The maximum speed of the cycle is 47.5 mph with an average speed of 15.3 mph. The High Speed and Low Speed Cruise modes reflect constant speed cycles at 65 mph and 55 mph, respectively, which are representative of drivers using cruise control whenever possible. The final rules include a new optional PTO test cycle in addition to the standard set of test cycles in order for manufacturers of advanced PTO systems to demonstrate in the laboratory environment fuel consumption reductions that would be realized from their systems in the real world. The composite PTO test cycle for utility and refuse trucks is described in greater detail in EPA/NHTSA (2011b, see Table 3-23).

TABLE F-1 Proposed Drive-Cycle Weightings (percent) for Hybrid Vehicles Without Power Take-off

Vehicle Category	Transient	55 mph	65 mph
Vocational vehicle	75	9	16
Day cab tractor	19	17	64
Sleeper cab tractor	5	9	86

Engine Dynamometer Evaluation

The engine test procedure involves exercising the conventional engine and the hybrid-engine system based on an engine testing strategy. Real-world functionality would need to be accurately represented. The testing would also need to recover vehicle kinetic energy. The agencies specify the use of the Heavy-Duty Engine FTP cycle for evaluation of hybrid vehicles, which is the same test cycle specified for engines used in vocational vehicles. Engine dynamometer evaluation may be undertaken in one of two ways:

1. Pre-transmission power-pack testing, which includes the engine and hybrid systems in a pre-transmission format, could utilize existing engine certification duty cycles. Changes to how the engine certification would be conducted to address energy capture and idle operation would need to be evaluated as a complete protocol is developed.
2. Post-transmission power-pack testing, which includes the transmission, would require a vehicle-like duty cycle, which provides the appropriate speeds and torques to match field operation.

Fuel-Consumption Credits

Heavy-duty hybrid vehicles and hybrid powertrains can be certified using an A to B test method. This concept entails testing the conventional vehicle or powertrain, identified as "A," and the hybrid version of the vehicle or powertrain, identified as "B." The benefit associated with the hybrid system for fuel consumption would be determined from the weighted fuel-consumption results from the tests of each vehicle or hybrid powertrain, as described below:

1. Improvement Factor = (Fuel Cons_A – Fuel Cons_B)/(Fuel Cons_A)
2. Gallons/1,000 ton-mile benefit = Improvement Factor × GEM Fuel Cons Result_B

Note in the above equations that the GEM (Greenhouse Gas Emissions Model) result would be calculated for the base vehicle or powertrain without hybridization, and the Improvement Factor would account for hybridization of the vehicle or powertrain.

The following equation for the credits (debits) accounts for the amount that the family emission limit is below (above) the standard, the payload tons, the production volume, and the useful life:

Fuel Consumption credit (deficit)(gallons) = (Std-FEL) × (Payload Tons) × (Volume) × (UL) × 10^3,

where:

Std = Standard associated with the regulatory category (gallons/1,000 ton-mile) (fuel consumption: Gallons/1,000 ton-mile)

Payload tons = Prescribed payload for each subcategory (12.5 tons for Class 7 tractors, 19 tons for Class 8 tractors, 2.85 tons for light heavy-duty [LHD] vocational, 5.6 tons for medium heavy-duty [MHD] vocational, 7.5 tons for heavy heavy-duty [HHD] vocational vehicles)

FEL = Family emission or fuel-consumption limit for the vehicle family (gallons/1,000 ton miles)

Volume = (Projected or actual) production volume of the vehicle family

UL = Useful life of the vehicle (435,000 miles for HHD, 185,000 miles for MHD, and 110,000 miles for LHD)

Early Credits

The final rules include an option for a manufacturer to obtain early credits by certifying a subcategory of vehicles at fuel-consumption levels below the standard prior to the model year the standard becomes effective. A 1.5 multiplier will be applied to early credits earned in model year 2013. Early credits provide an incentive for manufacturers to introduce more efficient engines and vehicles earlier than required by the standards.

Advanced Technology Credits

The final rules include a provision for obtaining credits for introducing advanced technologies to provide an incentive for their introduction. A 1.5 multiplier will be applied to these credits, but the total credits are capped in any model year. Hybrid powertrain designs that include energy storage systems are one of the advanced technologies defined by the agencies.

REFERENCES

EPA/NHTSA (Environmental Protection Agency [Office of Transportation and Air Quality]/National Highway Traffic Safety Administration [Department of Transportation]). 2010a. Greenhouse Gas Emissions Standards and Fuel Efficiency Standards for Medium- and Heavy-Duty Engines and Vehicles. Dockets No. EPA-HQ-OAR-2010-0162 and No. NHTSA-2010-0079, October 25, 2010. Available at http://www.regulations.gov.

EPA/NHTSA. 2010b. Draft Regulatory Impact Analysis, Proposed Rulemaking to Establish Greenhouse Gas Emissions Standards and Fuel Efficiency Standards for Medium- and Heavy-Duty Engines and Vehicles. EPA-420-D-10-901. October.

EPA/NHTSA. 2011a. Greenhouse Gas Emissions Standards and Fuel Efficiency Standards for Medium- and Heavy-Duty Engines and Vehicles, Final Rules. August 9.

EPA/NHTSA. 2011b. Final Rulemaking to Establish Greenhouse Gas Emissions Standards and Fuel Efficiency Standards for Medium- and Heavy-Duty Engines and Vehicles, Regulatory Impact Analysis. EPA-420-R-11-901. August.

G

History of Heavy-Duty Truck Aerodynamic and Tire Rolling Resistance Coefficients

The most recent 21st Century Truck Partnership (21CTP) white paper[1] adopted the Environmental Protection Agency's (EPA's) C_d and C_{rr} baselines from the proposed EPA/National Highway Traffic Safety Administration (NHTSA) greenhouse gas (GHG) rule that were to represent roughly the median of current in-use equipment for highway duty tractor-trailers with high-roof sleepers (identified in Table G-1) as "EPA GHG Rule Conventional High-Roof Average T-T." These coefficients are higher than those used in previous Partnership documents and prevailing industry documents, which represented circa 2005 to 2009 production available designs and components, identified in Table G-1 as "NRC 21CTP Review, Phase 2." See also footnote *e* in Table G-1.

Notice for example, that the currently available EPA SmartWay aerodynamics and tires (0.59 and 0.0063) provide lower fuel consumption than is achievable with the Average T-T base coefficients in the EPA/NHTSA GHG rule (0.69 and supposed combination weighted 0.0077), by about 7 percent and 6 percent, respectively; 13 percent when combined.

[1]Received by the committee on March 17, 2011.

TABLE G-1 History of Heavy-Duty Truck Aerodynamic and Tire Rolling Resistance Coefficients

		C_{rr}			
Truck Configuration Description	C_d	Steer	Drive	Trailer	Weighted Average[a]
1974 COE T-T[b]	1.02				0.0103[c]
1974 COE T-T, Closed Gap, Trailer Skirts, Boat-Tail[b]	0.46				
EPA Classic Conventional, Exterior Air Cleaners, Verticle Exhaust, Flat Bumber, No Aero Features T-T	0.77				
EPA GHG Rule (CE-CERT) 2000's Conventional High Roof T-T	0.74				
EPA GHG Rule Conventional High-Roof Average T-T	0.69	0.0078	0.0082	0.0072[d]	0.0077
NRC 21CTP, Phase 2, Mid-2000s Conventional High Roof, pre-S/W T-T[e]	0.63				0.0068
EPA GHG Rule Conventional High-Roof SmartWay T-T	0.59	0.0059	0.0074	0.0053	0.0063
EPA GHG Rule Conventional High-Roof Advanced SmartWay T-T	0.55				0.0055
NRC 21CTP Review, Phase 2, Conventional High Roof, 2015-2020 T-T	0.45				0.0045

NOTE: COE T-T: Cab Over Engine Tractor-Trailer. T-T: Tractor-Trailer. The C_ds' and the C_{rr}s' in the table are not derived from a consistent set of measurement or test procedures.

[a] Weightings: Steer (12), Drive (34), Trailer (34) reflect distribution of (80,000) lb Gross Vehicle Weight (GVW).

[b] K.R. Cooper, Commercial Vehicle Aerodynamic Drag Reduction: Historical Perspective as a Guide, The Aerodynamics of Heavy Vehicles: Trucks, Buses, and Trains, October 15, 2004. Springer.

[c] Based on bias ply tires.

[d] Assumed by National Research Council committee.

[e] Reducing Heavy-Duty Long Haul Combination Truck Fuel Consumption and CO_2 Emissions, NESCCAF/ICCT, Miller ed., October, 2009; and C_d same as the DOE 2006 21st Century Truck Roadmap and Technical White Papers. Doc No. 21CTP-003, December, Washington, D.C.

H

Bibliography for Long Combinations Vehicles in Canada, the United States, and Australia

LONG COMBINATION VEHICLE (LCVs) IN CANADA

- The British Columbia program overview, including equipment requirements and operational restrictions that improve safety. Available at http://www.th.gov.bc.ca/cvse/LCV/faqs.htm.
- A study of the experience with LCV use in Alberta by John Woodrooffe of the University of Michigan, with favorable outcomes for safety. Available at http://www.transportation.alberta.ca/Content/docType61/production/LCVEconomicEfficiencyReport.pdf.
- A 2005 study by the Canada Safety Council, showing that collision rates for LCVs are relatively low, given the numerous restrictions of operation. News summary: Available at http://archive.safety-council.org/info/traffic/LCVs.html. Full report: Available at http://archive.safety-council.org/info/traffic/LCVs.pdf.
- A study published by the *Canadian Journal of Civil Engineering* reviewed accident rates in Alberta for all types of articulated trucks over a period of several years, comparing LCVs to conventional semitrailer trucks. Turnpike doubles had the lowest overall accident rate. Available at http://www.nrcresearchpress.com/doi/abs/10.1139/L08-109.

LONG COMBINATION VEHICLES IN THE UNITED STATES

- Transportation Research Board (TRB). Crashes Involving Long Combination Vehicles: Data Quality Problems and Recommendations for Improvement, 2002. Available at http://trid.trb.org/view.aspx?id=717388.
- TRB. Western Uniformity Scenario Analysis, 2004. The U.S. Federal Highway Administration has released a report that assesses the impacts of lifting the LCV freeze and allowing harmonized LCV weights, dimensions, and routes among only those western states that currently allow LCVs. The report analyzes impacts of expanded LCV operations assuming that weights would be limited only by federal axle load limits and the federal bridge formula, with a maximum gross vehicle weight of 129,000 pounds. News summary: Available at: http://www.trb.org/Main/Blurbs/155006.aspx. Full report: Available at http://www.fhwa.dot.gov/policy/otps/truck/wusr/wusr.pdf.
- University of North Caroline (UNC). Highway Safety Research Center. Operational Characteristics of LCVs. LCVs handle and perform differently from tractor semitrailers or twin trailers because of their increased lengths and weights. These differences in handling and performance may jeopardize the safety of the LCV as well as other vehicles on the roadway. There is a clear need to conduct additional research to further evaluate LCV operations. News Summary: Available at http://trb.metapress.com/content/e7rj21x41635202g/?p=7796494a3f33446e875d54459b5ff39b&pi=11.
- Nevada. MTA Trucking big picture, asserts LCVs have better safety, 2011. Available at http://www.leg.state.nv.us/Session/76th2011/Exhibits/Senate/TRN/STRN282C.pdf.
- California Department of Transportation. Summary of LCV regulations in other states (not permitted in California), 2011. Per Intermodal Surface Transportation Efficiency Act (ISTEA), LCVs are allowed by a grandfathering clause only in states where they were in operation before June 1, 1991. Some form of LCV is currently allowed on designated routes in Alaska, Arizona, Colorado, Florida, Idaho, Indiana, Iowa, Kansas, Massachusetts, Michigan, Missouri, Montana, Nebraska, Nevada, New York, North Dakota, Ohio, Oklahoma, Oregon, South Dakota, Utah, Washington, and Wyoming. Oversize/overweight vehicles may be allowed by local jurisdictions in California for certain vehicles and loads. (20 states named).

- SmartWay Environmental Protection Agency (EPA). LCVs, A Glance at Clean Freight Strategies, 2010. LCVs generally have much better ton-mile fuel economy than other combination trucks. Increased productivity cuts fuel consumption and reduces greenhouse gas and air pollutant emissions. LCVs have inherent stability and control limitations because of their length and number of trailers. Therefore, it is important that only experienced drivers under safe conditions operate LCVs. Widespread use of LCVs could have an adverse affect on bridges and other transportation infrastructure. Available at http://www.epa.gov/smartwaylogistics/transport/documents/tech/longer-combination-vehicles.pdf.

LONG COMBINATION VEHICLES IN AUSTRALIA

National Transport Commission, 2005. Report that summarizes a 5-year, multi-stage project that investigated the behavior of long multi-combination vehicles transporting high center of gravity, heavy loads; found that spring versus air suspensions are more stable, and identified a new parameter that was used to characterize tracking misbehavior, Lateral Acceleration Gain. Available at http://www.ntc.gov.au/filemedia/Reports/RAG4OverarchingReportJan2005.pdf.

ADDITIONAL REFERENCES ON LCV SAFETY

1. Vehicle Weights and Dimensions Study.1986.13 volumes. Roads and Transportation Association of Canada, Ottawa.
2. Woodrooffe, J., P. Sweatman, D. Middleton, R. James, and J.R. Billing. 2010. Review of Canadian Experience with the Regulation of Large Commercial Motor Vehicles. National Academy of Sciences NCHRP Report 671. Washington, D.C.: The National Academies Press.
3. Ervin, R.D., and Y. Guy. 1986. The Influence of Weights and Dimensions on the Stability and Control of Heavy Trucks in Canada—Part 1. Prepared by the University of Michigan Transportation Research Institute for Roads and Transportation Association of Canada, Ottawa.
4. National Transportation Council. 2003. Performance Based Standards: Phase A– Standards and Measures. Regulatory Impact Statement. Melbourne, Australia.
5. York, J., and T. Maze. 1996. Applicability of Performance-Based Standards to Truck Size and Weight Regulation in the United States. Sesquicentennial Transportation Conference Proceedings. Iowa State University, Ames, Iowa.
6. U.S. Department of Transportation. 2000. Comprehensive Truck Size and Weight Study. FHWA-PL-00-029. Washington, D.C.: Federal Highway Administration.
7. U.S. Department of Transportation. 2004. Western Uniformity Analysis—A Regional Truck Size and Weight Scenario. Prepared for the Western Governors' Association. Washington, D.C.: Federal Highway Administration.
8. Woodrooffe, J., P.F. Sweatman, S. McFarlane, P. Dovile, M. Dunbabin, and D. Swenson. 1996. Dynamic Performance of Various Truck Configurations: Base Report. Prepared for the Federal Highway Administration Transportation Studies Division. Ottawa: Roaduser Research International.
9. Woodrooffe, J. 2002.Western Longer Combination Vehicle Scenario: Vehicle Operations and Safety Analysis. Prepared for the Federal Highway Administration, Office of Transportation Policy Studies. Ottawa: Woodrooffe & Associates, Inc.
10. Scopatz, R.A., and B.H. DeLucia. 2000. Longer Combination Vehicle Safety Data Collection. Washington, D.C.: AAA Foundation for Traffic Safety.
11. Government of Alberta Transportation. See http://www.transportation.albetaca/1179.htm.
12. British Columbia Ministry of Transportation and Infrastructure. See http://www.th.gov.bc.ca/cvse/LCV/faqs.htm.
13. New Brunswick Department of Transportation. 2010. Guidelines for Long Vehicle Combinations (LCVs) in the Province of New Brunswick. Version 4. Fredericton, New Brunswick.
14. Nova Scotia Department of Transportation, Infrastructure Renewal. 2009. Long Combination Vehicle Pilot Project. Halifax, Nova Scotia.
15. Ontario Ministry of Transportation. 2009. Ontario LCV Pilot Program Conditions. Toronto.
16. Council of Deputy Ministers Responsible for Transportation and Highway Safety. 2009. Task Force on Vehicle Weights and Dimensions Policy—Meeting Minutes. Ottawa.
17. Government of Saskatchewan, Department of Highways and Infrastructure. See http://www.highways.gov.sk.ca/trucking-programs/.
18. Woodrooffe, J., and L. Ash. 2001. Long Combination Vehicle (LCV) Safety Performance in Alberta 1995 to 1998. Edmonton, Alberta: Alberta Transportation.
19. Montufar, J., J. Regehr, G. Rempel, and R. McGregor. 2007. Long Combination Vehicle (LCV) Safety Performance in Alberta: 1999-2005. Edmonton, Alberta: Alberta Transportation.
20. Organization for Economic Cooperation Development. 2011. Moving Freight with Better Trucks: Improving Safety, Productivity and Sustainability. International Transport Forum. Paris: Organization for Economic Cooperation and Development.
21. Woodrooffe, J., K.P. Glaeser, P. Nordengen, M. Bereni, A. Germanchev, P. Eady, and B. Jacob. 2010. Safety,

Productivity, Infrastructure Wear, Fuel Use and Emissions Assessment of the International Truck Fleet—A Comparative Analysis. International Transport Forum. Paris: Organization for Economic Cooperation and Development.

22. Woodrooffe, J., K.P. Glaeser, and P. Nordengen, P. 2010. Truck productivity, efficiency, energy use and CO_2 output—International performance benchmarking. TRB Record.

I

Acronyms and Abbreviations

21CTP	21st Century Truck Partnership
ABS	antilock braking system
A/C	air conditioning
ACES	Advanced Collaborative Emission Study
AEC MOU	Advanced Engine Combustion Memorandum of Understanding
AH	ampere-hour
AHHPS	advanced heavy hybrid propulsion system
ALVW	adjusted loaded vehicle weight
ANL	Argonne National Laboratory
APBF	advanced petroleum based fuels
API	American Petroleum Institute
APU	auxiliary power unit
ARC	Automotive Research Center
ARRA	American Recovery and Reinvestment Act
ASTM	American Society for Testing and Materials
AT	aftertreatment
ATA	American Trucking Association
ATRI	American Transportation Research Institute
AVTA	Advanced Vehicle Testing Activity
BAC	blood alcohol content
bbl/day	barrels per day
bhp-h	brake horsepower-hour
BMEP	brake mean effective pressure
BSHR	brake specific heat rejection
BTE	brake thermal efficiency
BTL	biomass-to-liquids
BTU	British thermal unit
CAC	compressor after cooler
CARB	California Air Resources Board
CAT	Caterpillar
CDL	commercial driver's license
CERDEC	Communications-Electronics Research, Development and Engineering Center
CFD	computational fluid dynamics
CH_4	methane
CLEERS	Crosscut Lean Exhaust Emission Reduction Simulations
CLOSE	Collaborative Lubricating Oil Study on Emissions
CMV	commercial motor vehicle
CO	carbon monoxide
CO_2	carbon dioxide
CO_2-eq	CO_2-equivalent
CRADA	Cooperative Research and Development Agreement
CRC	Coordinating Research Council
CRS	Caterpillar Regeneration System
CTEI	Commercial Traffic Effects Institute
DCT	Diesel Crosscut Team
DDC	Detroit Diesel Corporation
DEC	Diesel Emission Control
DEER	Directions in Engine-Efficiency and Emissions Research (formerly Diesel Engine-Efficiency and Emissions Research)
DEF	Diesel Exhaust Fluid
DERA	Diesel Emission Reduction Act
DME	dimethyl ether
DOC	diesel oxidation catalyst
DOD	U.S. Department of Defense
DOE	U.S. Department of Energy
DOT	U.S. Department of Transportation
DPF	diesel particulate filter
DPIM	Dual Power Inverter Model
ECU	environmental control unit
EERE	Energy Efficiency and Renewable Energy (Office of)
EGR	exhaust gas recirculation
EIA	Energy Information Administration
EISA	Energy Independence and Security Act
EMA	Engine Manufacturers Association

EMD	engine manufacturer diagnostic
EPA	U.S. Environmental Protection Agency
EPAct 2005	Energy Policy Act of 2005
EPS	Electrified Parking Spaces
ESC	electronic stability control
EU	European Union
FACE	Fuels for Advanced Combustion Engines
FAME	fatty acid methyl esters
FCFP	FreedomCAR and Fuel Partnership
FCVT	FreedomCAR and Vehicle Technologies (Office of)
FHWA	Federal Highway Administration
FMCSA	Federal Motor Carrier Safety Administration
FMVSS	Federal Motor Vehicle Safety Standards
FOA	Funding Opportunity Announcement
FOH	fuel-operated heater
FOT	field operational test
F-T	Fischer-Tropsch
FTP	Federal Test Procedure
FY	fiscal year
GAC	granular activated carbon
gal	gallon
g/bhp-h	grams per brake horsepower-hour
GCW	gross combined weight
GES	General Estimates System
GFIC	ground-fault interrupter circuit
GHG	greenhouse gas
g/mi	grams per mile
gpm	gallons per mile
GPS	global positioning system
GREET	Greenhouse Gas Regulated Emissions, and Transportation
GTL	gas-to-liquid
GVW	gross vehicle weight
GVWR	Gross Vehicle Weight Rating
HC	hydrocarbon
HCCI	homogeneous-charge compression ignition
HD	heavy-duty
HECC	High Efficiency Clean Combustion
HEI	Health Effects Institute
HEMTT	Heavy Expanded Mobility Tactical Truck
HEV	hybrid electric vehicle
HFC	halogenated fluororcarbon
HFRR	high frequency reciprocating rig
HHDDE	heavy heavy-duty diesel engine
HHDDT	Heavy Heavy-Duty Diesel Truck
HHV	hybrid heavy-duty electric vehicle
HLA	hydraulic lift assist
HMMWV	High Mobility Multipurpose Wheeled Vehicle
HP	high pressure
hp	horsepower
HPV	high productivity vehicle (aka LCV)
HTHS	high temperature high shear
HTML	High Temperature Materials Laboratory
HVAC	heating, ventilation, and air conditioning
HVD	high-volatility diesel
HWFET	highway fuel economy test
IC	internal combustion
ICE	internal combustion engine
IGBT	insulated gate bipolar transistor
IRS	Internal Revenue Service
ISG	integrated starter-generator
ISTEA	Intermodal Surface Transportation Efficiency Act
ITS	Intelligent Transportation Systems
IVBSS	Integrated Vehicle-Based Safety Systems
JCAP	Japanese Clean Air Program
JCS	Johnson Controls-SAFT
JP-8	Jet Propellant 8
KOH	potassium hydroxide
kWh	kilowatt-hour
L	liter
LANL	Los Alamos National Laboratory
lb	pound
LCFS	Low Carbon Fuel Standard
LCV	Long Combination Vehicle (aka HPV)
LD	light-duty
LDT	light-duty truck
LDT1	light-duty truck 1
LDV	light-duty vehicle
LDW	lane departure warning
LEV	Low Emission Vehicle
LHDDE	light heavy-duty diesel engine
LI	lithium-ion
LLNL	Lawrence Livermore National Laboratory
LNT	lean NO_x trap
LP	low pressure
LSD	low sulfur diesel
LS/SF	long sleeper tractor and straight frame
LTC	low-temperature combustion
LVW	loaded vehicle weight
µm	micrometer
MBRC	Miles Between Road Call
MD/HD	medium-duty/heavy-duty
MDPV	medium-duty passenger vehicle
MDV	medium-duty vehicle
MET	More Electric Truck
MFC	Model Fund Consortium
MHDDE	medium heavy-duty diesel engine
MHDV	medium- and heav-duty vehicle
mi	mile

mpg	miles per gallon
mph	miles per hour
msec	millisecond
MYPP	Multi-Year Program Plan
N_2O	nitrous oxide
NAC	National Automotive Center
NAC NO_X	Absorber Catalysts
NASA	National Aeronautics and Space Administration
NCSU	North Caroline State University
NEC	National Electric Code
NETL	National Energy Technology Laboratory
NFAC	National Full-Scale Aerodynamics Complex
NHWBS	next-generation wide base single (tires)
NHTSA	National Highway Traffic Safety Administration
nm	nanometer
NMHC	nonmethane hydrocarbon
NO_X	oxides of nitrogen
NPBF	non-petroleum based fuels
NPRM	Notice of Proposed Rulemaking
NRC	National Research Council
NREL	National Renewable Energy Laboratory
NTE	Not-to-Exceed
NTP	National Toxicology Program
NTSB	National Transportation Safety Board
OBD	on-board diagnostic
OECD	Organization for Economic Co-operation and Development
OEM	original equipment manufacturer
OHVT	Office of Heavy Vehicles Technology
ORNL	Oak Ridge National Laboratory
PART	Program Assessment Rating Tool
PC	passenger car
PCCI	pre-mixed charge compression ignition
PEM	polymer electrolyte membrane
PHEV	plug-in hybrid electric vehicle
PM	particulate matter
$PM_{2.5}$	particulate matter smaller than 2.5 micrometers in diameter (also PM_{10})
PNGV	Partnership for a New Generation of Vehicles
PNNL	Pacific Northwest National Laboratory
ppm	parts per million
PSAT	Powertrain Systems Analysis Toolkit
psi	pounds per square inch
PTO	power take-off
PZEV	Partial Zero Emission Vehicle
R&D	research and development
RFS	Renewable Fuels Standards
rpm	revolutions per minute
SAE	Society of Automotive Engineers
SCR	selective catalytic reduction
SECA	Solid State Energy Conversion Alliance
SET	Supplemental Emission Test
SI	spark-ignited
SNL	Sandia National Laboratories
SNS	Spellation Neutron Source
SO_2	sulfur dioxide
SOFC	solid oxide fuel cell
STEP	Shorepower Truck Electrification Project
SULEV	Super Ultra Low Emission Vehicle
SUV	sport utility vehicle
21CTP	21st Century Truck Partnership
TACOM	U.S. Army Tank-Automotive Command
TARDEC	Tank-Automotive Research, Development and Engineering Center
TMA	Truck Manufacturers Association
TRB	Transportation Research Board
TSE	truck stop electrification
UDDS	Urban Dynamic Driving Schedule
UFP	ultra-fine particulates
ULEV	Ultra Low Emission Vehicle
ULSD	ultralow-sulfur diesel
USABC	U.S. Advanced Battery Consortium
USCAR	U.S. Council for Automotive Research
UTQGS	Uniform Tire Quality Grading System
V2I	vehicle to infrastructure
V2V	vehicle to vehicle
VG	variable geometry
VIUS	Vehicle Inventory and Use Survey
VMT	vehicle miles traveled
VSST	Vehicle Systems Simulation and Testing
VT	viscous transmission
VVA	variable valve actuation
WBS	wide base single
Wh/kg	watt-hours per kilogram
WHR	waste heat recovery
ZEV	Zero Emission Vehicle